22 Radio Receiver Projects for the Evil Genius

Evil Genius Series

Bionics for the Evil Genius: 25 Build-it-Yourself Projects

Electronic Circuits for the Evil Genius: 57 Lessons with Projects

Electronic Gadgets for the Evil Genius: 28 Build-it-Yourself Projects

Electronic Games for the Evil Genius

Electronic Sensors for the Evil Genius: 54 Electrifying Projects

50 Awesome Auto Projects for the Evil Genius

50 Model Rocket Projects for the Evil Genius

Mechatronics for the Evil Genius: 25 Build-it-Yourself Projects

MORE Electronic Gadgets for the Evil Genius: 40 NEW Build-it-Yourself Projects

101 Spy Gadgets for the Evil Genius

123 PIC® Microcontroller Experiments for the Evil Genius

123 Robotics Experiments for the Evil Genius

PC Mods for the Evil Genius

Solar Energy Projects for the Evil Genius

25 Home Automation Projects for the Evil Genius

51 High-Tech Practical Jokes for the Evil Genius

22 Radio Receiver Projects for the Evil Genius

TOM PETRUZZELLIS

New York   Chicago   San Francisco   Lisbon   London   Madrid
Mexico City   Milan   New Delhi   San Juan   Seoul
Singapore   Sydney   Toronto

The McGraw·Hill Companies

Library of Congress Cataloging-in-Publication Data on file with the Library of Congress.

Copyright © 2008 by The McGraw-Hill Companies, Inc. All rights reserved. Printed in the United States of America. Except as permitted under the United States Copyright Act of 1976, no part of this publication may be reproduced or distributed in any form or by any means, or stored in a data base or retrieval system, without the prior written permission of the publisher.

3   4   5   6   7   8   9   0   QPD/QPD   0   1   3   2   1   0

ISBN 978-0-07-148929-4
MHID 0-07-148929-0

The sponsoring editor for this book as Judy Bass, the editing supervisor was Maureen B. Walker, and the production supervisor was Pamela A. Pelton. It was set in Times New Roman by Keyword Group Ltd. The art director for the cover was Jeff Weeks.

Printer and bound by Quebecor/Dubuque.

This book is printer on acid-free paper.

McGraw-Hill books are available at special quantity discounts to use as premiums and sales promotions, or for use in corporate training programs. For more information, please write to the Director of Special Sales, McGraw-Hill Professional, Two Penn Plaza, New York, NY 10121-2298. Or contact your local bookstore.

Thomas Petruzzellis is an electronics engineer currently working at the geophysical laboratory at the State University of New York, Binghamton. Also an instructor at Binghamton, with 30 years' experience in electronics, he is a veteran author who has written extensively for industry publications, including *Electronics Now*, *Modern Electronics*, *QST, Microcomputer Journal*, and *Nuts & Volts*. Tom wrote five previous books, including an earlier volume in this series, *Electronic Sensors for the Evil Genius*. He is also the author of *Create Your Own Electronics Workshop*; *STAMP 2 Communications and Control Projects*; *Optoelectronics, Fiber Optics, and Laser Cookbook*; *Alarm, Sensor, and Security Circuit Cookbook*, all from McGraw-Hill. He lives in Vestal, New York.

Acknowledgments

I would like to thank the following people and companies listed below for their help in making this book possible. I would also like to thank senior editor Judy Bass and all the folks at McGraw-Hill publications who had a part in making this book possible. We hope the book will inspire both radio and electronics enthusiasts to build and enjoy the radio projects in this book.

Richard Flagg/RF Associates

Wes Greenman/University of Florida

Charles Higgins/Tennessee State University

Fat Quarters Software

Radio-Sky Publishing

Ramsey Electronics

Vectronics, Inc

Russell Clift

Todd Gale

Eric Vogel

Contents

Introduction

22 Radio Receiver Projects for the Evil Genius was created to inspire readers both young and old to build and enjoy radio and receiver projects, and perhaps propel interested experimenters into a career in radio, electronics or research. This book is for people who are interested in radio and electronics and those who enjoy building and experimenting as well as those who enjoy research.

Radio encompasses many different avenues for enthusiasts to explore, from simple crystal radios to sophisticated radio telescopes. This book is an attempt to show electronics and radio enthusiasts that there is a whole new world "out there" to explore.

Chapter 1 will present the history and background and elements of radio, such as modulation techniques, etc. Chapter 2 will help the newcomers to electronics, identifying components and how to look and understand schematics vs. pictorial diagrams. Next, Chapter 3 will show the readers how to install electronic components onto circuit boards and how to correctly solder before embarking on their new radio building adventure.

We will start our adventure with the simple "lowly" crystal radio in Chapter 4. Generally crystal radios are only thought of as simple AM radios which can only pickup local broadcast stations. But did you know that you can build crystal radios which can pickup long-distance stations as well as FM and shortwave broadcasts from around the world? You will learn how to build an AM, FM and shortwave crystal radio, in this chapter.

In Chapter 5, you will learn how AM radio is broadcast, from a radio station to a receiver in your home, and how to build your own TRF or Tuned Radio Frequency AM radio receiver. In Chapter 6, we will discover how FM radio works and how to build an FM radio with an SCA output for commercial free radio broadcasts.

Chapter 7 will present the exciting world of shortwave radio. Shortwave radio listening has a large following and encompasses an entire hobby in itself. You will be able to hear shortwave stations from around the world, including China, Russia, Italy, on your new shortwave broadcast receiver. Old time radio buffs will be interested in the single tube Doerle super-regenerative shortwave radio.

If you are interested in a portable shortwave receiver that you could take on a camping trip, then you may want to construct the multi-band integrated circuit shortwave radio receiver described in Chapter 8.

If you are interested in Amateur Radio or are thinking of learning Morse code or want to increase your code speed, you may want to consider building this 80 and 40-meter code receiver. This small lightweight portable receiver can be built in a small enclosure and taken on camping trips, etc.

In Chapter 10, you will learn how to build and use a WWW time code receiver, which can be used to pick up time signal broadcast from the National Institute of Standards and Technology (NIST) or

National Atomic Time Clock in Boulder CO. Time signal broadcasts present geophysical and propagation forecasts as well as marine and sea conditions. They will also help you set the time on your best chronometer.

With the VHF Public service receiver featured in Chapter 11, you will discover the high frequency "action bands" which cover the police, fire, taxis, highway departments and marine frequencies. You will be able to listen-in to all the exciting communications in your hometown.

The 6-meter and 2-meter dual VHF Amateur Radio receiver in Chapter 12 will permit to discover the interesting hobby of Amateur Radio. The 6-meter and 2-meter ham radio bands are two of the most popular VHF bands for technician class licensees. You may discover that you might just want to get you own ham radio license and talk to ham radio operators through local VHF repeaters or to the rest of the world.

Why not build an aircraft radio and listen-in to airline pilots talking from 747s to the control tower many miles away. You could also build the passive Air-band radio which you use to listen-in to your pilot during your own flight. Passive aircraft radios will not interfere with airborne radio so they are permitted on airplanes, without restriction. Check out these two receivers in Chapter 13.

Chapter 14 will also show you how to build an induction communication system, which will allow you to broadcast a signal around home or office using a loop of wire, to a special induction receiver. The induction loop broadcast system is a great aid to the hearing impaired, since it can broadcast to hearing aids as well.

The VLF, or "whistler" radio in Chapter 15, will pickup very very low frequency radio waves from around the world. You will be able to listen to low frequency beacon stations, submarine transmissions and "whistlers" or the radio waves created from electrical storms on the other side of the globe. This project is great for research projects where you can record and later analyze

your results by feeding your recorded signal into a sound card running an FFT program. Use your computer to record and analyze these interesting signals. There are many free programs available over the Internet. An FFT audio analyzer program can display the audio spectrum and show you where the signals plot out in respect to frequency.

If you are interested in weather, then you will appreciate the Lightening to Storm Receiver in Chapter 16, which will permit you to "see" the approaching storm berfore it actually arrives. This receiver will permit you to have advanced warning up to 50 miles or more away; it will warn you well in advance of an electrical storm, so you can disconnect any outdoor antennas.

The Ambient Power Module receiver project illustrated in Chapter 17, will allow to you pickup a broad spectrum of radio waves which get converted to DC power, and which can be used to power low current circuits around your home or office. This is a great project for experimentation and research. You can use it to charge cell phones, emergency lights, etc.

Our magnetometer project shown in Chapter 18 can be used to see the diurnal or daily changes in the Earth's magnetic field, and you can record the result to a data-logger or recording multi-meter.

If you are an avid amateur radio operator or shortwave listener, you many want to build a SIDs receiver shown in Chapter 19. A SIDs receiver can be used to determine when radio signals and/or propagation is disturbed by solar storms. This receiver will quickly alert you to unfavorable radio conditions. You can collect the receiver to your personal computer's sound card and use the data recorded to correlate radio propagation against storm conditions.

The Aurora receiver project in Chapter 20 will alert you, with both sound and meter display, when the Earth's magnetic increases just before an Aurora display is about to take place. UFO and Alien contact buffs can use this receiver to know when UFOs are close by.

For those interested in more earthly research projects, why not build your own ULF or ultra low frequently receiver, shown in Chapter 21, which can be utilized for detecting low frequency wave generated by earthquakes and fault lines. With this receiver you will be able to conduct your own research projects on monitoring the pulse of the Earth. You can connect your ELF receiver to a data-logger and record the signals over time to correlate your research with that of others.

You can explore the heavens by constructing your own radio telescope to monitor the radio signals generated from the planet Jupiter. This radio receiver, illustrated in Chapter 22, will pick up radio signals which indicate electrical and or magnetic storms on the Jovian planet. The Radio Jupiter receiver can be coupled to your personal computer, and can be used for a research project to record and analyze these radio storm signals.

Why not construct your own weather satellite receiving station, shown in Chapter 23. This receiver will allow you to receive APT polar satellites broadcasting while passing overhead. You can display the satellite weather maps on the computer's screen or save them later to show friends and relatives.

Chapter 24 discusses different analog to digital converters which you can use to collect and record data from the different receiver projects.

We hope you will find the *22 Radio Projects for the Evil Genius* a fun and thought-provoking book, that will find a permanent place on your electronics or radio bookshelf. Enjoy!

22 Radio Receiver Projects for the Evil Genius

Chapter 1

Radio Background and History

Electromagnetic energy encompasses an extremely wide frequency range. Radio frequency energy, both natural radio energy created by lightning and planetary storms as well as radio frequencies generated by man for communications, entertainment, radar, and television are the topic of this chapter. Radio frequency energy, or RF energy, covers the frequency range from the low end of the radio spectrum, around 10 to 25 kHz, which is used by high-power Navy stations that communicate with submerged nuclear submarines, through the familiar AM broadcast band from 550 to 1600 kHz. Next on the radio frequency spectrum are the shortwave bands from 2000 kHz to 30,000 kHz. The next band of frequencies are the very high frequency television channels covering 54 to 216 MHz, through the very popular frequency modulation FM band from 88 to 108 MHz. Following the FM broadcast band are aircraft frequencies on up through UHF television channels and then up through the radar frequency band of 1000 to 1500 MHz, and extending through approximately 300 gHz. See frequency spectrum chart in Figure 1-1. The radio frequency spectrum actually extends almost up to the lower limit of visible light frequencies.

Radio history

One of the more fascinating applications of electricity is in the generation of invisible ripples of energy called radio waves. Following Hans Oersted's accidental discovery of electromagnetism, it was realized that electricity and magnetism were related to each other. When an electric current was passed through a conductor, a magnetic field was generated perpendicular to the axis of flow. Likewise, if a conductor was exposed to a change in magnetic flux perpendicular to the conductor, a voltage was produced along the length of that conductor.

Joseph Henry, a Princeton University professor, and Michael Faraday, a British physicist, experimented separately with electromagnets in the early 1800s. They each arrived at the same observation: the theory that a current in one wire can produce a current in another wire, even at a distance. This phenomenon is called electromagnetic induction, or just induction. That is, one wire carrying a current induces a current in a second wire. So far, scientists knew that electricity and magnetism always seemed to affect each other at right angles. However, a major discovery lay hidden just beneath this seemingly simple concept of related perpendicularity, and its unveiling was one of the pivotal moments in modern science.

The man responsible for the next conceptual revolution was the Scottish physicist James Clerk Maxwell (1831–1879), who "unified" the study of electricity and magnetism in four relatively tidy equations. In essence, what he discovered was that electric and magnetic fields were intrinsically related to one another, with or without the presence of a conductive path for electrons to flow. Stated more formally, Maxwell's discovery was this: a changing electric field produces a perpendicular magnetic field, and a changing magnetic field produces a perpendicular electric field. All of this can take place in open space, the alternating electric and magnetic fields supporting each other as they travel through space at the speed of light. This dynamic structure of electric and magnetic fields propagating through space is better known as an electromagnetic wave.

Later, Heinrich Hertz, a German physicist, who is honored by our replacing the expression "cycles per second" with hertz (Hz), proved Maxwell's theory between the years 1886 and 1888. Shortly after that, in 1892, Eouard Branly, a French physicist, invented a device that could receive radio waves (as we know them today) and could cause them to ring an electric bell.

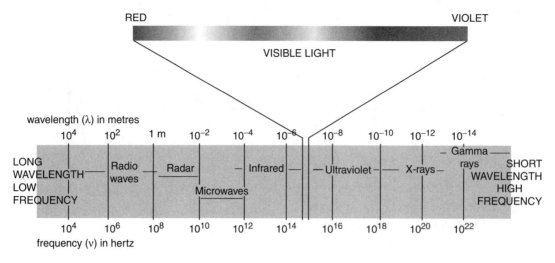

Figure 1-1 *Electromagnetic spectrum*

Note that at the time all the research being conducted in what was to become radio and later radio-electronics, was done by physicists.

In 1895, the father of modem radio, Guglielmo Marconi, of Italy, put all this together and developed the first wireless telegraph and was the first to commercially put radio into ships. The wire telegraph had already been in commercial use for a number of years in Europe. The potential of radio was finally realized through one of the most memorable events in history. With the sinking of the Titanic in 1912, communications between operators on the sinking ship and nearby vessels, and communications to shore stations listing the survivors brought radio to the public in a big way.

AM radio broadcasting began on November 2, 1920. Four pioneers: announcer Leo Rosenberg, engineer William Thomas, telephone line operator John Frazier and standby R.S. McClelland, made their way to a makeshift studio—actually a shack atop the Westinghouse "K" Building in East Pittsburgh—flipped a switch and began reporting election returns in the Harding vs. Cox Presidential race. At that moment, KDKA became the pioneer broadcasting station of the world.

Radio spread like wildfire to the homes of everyone in America in the 1920s. In a few short years, over 75 manufacturers began selling radio sets. Fledgling manufacturers literally came out of garages over-night. Many young radio enthusiasts rushed out to buy parts and radio kits which soon became available.

Radio experimenters discovered that an amplitude-modulated wave consists of a carrier and two identical sidebands which are both above and below the carrier wave. The Navy conducted experiments in which they attempted to pass one sideband and attenuate the other. These experiments indicated that one sideband contained all the necessary information for voice transmission, and these discoveries paved the way for development of the concept of single-sideband or SSB transmission and reception.

In 1923, a patent was granted to John R. Carson on his idea to suppress the carrier and one sideband. In that year the first trans-Atlantic radio telephone demonstration used SSB with pilot carrier on a frequency of 52 kc. Single sideband was used because of limited power capacity of the equipment and the narrow bandwidths of efficient antennas for those frequencies. By 1927, trans-Atlantic SSB radiotelephony was open for public service. In the following years, the use of SSB was limited to low-frequency and wire applications. Early developments in FM transmission suggested that this new mode might prove to be the ultimate in voice communication. The resulting slow development of SSB technology precluded practical SSB transmission and reception at high frequencies. Amateur radio SSB activity followed very much the same pattern. It wasn't until 1948 that amateurs began seriously experimenting with SSB, likely delayed by the wartime blackouts.

The breakthroughs in the war years, and those following the war, were important to the development of HF-SSB communication. Continued advances in technology made SSB the dominant mode of HF radio communication.

The radio-frequency spectrum, once thought to be adequate for all needs, has become very crowded. As the world's technical sophistication progresses, the requirements for rapid and dependable radio communications increase. The competition for available radio spectrum space has increased dramatically. Research and development in modern radio systems has moved to digital compression and narrow bandwidth with highly developed modulation schemes and satellite transmission.

The inventor most responsible for the modern day advances in radio systems was Edwin H. Armstrong. He was responsible for the Regenerative circuit in 1912, the Superheterodyne radio circuit in 1918, the Superregenerative radio circuit design in 1922 and the complete FM radio system in 1933. His inventions and developments form the backbone of radio communications as we know it today. The majority of all radio sets sold are FM radios, all microwave relay links are FM, and FM is the accepted system in all space communications. Unfortunately, Armstrong committed suicide while still embittered in patent lawsuits: later vindicated, his widow received a windfall.

Sony introduced their first transistorized radio in 1960, small enough to fit in a vest pocket, and able to be powered by a small battery. It was durable, because there were no tubes to burn out. Over the next 20 years, transistors displaced tubes almost completely except for very high power, or very high frequency, uses. In the 1970s; LORAN became the standard for radio navigation system, and soon, the US Navy experimented with satellite navigation. Then in 1987, the GPS constellation of satellites was launched and navigation by radio in the sky had a new dimension. Amateur radio operators began experimenting with digital techniques and started to send pictures, faxes and teletype via the personal computer through radio. By the late 1990s, digital transmissions began to be applied to radio broadcasting.

Types of radio waves

There are many kinds of natural radiative energy composed of electromagnetic waves. Even light is electromagnetic in nature. So are shortwaves, X-rays and "gamma" ray radiation. The only difference between these kinds of electromagnetic radiation is the frequency of their oscillation (alternation of the electric and magnetic fields back and forth in polarity). By using a source of AC voltage and a device called an antenna, we can create electromagnetic waves.

It was discovered that high frequency electromagnetic currents in an antenna wire, which in turn result in a high frequency electromagnetic field around the antenna, will result in electromagnetic radiation which will move away from the antenna into free space at the velocity of light, which is approximately 300,000,000 meters per second.

In radio transmission, a radiating antenna is used to convert a time-varying electric current into an electromagnetic wave, which freely propagates through a nonconducting medium such as air or space. An antenna is nothing more than a device built to produce a dispersing electric or magnetic field. An electromagnetic wave, with its electric and magnetic components, is shown in Figure 1-2.

When attached to a source of radio frequency signal generator, or transmitter, an antenna acts as a transmitting device, converting AC voltage and current into electromagnetic wave energy. Antennas also have the ability to intercept electromagnetic waves and convert their energy into AC voltage and current. In this mode, an antenna acts as a receiving device.

Radio frequencies spectrum

Radio frequency energy is generated by man for communications, entertainment, radar, television,

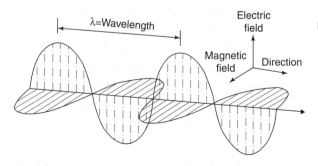

Figure 1-2   *Magnetic vs. electric wave*

navigation, etc. This radio frequency or (RF) energy covers quite a large range of radio frequencies from the low end of the radio spectrum from l0 to 25 kHz, which is the domain occupied by the high-power Navy stations that communicate with submerged nuclear submarines: these frequencies are called Very Low Frequency waves or VLF. Above the VLF frequencies are the medium wave frequencies or (MW), i.e. the AM radio broadcast band from 550 to 1600 kHz. The shortwave bands or High Frequency or (HF) bands cover from 2000 kHz to 30,000 kHz and make use of multiple reflections from the ionosphere which surrounds the Earth, in order to propagate the signals to all parts of the Earth. The Very High Frequencies or VHF bands begin around 30 MHz; these lower VHF frequencies are called low-band VHF. Mid-band VHF frequencies begin around 50 MHz which cover the lowest TV channel 2. Low-band television channels 2 through 13 cover the 54 to 216 MHz range. The popular frequency modulation or (FM) broadcast band covers the range from 88 to l08 MHz, which is followed by low-band Air-band frequencies from 118 to 136 MHz. High-band VHF frequencies around 144 are reserved for amateur radio, public service around 150 MHz, with marine frequencies around 156 MHz. UHF frequencies begin around 300 MHz and go up through the radar frequency band of 1000 to 1500 MHz, and extending through approximately 300 gHz. Television channels 14 through 70 are placed between 470 and 800 MHz. American cell phone carriers have cell phone communications around 850 MHz. Geosynchronous weather satellites signals are placed around 1.6 GHz, and PCS phone devices are centered around 1.8 GHz. The Super-high frequency (SHF) bands range from 3 to 30 GHz, with C-band microwave frequencies around 3.8 GHz, then X-band, from 7.25 to 8.4 GHz, followed by the KA and KU-band microwave bands.

Table 1-1 illustrates the division of radio frequencies. The radio frequency spectrum extends almost up to the lower limit of visible light frequencies, with just the infrared frequencies lying in between it and visible light. The radio frequency spectrum is a finite resource which must be used and shared with many people and agencies around the world, so cooperation is very important.

So how does a radio work? As previously mentioned, radio waves are part of a general class of waves known as electromagnetic waves. In essence, they are electrical and magnetic energy which travels through space in the form of a wave. They are different from sound waves (which are pressure waves that travel through air or water, as an example) or ocean waves (similar to sound waves in water, but much lower in frequency and are much larger). The wave part is similar, but the energy involved is electrical and magnetic, not mechanical.

Electromagnetic waves show up as many things. At certain frequencies, they show up as radio waves. At much higher frequencies, we call them infrared light. Still higher frequencies make up the spectrum known as visible light. This goes on up into ultraviolet light, and X-rays, things that radio engineers rarely have to worry about. For our discussions, we'll leave light to the physicists, and concentrate on radio waves.

Radio waves have two important characteristics that change. One is the amplitude, or strength of the wave. This is similar to how high the waves are coming into shore from the ocean. The bigger wave has a higher amplitude. The other thing is frequency. Frequency is how often the wave occurs at any point. The faster the wave repeats itself, the higher the frequency. Frequency is measured by the number of times in a second that the wave repeats itself. Old timers remember when frequency was described in units of cycles per second. In more recent times we have taken to using the simplified term of hertz (named after the guy who discovered radio waves). Metric prefixes are often used, so that 1000 hertz is a kilohertz, one million hertz is a megahertz, and so on.

A typical radio transmitter, for example, takes an audio input signal, such as voice or music and amplifies it. The amplified audio is in turn sent to a modulator and an RF exciter which comprises the radio frequency transmitter. The exciter in the transmitter generates a main carrier wave. The RF signal from the exciter is further amplified by a power amplifier and then the RF signal is sent out to an antenna which radiates the signal into the sky and out into the ionosphere. Depending upon the type of transmitter used the modulation technique can be either AM, FM, SSB signal sideband, CW, or digital modulation, etc.

AM modulation

Amplitude modulation (AM) is a technique used in electronic communication, most commonly for transmitting information via a carrier wave wirelessly.

Table 1-1

Radio frequency spectrum chart

Frequency range

Extremely Low Frequency (ELF)	0	to	3 kHz
Very Low Frequency (VLF)	3 kHz	to	30 kHz
Radio Navigation & maritime/aeronautical mobile	9 kHz	to	540 kHz
Low Frequency (LF)	30 kHz	to	300 kHz
Medium Frequency (MF)	300 kHz	to	3000 kHz
AM Radio Broadcast	540 kHz	to	1630 kHz
Travellers Information Service	1610 kHz		
High Frequency (HF)	3 MHz	to	30 MHz
Shortwave Broadcast Radio	5.95 MHz	to	26.1 MHz
Very High Frequency (VHF)	30 MHz	to	300 MHz
Low Band: TV Band 1 - Channels 2-6	54 MHz	to	88 MHz
Mid Band: FM Radio Broadcast	88 MHz	to	174 MHz
High Band: TV Band 2 - Channels 7-13	174 MHz	to	216 MHz
Super Band (mobile/fixed radio & TV)	216 MHz	to	600 MHz
Ultra-High Frequency (UHF)	300 MHz	to	3000 MHz
Channels 14-70	470 MHz	to	806 MHz
L-band:	500 MHz	to	1500 MHz
Personal Communications Services (PCS)	1850 MHz	to	1990 MHz
Unlicensed PCS Devices	1910 MHz	to	1930 MHz
Superhigh Frequencies (SHF)			
(Microwave)	3 GHz	to	30.0 GHz
C-band	3600 MHz	to	7025 MHz
X-band	7.25 GHz	to	8.4 GHz
Ku-band	10.7 GHz	to	14.5 GHz
Ka-band	17.3 GHz	to	31.0 GHz
Extremely High Frequencies (EHF)			
(Millimeter Wave Signals)	30.0 GHz	to	300 GHz
Additional Fixed Satellite	38.6 GHz	to	275 GHz
Infrared Radiation	300 GHz	to	430 THz
Visible Light	430 THz	to	750 THz
Ultraviolet Radiation	1.62 PHz	to	30 PHz
X-Rays	0.30 PHz	to	30 EHz
Gamma Rays	0.30 EHz	to	3000 EHz

It works by varying the strength of the transmitted signal in relation to the information being sent.

In the mid-1870s, a form of amplitude modulation was the first method to successfully produce quality audio over telephone lines. Beginning in the early 1900s, it was also the original method used for audio radio transmissions, and remains in use by some forms of radio communication—"AM" is often used to refer to the medium-wave broadcast band (see AM Radio–Chapter 5).

Amplitude modulation (AM) is a type of modulation technique used in communication. It works by varying

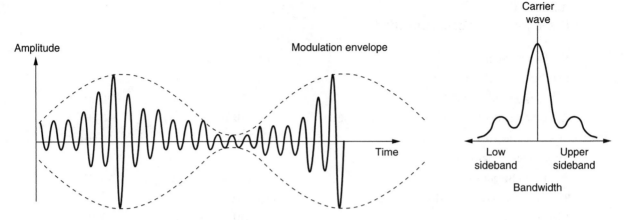

Figure 1-3 *Amplitude modulation waveform*

the strength of the transmitted signal in relation to the information being sent, for example, changes in the signal strength can be used to reflect sounds being reproduced in the speaker. This type of modulation technique creates two sidebands with the carrier wave signal placed in the center between the two sidebands. The transmission bandwidth of AM is twice the signal's original (baseband) bandwidth—since both the positive and negative sidebands are 'copied' up to the carrier frequency, but only the positive sideband is present originally. See Figure 1-3. Thus, double-sideband AM (DSB-AM) is spectrally inefficient. The power consumption of AM reveals that DSB-AM with its carrier has an efficiency of about 33% which is too efficient. The benefit of this system is that receivers are cheaper to produce. The forms of AM with suppressed carriers are found to be 100% power efficient, since no power is wasted on the carrier signal which conveys no information. Amplitude modulation is used primarily in the medium wave band or AM radio band which covers 520 to 1710 kHz. AM modulation is also used by shortwave broadcasters in the SW bands from between 5 MHz and 24 MHz, and in the aircraft band which covers 188 to 136 MHz.

FM modulation

Frequency modulation (FM) is a form of modulation which represents information as variations in the instantaneous frequency of a carrier wave. Contrast this with amplitude modulation, in which the amplitude

of the carrier is varied while its frequency remains constant. In analog applications, the carrier frequency is varied in direct proportion to changes in the amplitude of an input signal. Digital data can be represented by shifting the carrier frequency among a set of discrete values, a technique known as frequency-shift keying. The diagram in Figure 1-4, illustrates the FM modulation scheme, the RF frequency is varied with the sound input rather than the amplitude.

FM is commonly used at VHF radio frequencies for high-fidelity broadcasts of music and speech, as in FM broadcasting. Normal (analog) TV sound is also broadcast using FM. A narrowband form is used for voice communications in commercial and amateur radio settings. The type of FM used in broadcast is generally called wide-FM, or W-FM. In two-way radio, narrowband narrow-FM (N-FM) is used to conserve bandwidth. In addition, it is used to send signals into space.

Wideband FM (W-FM) requires a wider bandwidth than amplitude modulation by an equivalent modulating signal, but this also makes the signal more robust against noise and interference. Frequency modulation is also more robust against simple signal amplitude fading phenomena. As a result, FM was chosen as the modulation standard for high frequency, high fidelity radio transmission: hence the term "FM radio." FM broadcasting uses a well-known part of the VHF band between 88 and 108 MHz in the USA.

FM receivers inherently exhibit a phenomenon called capture, where the tuner is able to clearly receive the stronger of two stations being broadcast on

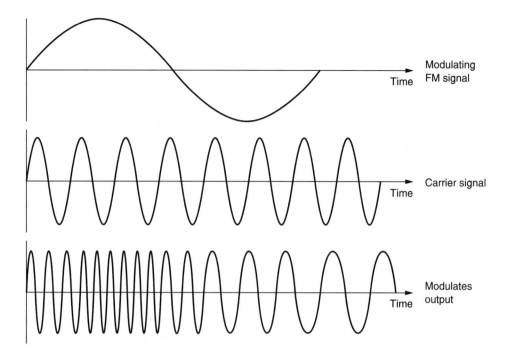

Figure 1-4 *FM modulation waveform*

the same frequency. Problematically, however, frequency drift or lack of selectivity may cause one station or signal to be suddenly overtaken by another on an adjacent channel. Frequency drift typically constituted a problem on very old or inexpensive receivers, while inadequate selectivity may plague any tuner. Frequency modulation is used on the FM broadcast band between 88 and 108 MHz as well as in the VHF and UHF bands for both public service and amateur radio operators.

Single sideband (SSB) modulation

Single sideband modulation (SSB) is a refinement upon amplitude modulation, which was designed to be more efficient in its use of electrical power and spectrum bandwidth. Single sideband modulation avoids this bandwidth doubling, and the power wasted on a carrier, but the cost of some added complexity.

The balanced modulator is the most popular method of producing a single sideband modulated signal. The balanced modulator provides the "sidebands" of energy that exist on either side of the carrier frequency but eliminates the RF carrier, see Figure 1-5. The carrier is

removed because it is the sidebands that provide the actual meaningful content of material, within the modulation envelope. In order to make SSB even more efficient, one of these two sidebands is removed by a bandpass. So the intelligence is preserved with SSB and it becomes a more efficient use of radio spectrum energy. It provides almost 9 Decibels (dBs) of signal gain over an amplitude modulated signal that includes an RF "carrier" of the same power level! As the final RF amplification is now concentrated in a single sideband, the effective power output is greater than in normal AM (the carrier and redundant sideband account for well over half of the power output of an AM transmitter). Though SSB uses substantially less bandwidth and power, it cannot be demodulated by a simple envelope detector like standard AM.

SSB was pioneered by telephone companies in the 1930s for use over long-distance lines, as part of a technique known as frequency-division multiplexing (FDM). This enabled many voice channels to be sent down a single physical circuit. The use of SSB meant that the channels could be spaced (usually) just 4000 Hz apart, while offering a speech bandwidth of nominally 300–3400 Hz. Amateur radio operators began to experiment with the method seriously after World War II. It has become a *de facto* standard for long-distance voice radio transmissions since then.

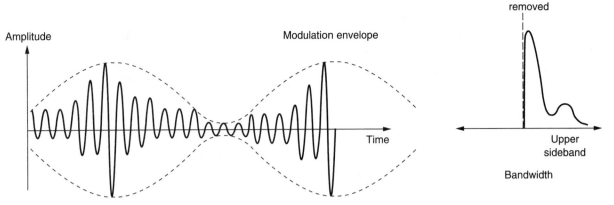

Figure 1-5 *Single sideband modulation waveform*

Single Sideband Suppressed Carrier (SSB-SC) modulation was the basis for all long-distance telephone communications up until the last decade. It was called "L carrier." It consisted of groups of telephone conversations modulated on upper and/or lower sidebands of contiguous suppressed carriers. The groupings and sideband orientations (USB, LSB) supported hundreds and thousands of individual telephone conversations. Single sideband communications are used by amateur radio operators and government, and utility stations primarily in the shortwave bands for long-distance communications.

Shortwave radio

Shortwave radio operates between the frequencies of 1.80 MHz and 30 MHz and came to be referred to as such in the early days of radio because the wavelengths associated with this frequency range were shorter than those commonly in use at that time. An alternate name is HF or high frequency radio. Short wavelengths are associated with high frequencies because there is an inverse relationship between frequency and wavelength. Shortwave frequencies are capable of reaching the other side of the Earth, because these waves can be refracted by the ionosphere, by a phenomenon known as Skywave propagation. High-frequency propagation is dependent upon a number of different factors, such as season of the year, solar conditions, including the number of sunspots, solar flares, and overall solar activity. Solar flares can prevent the ionosphere from reflecting or refracting radio waves.

Another factor which determines radio propagation is the time of the day; this is due to a particular transient atmosphere ionized layer forming only during day when atoms are broken up into ions by sun photons. This layer is responsible for partial or total absorption of particular frequencies. During the day, higher shortwave frequencies (i.e., above 10 MHz) can travel longer distances than lower ones; at night, this property is reversed.

Different types of modulation techniques are used on the shortwave frequencies in addition to AM and FM. AM, amplitude modulation, is generally used for shortwave broadcasting, and some aeronautical communications, while Narrow-band frequency modulation (NFM) is used at the higher HF frequencies. Single sideband or (SSB), is used for long-range communications by ships and aircraft, for voice transmissions by amateur radio operators. CW, Continuous Carrier Wave or (CW), is used for Morse code communications. Various other types of digital communications such as radioteletype, fax, digital, SSTV and other systems require special hardware and software to decode. A new broadcasting technique called Digital Radio Mondiale or (DRM) is a digital modulation scheme used on bands below 30 MHz.

Shortwave listening

Many hobbyists listen to shortwave broadcasters and for some listeners the goal is to hear as many stations from as many countries as possible (DXing); others listen to specialized shortwave utility, or "UTE," transmissions such as maritime, naval, aviation, or military signals. Others focus on intelligence signals. Many, though, tune the shortwave bands for the program content of shortwave broadcast stations, aimed to a general audience (such as the Voice of America, BBC World Service, Radio Australia, etc.). Some even listen to two-way communications by amateur radio operators. Nowadays, as the Internet evolves, the hobbyist can listen to shortwave signals via remotely controlled shortwave receivers around the world, even without owning a shortwave radio (see for example http://www.dxtuners.com). Alternatively, many international broadcasters (such as the BBC) offer live streaming audio on their web-sites. Table 1-2, lists some of the popular shortwave broadcast bands.

Shortwave listeners, or SWLs, can obtain QSL cards from broadcasters, utility stations or amateur radio operators as trophies of the hobby. Some stations even give out special certificates, pennants, stickers and other tokens and promotional materials to shortwave listeners.

Major users of the shortwave radio bands include domestic broadcasting in countries with a widely dispersed population with few long-wave, medium-wave, or FM stations serving them. International broadcasting stations beamed radio broadcasts to foreign audiences.

Table 1-2
Shortwave broadcast chart

Band	Megahertz Band (MHz)	Kilohertz (KHz)
120 Meter	2.3–2.5 MHz	2300–2500 KHz
90 Meter	3.2–3.40 MHZ	3200–3400 KHz
75 Meter	3.90–4.00 MHZ	3900–4000 KHz
60 Meter	4.750–5.060 MHz	4750–5060 KHz
49 Meter	5.950–6.20 MHz	5950–6200 KHz
41 Meter	7.10–7.60 MHz	7100–7600 KHz
31 Meter	9.20–9.90 MHz	9500–9900 KHz
25 Meter	11.60–12.200 MHz	11600–12200 KHz
22 Meter	13.570–13.870 MHZ	13570–13870 KHz
19 Meter	15.10–15.800 MHz	15100–15800 KHz
16 Meter	17.480–17.900 MHz	17480–17900 KHz
13 Meter	21.450–21.850 MHz	21450–21850 KHz
11 Meter	25.60–26.100 MHz	25600–26100 KHz

Speciality political, religious, and conspiracy theory radio networks, individual commercial and non-commercial paid broadcasts for the north American and other markets. Utility stations transmitting messages not intended for a general public, such as aircraft flying between continents, encoded or ciphered diplomatic messages, weather reporting, or ships at sea. Amateur radio operators have rights to use many frequencies in the shortwave bands; you can hear their communications using different modulation techniques and even obtain a license to communicate in these bands yourself. Contact the Amateur Radio Relay League for more information. Table 1-3 illustrates the amateur radio frequencies and how they are divided between the different license classes. On the shortwave band you will also encounter time signal stations and number stations, thought to be spy stations operating on the shortwave bands.

Types of receivers

A radio signal is transmitted through the ionosphere and is picked up by the antenna in your radio receiver. The antenna is fed to an RF amplifier and usually an intermediate amplifier or IF amplifier and then on to a detector of some sort depending upon the type of receiver you are using. From the detector, the resultant audio signal is amplified and sent to a loudspeaker for listening. Figure 1-6 illustrates a block diagram of a typical AM radio receiver. The antenna is sent to the RF amplifier. The mixer is fed by both the local oscillator and the RF amplifier. The signal from the mixer is sent to a bandpass filter and then on to the first IF amplifier. The first IF amplifier is next sent to the detector and then on to the final audio amplifier stage which drives the speaker. The illustration depicted in Figure 1-7 shows a typical FM receiver block diagram. The antenna feeds the RF amplifier stage. Both the RF amplifier and local oscillator are fed into the mixer. The signal from the mixer is next sent to the IF amplifier stage. From the IF amplifier stage the signal is next sent to the FM demodulator section, which feeds the signal to the voltage amplifier and then the signal is fed to the final audio amplifier stage and on to the speaker. Note the feedback path between, i.e. the AGC or automatic frequency control from the FM demodulator back to

Table 1-3
US amateur radio bands

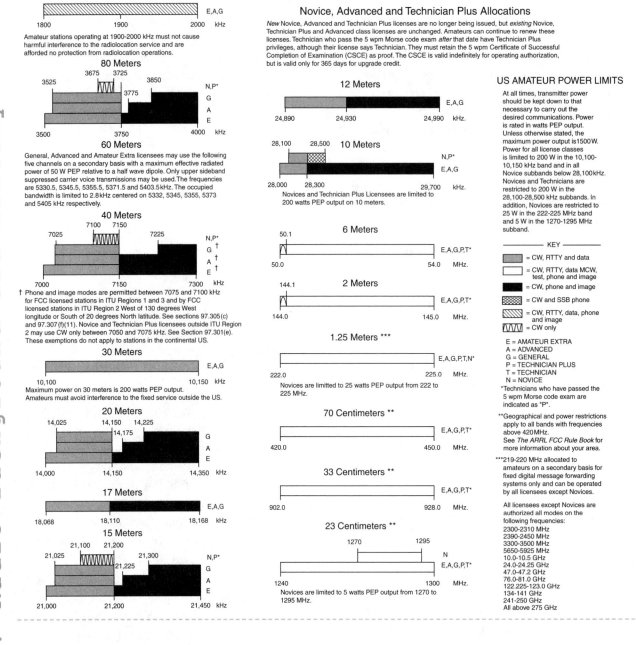

80 Meters

60 Meters

General, Advanced and Amateur Extra licensees may use the following five channels on a secondary basis with a maximum effective radiated power of 50 W PEP relative to a half wave dipole. Only upper sideband suppressed carrier voice transmissions may be used. The frequencies are 5330.5, 5345.5, 5355.5, 5371.5 and 5403.5 kHz. The occupied bandwidth is limited to 2.8 kHz centered on 5332, 5345, 5355, 5373 and 5405 kHz respectively.

40 Meters

† Phone and image modes are permitted between 7075 and 7100 kHz for FCC licensed stations in ITU Regions 1 and 3 and by FCC licensed stations in ITU Region 2 West of 130 degrees West longitude or South of 20 degrees North latitude. See sections 97.305(c) and 97.307(f)(11). Novice and Technician Plus licensees outside ITU Region 2 may use CW only between 7050 and 7075 kHz. See Section 97.301(e). These exemptions do not apply to stations in the continental US.

30 Meters

Maximum power on 30 meters is 200 watts PEP output. Amateurs must avoid interference to the fixed service outside the US.

20 Meters

17 Meters

15 Meters

Amateur stations operating at 1900-2000 kHz must not cause harmful interference to the radiolocation service and are afforded no protection from radiolocation operations.

Novice, Advanced and Technician Plus Allocations

New Novice, Advanced and Technician Plus licenses are no longer being issued, but *existing* Novice, Technician Plus and Advanced class licenses are unchanged. Amateurs can continue to renew these licenses. Technician who pass the 5 wpm Morse code exam *after* that date have Technician Plus privileges, although their license says Technician. They must retain the 5 wpm Certificate of Successful Completion of Examination (CSCE) as proof. The CSCE is valid indefinitely for operating authorization, but is valid only for 365 days for upgrade credit.

12 Meters

10 Meters

Novices and Technician Plus Licensees are limited to 200 watts PEP output on 10 meters.

6 Meters

2 Meters

1.25 Meters ***

Novices are limitted to 25 watts PEP output from 222 to 225 MHz.

70 Centimeters **

33 Centimeters **

23 Centimeters **

Novices are limited to 5 watts PEP output from 1270 to 1295 MHz.

US AMATEUR POWER LIMITS

At all times, transmitter power should be kept down to that necessary to carry out the desired communications. Power is rated in watts PEP output. Unless otherwise stated, the maximum power output is 1500 W. Power for all license classes is limited to 200 W in the 10,100-10,150 kHz band and in all Novice subbands below 28,100 kHz. Novices and Technicians are restricted to 200 W in the 28,100-28,500 kHz subbands. In addition, Novices are restricted to 25 W in the 222-225 MHz band and 5 W in the 1270-1295 MHz subband.

— KEY —

= CW, RTTY and data

= CW, RTTY, data MCW, test, phone and image

= CW, phone and image

= CW and SSB phone

= CW, RTTY, data, phone and image

= CW only

E = AMATEUR EXTRA
A = ADVANCED
G = GENERAL
P = TECHNICIAN PLUS
T = TECHNICIAN
N = NOVICE

*Technicians who have passed the 5 wpm Morse code exam are indicated as "P".

**Geographical and power restrictions apply to all bands with frequencies above 420 MHz.
See *The ARRL FCC Rule Book* for more information about your area.

***219-220 MHz allocated to amateurs on a secondary basis for fixed digital message forwarding systems only and can be operated by all licensees except Novices.

All licensees except Novices are authorized all modes on the following frequencies:
2300-2310 MHz
2390-2450 MHz
3300-3500 MHz
5650-5925 MHz
10.0-10.5 GHz
24.0-24.25 GHz
47.0-47.2 GHz
76.0-81.0 GHz
122.225-123.0 GHz
134-141 GHz
241-250 GHz
All above 275 GHz

the local oscillator. Finally, the shortwave radio block diagram is illustrated in the diagram in Figure 1-8. The antenna line is fed to the RF amplifier section. Both the local oscillator and the RF amplifier are fed to a filter section, which is in turn sent to the IF amplifier section. The output signal from the IF amplifier section is next sent to the product detector. A BFO or beat frequency oscillator signal is sent to the product detector, this is what permits SSB reception. The signal from the product detector is next sent to the audio amplifier and

then on to the speaker. The receivers shown are the most common types of receivers. There are in fact many different variations in receiver designs including receivers made to receive special digital signals, which we will not discuss here.

Next, we will move our discussion to identifying electronics components and reading schematics and learning how to solder before we forge ahead and begin building some fun radio receiver projects.

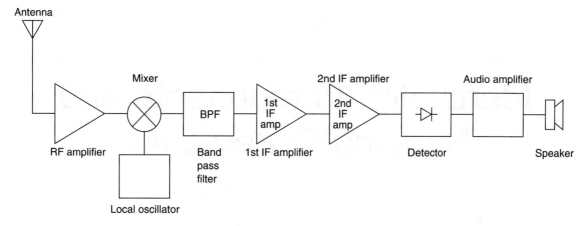

Figure 1-6 *AM radio block diagram*

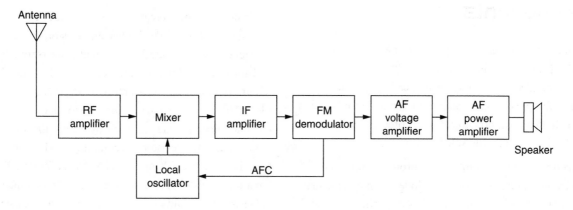

Figure 1-7 *FM radio block diagram*

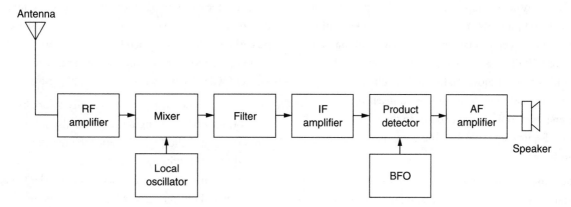

Figure 1-8 *SSB shortwave receiver block diagram*

Chapter 2

Identifying Components and Reading Schematics

Identifying electronic components

If you are a beginner to electronics or radio, you may want to take a few minutes to learn a little about identifying electronic components, reading schematics, and installing electronic components on a circuit board. You will also learn how to solder, in order to make long-lasting and reliable solder joints.

Electronic circuits comprise electronic components such as resistors and capacitors, diodes, semiconductors and LEDs, etc. Each component has a specific purpose that it accomplishes in a particular circuit. In order to understand and construct electronic circuits it is necessary to be familiar with the different types of components, and how they are used. You should also know how to read resistor and capacitor color codes, recognize physical components and their representative diagrams and pin-outs. You will also want to know the difference between a schematic and a pictorial diagram. First, we will discuss the actual components and their functions and then move on to reading schematics, then we will help you to learn how to insert the components into the circuit board. In the next chapter we will discuss how to solder the components to the circuit board.

The diagrams shown in Figures 2-1, 2-2 and 2-3 illustrate many of the electronic components that we will be using in the projects presented in this book.

Types of resistors

Resistors are used to regulate the amount of current flowing in a circuit. The higher the resistor's value or resistance, the less current flows and conversely a lower resistor value will permit more current to flow in a circuit. Resistors are measured in ohms (Ω) and are identified by color bands on the resistor body. The first band at one end is the resistor's first digit, the second color band is the resistor's second digit and the third band is the resistors's multiplier value. A fourth color band on a resistor represents the resistor's tolerance value. A silver band denotes a 10% tolerance resistor, while a gold band denotes a 5% resistor tolerance. No fourth band denotes that a resistor has a 20% tolerance. As an example, a resistor with a brown, black, and red band will represent the digit (1), the digit (0), with a multiplier value of (00) or 1000, so the resistor will have a value of 1k or 1000 ohms. There are a number of different styles and sizes of resistor. Small resistors can be carbon, thin film or metal. Larger resistors are made to dissipate more power and they generally have an element wound from wire.

A potentiometer (or pot) is basically a variable resistor, generally having three terminals and fitted with a rotary control shaft which varies the resistance as it is rotated. A metal wiping contact rests against a circular carbon or wire wound resistance track. As the wiper arm moves about the circular resistance, the resistance to the output terminals changes. Potentiometers are

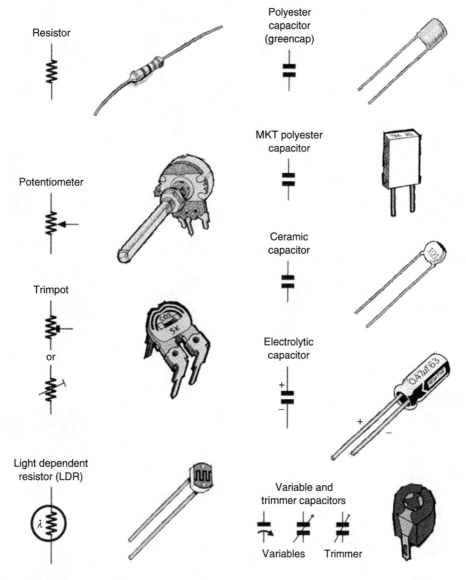

Figure 2-1   *Electronic components 1*

commonly used as volume controls in amplifiers and radio receivers.

A trimpot is a special type of potentiometer which, while variable, is intended to be adjusted once or only occasionally. For this reason a control shaft is not provided but a small slot is provided in the center of the control arm. Trimpots are generally used on printed circuit boards.

A light-dependent resistor (LDR) is a special type of resistor that varies its resistance value according to the amount of light falling on it. When it is in the dark, an LDR will typically have a very high resistance,

i.e. millions of ohms. When light falls on the LDR the resistance drops to a few hundred ohms.

Types of capacitors

Capacitors block DC current while allowing varying or AC current signals to pass. They are commonly used for coupling signals from part of a circuit to another part of a circuit, they are also used in timing circuits. There are a number of different types of capacitor as described below.

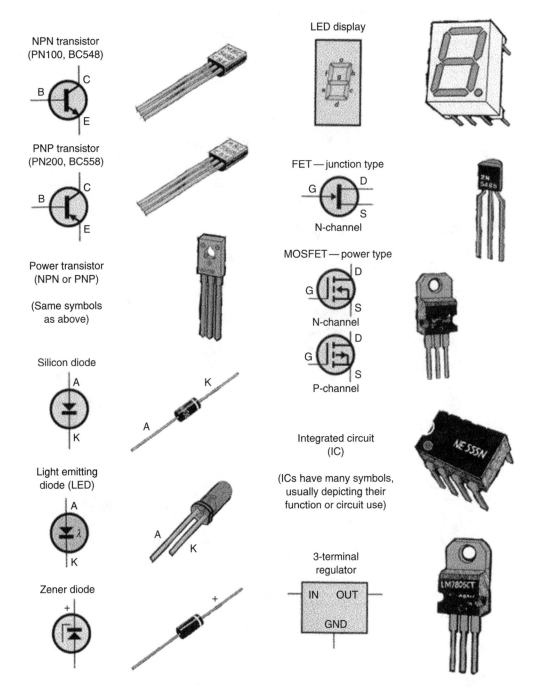

NPN transistor
(PN100, BC548)

PNP transistor
(PN200, BC558)

Power transistor
(NPN or PNP)

(Same symbols
as above)

Silicon diode

Light emitting
diode (LED)

Zener diode

LED display

FET — junction type
N-channel

MOSFET — power type
N-channel

P-channel

Integrated circuit
(IC)

(ICs have many symbols,
usually depicting their
function or circuit use)

3-terminal
regulator

Figure 2-2 *Electronic components 2*

Polyester capacitors use polyester plastic film as their insulating dielectric. Some polyester capacitors are called greencaps because they are coated with a green or brown color coating on the outside of the component. Their values are specified in microfarads or (μF), nanofarads, (nF), or picofarads (pF) and range from 1 nF up to about 10 μF. These capacitors do not have polarity and have fixed values.

MKT capacitors are another type of capacitor, but they are rectangular or (block) in shape and are usually yellow in color. One of the major advantages of these capacitors is a more standardized lead spacing, making them more useful for PC board projects. The components can generally be substituted for polyester types.

Ceramic capacitors use a tiny disk of ceramic or porcelain material in their construction for a dielectric

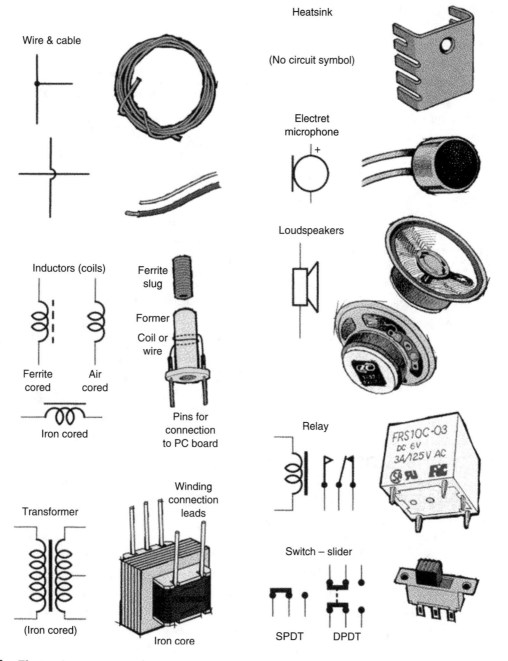

Figure 2-3 *Electronic components 3*

and they range in value from 1 pF up to 2.2 μF. Those with values above 1 nF are often made with multiple layers of metal electrodes and dielectric, to allow higher capacitance values in smaller bodies. These capacitors are usually called 'multilayer monolithics' and distinguished from lower value disk ceramic types. Ceramic capacitors are often used in RF radio circuits and filter circuits.

Electrolytic capacitors use very thin film of metal oxide as their dielectric, which allows them to provide a large amount of capacitance in a very small volume. They range in value from 100 nF up to hundreds and thousands of microfarads (μF). They are commonly used to filter power supply circuits, coupling audio circuits and in timing circuits. Electrolytic capacitors have polarity and must be installed with respect to these polarity marking. The capacitor will have either a white or black band denoting polarity with a plus (+) or minus (−) marking next to the color band.

Variable capacitors are used in circuits for (trimming) or adjustment, i.e. for setting a frequency. A variable capacitor has one set of fixed plates and one set of plates which can be moved by turning a knob. The dielectric between the plates is usually a thin plastic film. Most variable capacitors have low values up to a few tens of picofarads (pF) and a few hundreds of microfarads for larger variable capacitors.

Diodes

A diode is a semiconductor device which can pass current in one direction only. In order for current to flow the anode (A) must be positive with respect to the cathode (K). In this condition, the diode is said to be forward biased and a voltage drop of about .6 volt appears across its terminals. If the anode is less than .6 volt positive with respect to the cathode, negligible current will flow and the diode behaves as an open circuit.

Types of transistors

Transistors are semiconductor devices that can be used either as electronic switches or to amplify signals. They have three leads, called the Collector, Base, and Emitter. A small current flowing between base and emitter (junction) causes a much larger current to flow between the emitter and collector (junction). There a two basic types of transistors, PNP and NPN styles.

Field Effect Transistors, or FETs, are a different type of transistor, which usually still have three terminals but work in a different way. Here the control element is the "gate" rather than the base, and it is the "gate" voltage which controls the current flowing in the "channel" between the other terminals—the "source" and the "drain." Like ordinary transistors FETs can be used either as electronic switches or as amplifiers; they also come in P-channel and N-channel types, and are available in small signal types as well as power FETs.

Power transistors are usually larger than the smaller signal type transistors. Power transistors are capable of handling larger currents and voltages. Often metal tabs and heatsinks are used to remove excess heat from the part.

These devices are usually bolted to the chassis and are used for amplifying RF or audio energy.

Integrated circuits

Integrated circuits, or ICs, contain all, or most, of the components necessary for a particular circuit function, in one package. Integrated circuits contain as few as 10 transistors or many millions of transistors, plus many resistors, diodes and other components. There are many shapes, styles and sizes of integrated circuits: in this book we will use the dual-in-line style IC, either 8, 14 or 16 pin devices.

Three-terminal regulators are special types of integrated circuits, which supply a regulated or constant and accurate voltage from output regardless (within limits) of the voltage applied to input. They are most often used in power supplies. Most regulators are designed to give specific output voltages, i.e. a 'LM7805" regulator provides a 5 volt output, but some IC regulators can provide adjustable output based on an external potentiometer which can vary the output voltage.

Heatsinks

Many electronic components generate heat when they are operating. Generally heatsinks are used on semiconductors like transistors to remove heat. Overheating can damage a particular component or the entire circuit. The heatsink cools the transistor and ensures a long circuit life by removing the excess heat from the circuit area.

Light-emitting diodes

Light-emitting diodes, or LEDs, are special diodes which have a plastic translucent body (usually clear, red, yellow, green or blue in color) and a small semiconductor element which emits light when the diode passes a small current. Unlike an incandescent lamp, an LED does not need to get hot to produce light. LEDs must always be forward biased to operate. Special LEDs can also produce infrared light.

LED displays consist of a number of LEDs together in a single package. The most common type has seven elongated LEDs arranged in an "8" pattern. By choosing which combinations of LEDs are lit, any number of digits from "0" through "9" can be displayed. Most of these "7-segment" displays also contain another small round LED which is used as a decimal point.

Types of inductors

Inductors or "coils" are basically a length of wire, wound into a cylindrical spiral (or layers of spirals) in order to increase their inductance. Inductance is the ability to store energy in a magnetic field. Many coils are wound on a former of insulating material, which may also have connection pins to act as the coil's terminals. The former may also be internally threaded to accept a small core or "slug" of ferrite, which can be adjusted in position relative to the coil itself to vary the coil inductance.

A transformer consists of a number of coils of windings of wire wound on a common former, which is also inside a core of iron alloy, ferrite of other magnetic material. When an alternating current is passed through one of the windings (primary), it produces an alternating magnetic field in the core and this in turn induces AC voltages in the other (secondary) windings. The voltages produced in the other winding depend on the number of turns in those windings, compared with the turns in the primary winding. If a secondary winding has fewer turns than the primary, it will produce a lower voltage, and be called a step-down transformer. If the secondary winding has more windings than the primary, then the transformer will produce a higher voltage and it will be a step-up transformer. Transformers can be used to change the voltage levels of AC power and they are available in many different sizes and power handling capabilities.

Microphones

A microphone converts audible sound waves into electrical signals which can be then amplified. In an electret microphone, the sound waves vibrate a circular

diaphragm made from very thin plastic material which has a permanent charge in it. Metal films coated on each side form a capacitor, which produces a very small AC voltage when the diaphragm vibrates. All electret microphones also contains FET which amplifies the very small AC signals. To power an FET amplifier, the microphone must be supplied with a small DC voltage.

Loudspeakers

A loudspeaker converts electrical signals into sound waves that we can hear. It has two terminals which go to a voice coil, attached to a circular cone made of either cardboard or thin plastic. When electrical signals are applied to the voice coil, its creates a varying magnetic field from a permanent magnet at the back of the speaker. As a result the cone vibrates in sympathy with the applied signal to produce sound waves.

Relays

Many electronic components are not capable of switching higher currents or voltages, so a device called a relay is used. A relay has a coil which forms an electromagnet, attracting a steel "armature" which itself pushes on one or more sets of switching contacts. When a current is passed through the coil to energize it, the moving contacts disconnect from one set of contacts to another, and when the coil is de-energized the contacts go back to their original position. In most cases, a relay needs a diode across the coil to prevent damage to the semiconductor driving the coil.

Switches

A switch is a device with one or more sets of switching contacts, which are used to control the flow of current in a circuit. The switch allows the contacts to be controlled by a physical actuator of some kind—such as a press-button toggle lever, rotary or knob, etc. As the name denotes, this type of switch has an actuator bar which slides back and forth between the various contact positions. In a single-pole, double throw, or "SPDT" slider switch, a moving contacte links the center contact to either of the two end contacts. In contrast, a double-pole double throw (DPDT) slider switch has two of these sets of contacts, with their moving contacts operating in tandem when the slider is actuated.

Wire

A wire is simply a length of metal conductor, usually made from copper since its conductivity is good, which means its resistance is low. When there is a risk of a wire touching another wire and causing a short, the copper wire is insulated or covered with a plastic coating which acts as an insulating material. Plain copper wire is not usually used since it will quickly oxidize or tarnish in the presence of air. A thin metal alloy coating is often applied to the copper wire; usually an alloy of tin or lead is used.

Single or multi-strand wire is covered in colored PVC plastic insulation and is used quite often in electronic applications to connect circuits or components together. This wire is often called "hooku" wire. On a circuit diagram, a solid dot indicates that the wires or PC board tracks are connected together or joined, while a "loop-over" indicates that they are not joined and must be insulated. A number of insulated wires enclosed in an outer jacket is called an electrical cable. Some electrical cables can have many insulated wires in them.

Semiconductor substitution

There are often times when building an electronic circuit, it is difficult or impossible to find or locate the original transistor or integrated circuit. There are a number of circuits shown in this book which feature transistors, SCRs, UJTs, and FETs that are specified but cannot easily be found. Where possible many of these foreign components are converted to substitute values, either with a direct replacement or close substitution. Many foreign parts can be easily converted directly to a commonly used transistor or component. Occasionally an outdated component has no direct common

replacement, so the closest specifications of that component are attempted. In some instances we have specified replacement components with substitution components from the NTE brand or replacements. Most of the components for the projects used in this book are quite common and easily located or substituted without difficulty.

When substituting components in the circuit, make sure that the pin-outs match the original components. Sometimes, for example, a transistor may have bottom view drawing, while the substituted value may have a drawing with a top view. Also be sure to check the pin-outs or the original components versus the replacement. As an example, some transistors will have EBC versus ECB pin-outs, so be sure to look closely at possible differences which may occur.

Reading electronic schematics

The heart of all radio communication devices, both transmitters and receivers, all revolve around some type of oscillator. In this section we will take a look at what is perhaps the most important part of any receiver, and that is the oscillator. Communication transmitters, receivers, frequency standards and synthesizers all use some type of oscillators circuits. Transmitters need oscillators for their exciters, while receivers most often use local oscillators to mix signals. In this section you will see how specific electronics components are utilized to form oscillator circuits. Let's examine a few of the more common types of oscillator designs and their building considerations. In this section, you will also tell the difference between a schematic diagram and a pictorial diagram. A schematic diagram illustrates the electronic symbols and how the components connect to one another, it is the circuit blue-print and pretty much universal among electronic enthusiasts. A pictorial diagram, on the other hand, is a "picture" of how the components might appear in an actual circuit on a circuit board of one type or another. Take a close look at the difference between the two types of diagram, and they will help you later when building actual circuits. Our first type of oscillator shown below is The Hartley.

Hartley oscillator

The Hartley RF oscillator, illustrated in Figure 2-4, is centered around the commonly available 2N4416A FET transistor. This general purpose VFO oscillator operates around 5100 kHz. The frequency determining components are L1, C1, a 10 pF trimmer and capacitors C2, C3, C4, C5 and C6. Note capacitor C6 is a 10-100 pF variable trimmer type. Capacitor C7 is to reduce the loading on the tuned circuit components. Its value can be small but be able to provide sufficient drive to the succeeding buffer amplifier stage. You can experiment using a small viable capacitor trimmer, such as a 5-25 pF.

The other components, such as the two resistors, silicon diode at D1 are standard types, nothing particularly special. The Zener diode at D2 is a 6.2 volt type. Capacitor C8 can be selected to give higher/lower output to the buffer amplifier. Smaller C6 values give lower output and conversely higher values give larger output.

In order to get the circuit to work properly, you need to have an inductive reactance for L1 of around about 180 ohms. At 5 MHz this works out at about 5.7 uH. The important consideration, is that the feedback point from the source of the JFET connects to about 25% of the windings of L1 from the ground end. An air cored inductor is shown in the diagram. It could be, for example, 18-19 turns of #20 gauge wire on a 25.4 mm (1″) diameter form spread evenly over a length of about 25.4 mm (1″). The tap would be at about 4½ turns. Alternatively, with degraded performance,

you could use a T50-6 toroid and wind say 37 turns of #24 wire (5.48 µH) tapping at 9 turns.

So to have the oscillator operate at around 5 MHz, we know the LC is 1013 and if L is say 5.7 µH then total C for resonance (just like LC Filters eh!) is about 177 pF. We want to be able to tune from 5000 to 5100 kHz, a tuning ratio of 1.02, which means a capacitance ratio of 1.04 (min to max).

Colpitts oscillators

Colpitts oscillators are similar to the shunt fed Hartley oscillator circuit except the Colpitts oscillator, instead of having a tapped inductor, utilizes two series capacitors in its LC circuit. With the Colpitts oscillator the connection between these two capacitors is used as the center tap for the circuit. A Colpitts oscillator circuit is shown at Figure 2-5, and you will see some similarities with the Hartley oscillator.

The simplest Colpitts oscillator to construct and get running is the "series tuned" version, more often referred to as the "Clapp Oscillator." Because there is no load on the inductor, a high "Q" circuit results with a high L/C ratio and of course much less circulating current. This aids drift reduction. Because larger inductances are required, stray inductances do not have as much impact as perhaps in other circuits.

The total capacitive reactance of the parallel combination of capacitors depicted as series tuning below the inductor in a series tuned Colpitts oscillator or "Clapp oscillator" should have a total reactance of

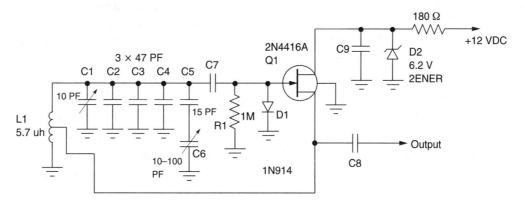

Figure 2-4 *Hartley oscillator circuit*

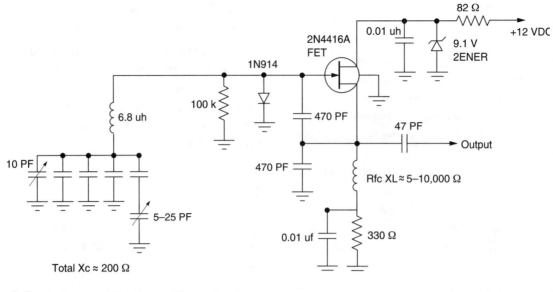

Figure 2-5 *Series tuned Colpitts oscillator circuit*

around 200 ohms. Not all capacitors may be required in your particular application. Effectively all the capacitors are in series in a Colpitts oscillator, i.e. they appear as parallel connected but their actual values are in fact in series.

Ideally, your frequency determining components L1 and the parallel capacitors should be in a grounded metal shield. The FET used in the Colpitts oscillator is the readily available 2N4416A. Note, the metal FET case is connected to the circuit ground. The output from the Colpitts oscillator is through output capacitor 47 pF; this should be the smallest of values possible, consistent with continued reliable operation into the next buffer amplifier stage.

Crystal oscillators

Crystal oscillators are oscillators where the primary frequency determining element is a quartz crystal. Because of the inherent characteristics of the quartz crystal the crystal oscillator may be held to extreme accuracy of frequency stability.

Crystal oscillators are, usually, fixed frequency oscillators where stability and accuracy are the primary considerations. For example, it is almost impossible to design a stable and accurate LC oscillator for the upper HF and higher frequencies without resorting to some sort of crystal control. Hence the reason for crystal oscillators.

The crystal oscillator depicted at Figure 2-6 is a typical example of an RF or radio frequency crystal oscillators which may be used for exciters or RF converters. Transistor Q1, can be any transistor which will operate up to 150 MHz, such as a 2N2222A.

The turns ratio on the tuned circuit depicts an anticipated nominal load of 50 ohms. This allows a theoretical 2.5k ohms on the collector, so typically a 7:1 turns ratio for T1 would be used. Use the: $L * C = 25330 / F0^2$ formula for determining L and C in the tuned circuits of crystal oscillator. Personally I would make L a reactance of around 250 ohms. In this case, I'd make C1 a smaller trimmer capacitor. The crystal at X1 could be an overtone type crystal for the crystal, selecting a $L * C$ for the odd particular multiple of overtone wanted in your crystal oscillators. Typically the output of the crystal oscillator would be followed by a buffer circuit. A pictorial diagram of the crystal controlled oscillator circuit is shown in Figure 2-7. Note that the components in this diagram are illustrated as component blocks as they might actually look placed on the circuit, rather than as electronic symbols in a circuit diagram or schematic.

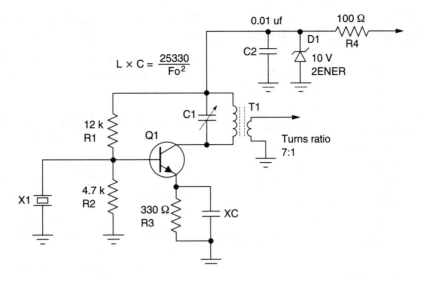

$$L \times C = \frac{25330}{Fo^2}$$

Figure 2-6 *Crystal oscillator circuit*

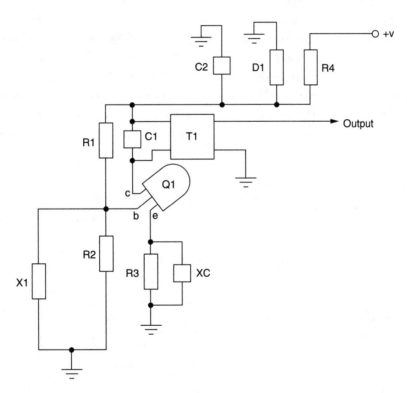

Figure 2-7 *Crystal oscillator pictorial diagram*

Ceramic resonator VFO oscillator

The ceramic resonator VFO, or variable frequency oscillator circuit shown in Figure 2-8, illustrates a 7MH oscillator with a variable crystal oscillator (VXO). The VXO oscillator is extremely stable, but allows only a small variation in frequency, as compared with a conventional VFO. In contrast, a VFO with an LC resonant circuit can be tuned over a range of several hundred kHz, but its frequency stability will depend upon its construction, and is never as good as a crystal oscillator. The use of a ceramic resonator as a frequency determining component fulfills both requirements. The VXO oscillator is very stable yet it can vary the frequency, so the oscillator can be tuned. The range of a VXO oscillator circuit is not as wide as an LC oscillator but it offers a tuning range of 35 kHz with good frequency stability. The somewhat unusual resonant LC circuit at the collector of Q1 has two functions. It improves the shape of the output signal and also compensates for the amplitude drop starting at approx. 7020 kHz. Transistor Q1 is a readily obtainable 2N3904 and the ceramic resonator is a Murata SFE 7.02 M2C type or equivalent. Inductor L1 consists of two coils on a T50-2 powered iron toroid. The primary coil is 8-turns, while the secondary coil is 2-turns.

The diagram shown in Figure 2-9 depicts the ceramic resonator as a pictorial diagram, where the components look as a components block that might be placed in a circuit rather than an actual schematic diagram.

Voltage controlled oscillator (VCO)

A voltage controlled oscillator, or VCO, is an oscillator where the main variable tuning capacitor is a varactor diode. The voltage controlled oscillator is tuned across the "band" by a well regulated Dc voltage applied to the varactor diode, which varies the net capacitance applied to the tuned circuit. The voltage controlled oscillator, shown in Figure 2-10, illustrates a VCO which operates in the amateur radio band between 1.8 and 2.0 MHz.

Buying quality variable capacitors today is often quite expensive, so VCOs are an extremely attractive alternative. As an alternative, all you need is an extremely stable and clean source of Dc power, a varactor diode and a high quality potentiometer—usually a 10 turn type. Note that circuit "Q" tends to be somewhat degraded by using varactor diodes instead of variable capacitors.

When a reverse voltage is applied to a diode, it exhibits the characteristics of a capacitor. Altering the voltage alters the capacitance. Common diodes such as

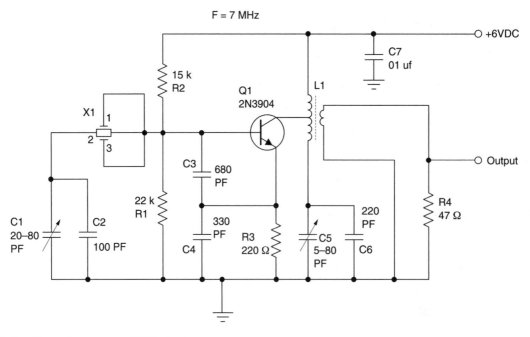

Figure 2-8 *Ceramic resonator VFO circuit*

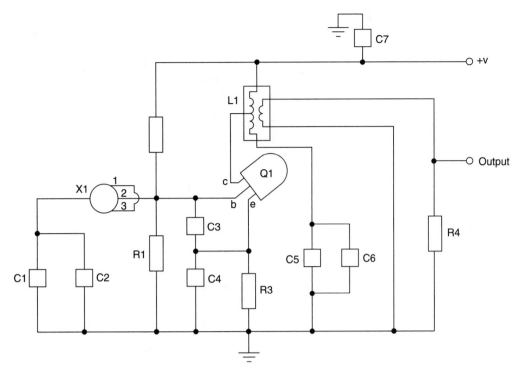

Figure 2-9 *Ceramic resonator VFO pictorial circuit diagram*

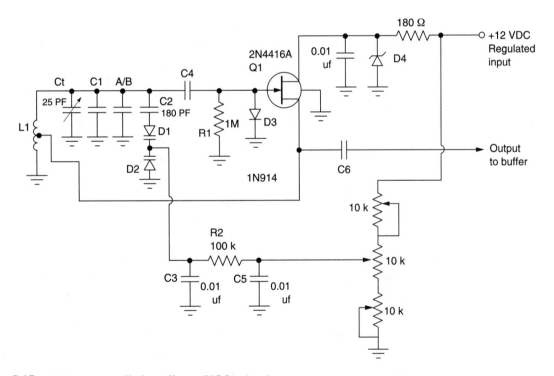

Figure 2-10 *Voltage controlled oscillator (VCO) circuit*

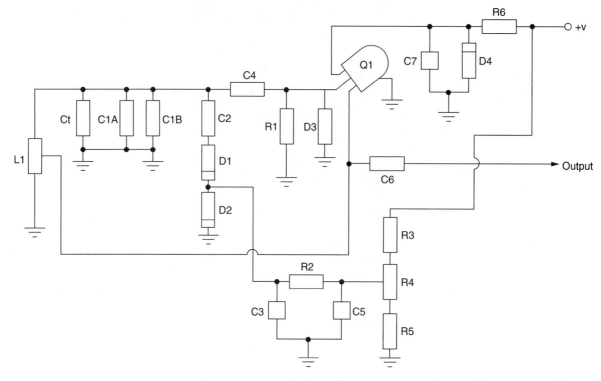

Figure 2-11 *Voltage controlled oscillator (VCO) pictorial diagram*

1N914 and 1N4004 can be used, but more commonly a special diode known as a varactor diode is specifically manufactured for VCO use, such as Motorola's MVAM115.

The variable capacitor in this circuit is replaced by a varactor diode as a tuning diode, in series with capacitor C2. Note, the two, diodes back-to-back in series. Although this in effect divides total varactor diode capacitance by two, it eliminates RF energy present in the tank circuit driving a single diode into conduction on peaks which will increase the bias voltage; this also gives rise to harmonics.

Inductor L is wound on a toroid, an Amidon T50-2 type which would require about 55 turns of #26 wire, or even the T68-2 type requiring about 51 turns of #24 wire. Both gauges mentioned are those which will conveniently fit around the core. The inductor should have a value of about 15 µH. Capacitor C2 should have around180 pF, and the total capacitance of C1 a/b should be 390 pF. C4 is used to reduce the loading of the tuned circuit components. Its value can be small but be able to provide sufficient drive to the succeeding

buffer amplifier stage. You can experiment using a small viable capacitor trimmer, such as a 5-25 pF.

The 10K potentiometer is a 10 turn "quality" potentiometer for tuning; the upper and lower trim pots (set and forget) allow you to adjust the voltage range of your choice that your tuning potentiometer will see. Use "quality" trimpots for these potentiometers, since they will affect the stability of the tuning of the circuit. The 100K resistor and the two 0.1 µF capacitors are further filtering. The diagram in Figure 2-11 depicts a pictorial diagram of the voltage controlled oscillator circuit. The circuit is shown as the components might appear rather than as symbols in a schematic diagram. The pictorial diagram often reflects an actual layout or parts placement and how the parts may appear on a circuit board.

Now that you can identify electronics components and you understand how they function in an electronics circuit, as well how electronics circuits are drawn, we can move on to the how-to-soldering section. In the next chapter, you will learn how to install components and how to solder the components onto a circuit board.

Chapter 3

Electronic parts installation and soldering

Before we discuss mounting the components to the printed circuit board we should take a few minutes to discuss the resistor and capacitor codes which are used to help identify and install these components. Table 3-1 lists the resistor color code information. Resistors generally have three or four color bands which help identify the resistor values. The color band will start from one edge of the resistor body; this is the first color code which represents the first digit, and the second color band depicts the second digit value, while the third color band is the resistor's multiplier value. The fourth color band represents the resistor tolerance value. If there is no fourth band then the resistor has a 20% tolerance; if the fourth band is silver, then the resistor has a 10% tolerance value, and if the fourth band is gold then the resistor has a 5% tolerance value. Therefore, if you have a resistor with the first band having a brown color, the second band with a black color and a third band with an orange color and a fourth with gold, then one (1) times zero (0) multiplied by a value of (1000), so the resistor will have a 10,000 ohm or 10k ohm value with a 5% tolerance.

Now let's move to identifying various capacitors. When identifying and installing the ceramic, mylar or poly capacitors, you will need to refer to Table 3-2, which lists capacitor (code) values vs. actual capacitor values. Most large capacitors, such as electrolytic types, will have their actual value printed on the body of the capacitor. But often with small capacitors they are just too small to have the actual capacitor values printed on them. An abbreviated code was devised to mark the capacitor, as shown in the table. A code marking of (104) would therefore denote a capacitor with a value of .1 µF (microfarads) or 100 nF (nanofarads). Once you understand how these component codes work, you can easily identify the resistors and capacitor and proceed to mount them on the printed circuit board.

Before we begin building actual electronic radio receiver circuits, you will need to understand how to prepare the electronic components before installing them onto the circuit board. First, you will need to know how to "dress" or prepare resistor leads, which is shown in Figure 3-1. Resistor leads are first bent to fit the PC board component holes, then the resistor is mounted to the circuit board. Figure 3-2 depicts how capacitor leads are bent to fit the PC board holes. Once, the capacitor leads are prepared, the capacitor can be installed onto the circuit board. The diagram shown in Figure 3-3 illustrates how to prepare diode leads before mounting them to the PB board. Finally, Figure 3-4 depicts transistor lead preparation, while Figure 3-5 shows how integrated circuit leads are prepared for installation.

Learning how to solder

Everyone working in electronics needs to know how to solder well. Before you begin working on a circuit, carefully read this chapter on soldering. In this section you will learn how to make good solder joints when soldering point-to-point wiring connections as well as for PC board soldering connections.

In all electronics work, the wiring connections must be absolutely secure. A loose connection in a radio results in noise, scratching sounds, or no sound at all. In a TV, poor connections can disrupt the sound or picture. The safe operation of airplanes and the lives of astronauts in flight depend on secure electronics connections.

Table 3-1

Resistor color code chart

Color Band	1st Digit	2nd Digit	Multiplier	Tolerance
Black	0	0	1	
Brown	1	1	10	1%
Red	2	2	100	2%
Orange	3	3	1,000 (K)	3%
Yellow	4	4	10,000	4%
Green	5	5	100,000	
Blue	6	6	1,000,000 (M)	
Violet	7	7	10,000,000	
Gray	8	8	100,000,000	
White	9	9	1,000,000,000	
Gold			0.1	5%
Silver			0.01	10%
No color				20%

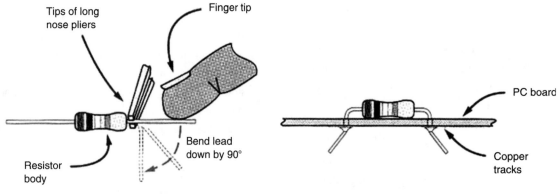

(a) Grip leads with pliers near body and bend down free ends

(c) Leads passed through PC board holes, bent and soldered to pads

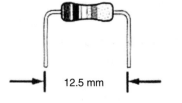

(b) Leads bent down and parallel spaced to match board holes

(d) Excess leads trimmed off with sidecutters

Figure 3-1 *Preparing resistor leads for horizontal mounting*

Table 3-2

Capacitor code identification information

This table is designed to provide the value of alphanumeric coded ceramic, mylar and mica capacitors in general. They come in many sizes, shapes, values and ratings; many different manufacturers worldwide produce them and not all play by the same rules. Most capacitors actually have the numeric values stamped on them; however, some are color coded and some have alphanumeric codes. The capacitor's first and second significant number IDs are the first and second values, followed by the multiplier number code, followed by the percentage tolerance letter code. Usually the first two digits of the code represent the significant part of the value, while the third digit, called the multiplier, corresponds to the number of zeros to be added to the first two digits.

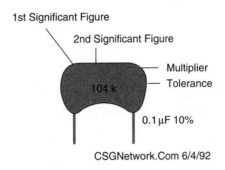

CSGNetwork.Com 6/4/92

Value	Type	Code	Value	Type	Code
1.5 pF	Ceramic		1,000 pF /.001 µF	Ceramic / Mylar	102
3.3 pF	Ceramic		1,500 pF /.0015 µF	Ceramic / Mylar	152
10 pF	Ceramic		2,000 pF /.002 µF	Ceramic / Mylar	202
15 pF	Ceramic		2,200 pF /.0022 µF	Ceramic / Mylar	222
20 pF	Ceramic		4,700 pF /.0047 µF	Ceramic / Mylar	472
30 pF	Ceramic		5,000 pF /.005 µF	Ceramic / Mylar	502
33 pF	Ceramic		5,600 pF /.0056 µF	Ceramic / Mylar	562
47 pF	Ceramic		6,800 pF /.0068 µF	Ceramic / Mylar	682
56 pF	Ceramic		.01	Ceramic / Mylar	103
68 pF	Ceramic		.015	Mylar	
75 pF	Ceramic		.02	Mylar	203
82 pF	Ceramic		.022	Mylar	223
91 pF	Ceramic		.033	Mylar	333
100 pF	Ceramic	101	.047	Mylar	473
120 pF	Ceramic	121	.05	Mylar	503
130 pF	Ceramic	131	.056	Mylar	563
150 pF	Ceramic	151	.068	Mylar	683
180 pF	Ceramic	181	.1	Mylar	104
220 pF	Ceramic	221	.2	Mylar	204
330 pF	Ceramic	331	.22	Mylar	224
470 pF	Ceramic	471	.33	Mylar	334
560 pF	Ceramic	561	.47	Mylar	474
680 pF	Ceramic	681	.56	Mylar	564
750 pF	Ceramic	751	1	Mylar	105
820 pF	Ceramic	821	2	Mylar	205

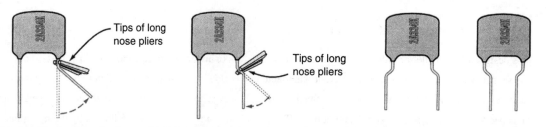

(a) Grip leads with pliers near body and bend out or in by 45°

(b) Grip now about 2 mm down and bend down by 45°, parallel again

(c) Both leads cranked in (or out) if necessary to match PC board hole spacing

Figure 3-2 *"Dressing" capacitor component leads*

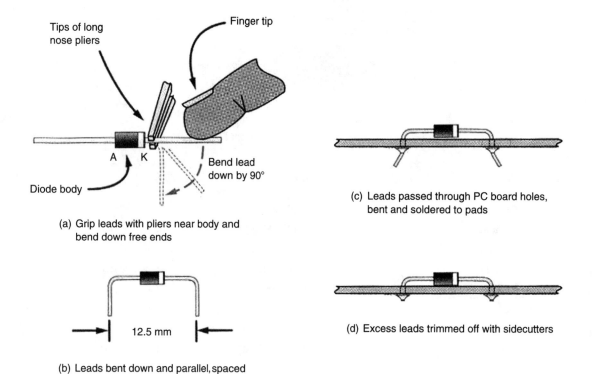

(a) Grip leads with pliers near body and bend down free ends

12.5 mm

(b) Leads bent down and parallel, spaced to match board holes

(c) Leads passed through PC board holes, bent and soldered to pads

(d) Excess leads trimmed off with sidecutters

Figure 3-3 *Preparing diode leads for horizontal mounting*

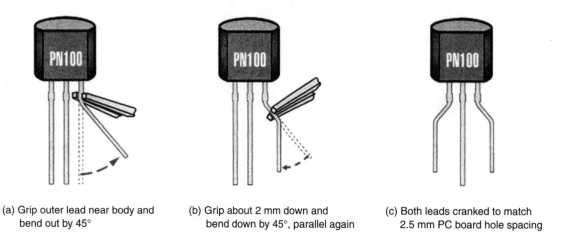

(a) Grip outer lead near body and bend out by 45°

(b) Grip about 2 mm down and bend down by 45°, parallel again

(c) Both leads cranked to match 2.5 mm PC board hole spacing

Figure 3-4 *"Dressing" TO-92 transistor leads*

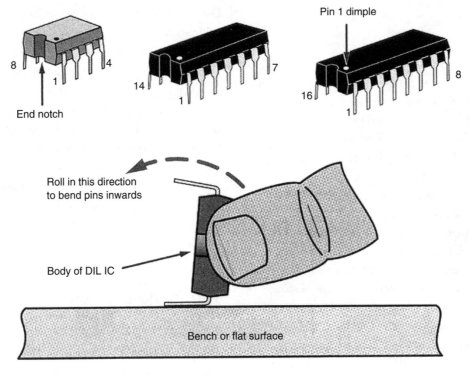

Figure 3-5 *DIL IC packages and preparing them for insertion*

Soldering joins two pieces of metal, such as electrical wires, by melting them together with another metal to form a strong, chemical bond. Done correctly, it unites the metals so that electrically they act as one piece of metal. Soldering is not just gluing metals together. Soldering is tricky and intimidating in practice, but easy to understand in theory. Basic supplies include a soldering iron, which is a prong of metal that heats to a specific temperature through electricity, like a regular iron. The solder, or soldering wire, often an alloy of aluminum and lead, needs a lower melting point than the metal you're joining. Finally, you need a cleaning resin called flux that ensures the joining pieces are incredibly clean. Flux removes all the oxides on the surface of the metal that would interfere with the molecular bonding, allowing the solder to flow into the joint smoothly. You also need two things to solder together.

The first step in soldering is cleaning the surfaces, initially with sandpaper or steel wool and then by melting flux onto the parts. Sometimes, flux is part of the alloy of the soldering wire, in an easy-to-use mixture. Then, the pieces are both heated above the melting point of the solder (but below their own melting point) with the soldering iron. When touched to the joint, this precise heating causes the solder to "flow" to the place of highest temperature and makes a chemical bond. The solder shouldn't drip or blob, but spread smoothly, coating the entire joint. When the solder cools, you should have a clean, sturdy connection.

Many people use soldering in their field, from electrical engineering and plumbing to jewelry and crafts. In a delicate procedure, a special material, called solder, flows over two pre-heated pieces and attaches them through a process similar to welding or brazing. Various metals can be soldered together, such as gold and sterling silver in jewelry, brass in watches and clocks, copper in water pipes, or iron in leaded glass stained windows. All these metals have different melting points, and therefore use different solder. Some "soft" solder, with a low melting point, is perfect for wiring a circuit board. Other "hard" solder, such as for making a bracelet, needs a torch rather than a soldering iron to get a hot enough temperature. Electrical engineers and hobbyists alike can benefit from learning the art and science of soldering.

Solder

The best solder for electronics work is 60/40 rosin-core solder. It is made of 60% tin and 40% lead. This mixture melts at a lower temperature than either lead or tin alone. It makes soldering easy and provides good connections. The rosin keeps the joint clean as it is being soldered. The heat of the iron often causes a tarnish or oxide to form on the surface. The rosin dissolves the tarnish to make the solder cling tightly. Solders have different melting points, depending on the ratio of tin to lead. Tin melts at 450°F and lead at 621°F. Solder made from 63% tin and 37% lead melts at 361°F, the lowest melting point for a tin and lead mixture. Called 63-37 (or eutectic), this type of solder also provides the most rapid solid-to-liquid transition and the best stress resistance. Solders made with different lead/tin ratios have a plastic state at some temperatures. If the solder is deformed while it is in the plastic state, the deformation remains when the solder freezes into the solid state. Any stress or motion applied to "plastic solder" causes a poor solder joint.

The 60-40 solder has the best wetting qualities. Wetting is the ability to spread rapidly and bond materials uniformly; 60-40 solder also has a low melting point. These factors make it the most commonly used solder in electronics.

Some connections that carry high current can't be made with ordinary tin-lead solder because the heat generated by the current would melt the solder. Automotive starter brushes and transmitter tank circuits are two examples. Silver-bearing solders have higher melting points, and so prevent this problem. High-temperature silver alloys become liquid in the 1100°F to 1200°F range, and a silver-manganese (85-15) alloy requires almost 1800°F.

Because silver dissolves easily in tin, tin bearing solders can leach silver plating from components. This problem can greatly reduced by partially saturating the tin in the solder with silver or by eliminating the tin. Tin-silver or tin-lead-silver alloys become liquid at temperatures from 430°F for 96.5-3.5 (tin-silver), to 588°F for 1.0-97.5-1.5 (tin-lead-silver). A 15.080.0-5.0 alloy of lead-indium-silver melts at 314°F.

Never use acid-core solder for electrical work. It should be used only for plumbing or chassis work.

For circuit construction, only use fluxes or solder-flux combinations that are labeled for electronic soldering.

The rosin or the acid is a flux. Flux removes oxide by suspending it in solution and floating it to the top. Flux is not a cleaning agent! Always clean the work before soldering. Flux is not a part of a soldered connection— it merely aids the soldering process. After soldering, remove any remaining flux. Rosin flux can be removed with isopropyl or denatured alcohol. A cotton swab is a good tool for applying the alcohol and scrubbing the excess flux away. Commercial flux-removal sprays are available at most electronic-part distributors.

The soldering iron

Soldering is used in nearly every phase of electronic construction so you'll need soldering tools. A soldering tool must be hot enough to do the job and lightweight enough for agility and comfort. A temperature controlled iron works well, although the cost is not justified for occasional projects. Get an iron with a small conical or chisel tip. Soldering is not like gluing; solder does more than bind metal together and provide an electrically conductive path between them. Soldered metals and the solder combine to form an alloy.

You may need an assortment of soldering irons to do a wide variety of soldering tasks. They range in size from a small 25-watt iron for delicate printed-circuit work to larger 100 to 300-watt sizes used to solder large surfaces. If you could only afford a single soldering tool when initially setting up your electronics workbench than, an inexpensive to moderately priced pencil-type soldering iron with between 25 and 40-watt capacity is the best for PC board electronics work. A 100-watt soldering gun is overkill for printed-circuit work, since it often gets too hot, cooking solder into a brittle mess or damaging small parts of a circuit. Soldering guns are best used for point-to-point soldering jobs, for large mass soldering joints or large components. Small "pencil" butane torches are also available, with optional soldering-iron tips. A small butane torch is available from the Solder-It Company. Butane soldering irons are ideal for field service problems and will allow you to solder where there is no 110 volt power source. This company also sells a soldering kit that contains paste solders (in syringes) for electronics, pot metal and plumbing. See the Appendix for the address information.

Keep soldering tools in good condition by keeping the tips well tinned with solder. Do not run them at full temperature for long periods when not in use. After each period of use, remove the tip and clean off any scale that may have accumulated. Clean an oxidized tip by dipping the hot tip in sal ammoniac (ammonium chloride) and then wiping it clean with a rag. Sal ammoniac is somewhat corrosive, so if you don't wipe the tip thoroughly, it can contaminate electronic soldering. You can also purchase a small jar of "Tip Tinner," a soldering iron tip dresser, from your local Radio Shack store. Place the tip of the soldering iron into the "Tip Tinner" after every few solder joints.

If a copper tip becomes pitted, file it smooth and bright and then tin it immediately with solder. Modern soldering iron tips are nickel or iron clad and should not be filed. The secret of good soldering is to use the right amount of heat. Many people who will have not soldered before use too little heat dabbing at the joint to be soldered and making little solder blobs that cause unintended short circuits. Always use caution when soldering. A hot soldering iron can burn your hand badly or ruin a tabletop. It's a good idea to buy or make a soldering iron holder.

Soldering station

Often when building or repairing a circuit, your soldering iron is kept switched "on" for unnecessarily long periods, consuming energy and allowing the soldering iron tip to burn and develop a buildup of oxide. Using this soldering-iron temperature controller, you will avoid destroying sensitive components when soldering.

Buying a lower wattage iron may solve some of the problems, but new problems arise when you want to solder some heavy-duty component, setting the stage for creating a "cold" connections. If you've ever tried to troubleshoot some instrument in which a cold solder joint was at the root of the problem, you know how difficult such defects are to locate. Therefore, the best way to satisfy all your needs is to buy a temperature controller electronics workbench.

A soldering station usually consists of a temperature controlled soldering iron with an adjustable heat or temperature control and a soldering iron holder and cleaning pad. If you are serious about your electronics hobby or if you have been involved with electronics building and repair for any length of time, you will eventually want to invest in a soldering station at some point in time. There are real low cost soldering station for hobbyists, for under $30, but it makes more sense to purchase a moderately price soldering station such as the quality Weller series. A typical soldering station is shown in Figure 3-6.

Soldering gun

An electronics workbench would not be complete without a soldering gun. Soldering guns are useful for soldering large components to terminal strips, or splicing wires together or when putting connectors on coax cable. There are many instances where more heat is needed than a soldering iron can supply. For example, a large connector mass cannot be heated with a small soldering iron, so you would never be able to "tin" a connector with a small wattage soldering iron. A soldering gun is a heavy duty soldering device which does in fact look like a gun. Numerous tips are available for a soldering gun and they are easily replaceable using two small nuts on the side arm of the soldering gun. Soldering guns are available in two main heat ranges.

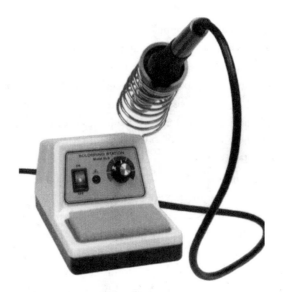

Figure 3-6 *Temperature controlled soldering station*

Most soldering guns have a two-step "trigger" switch which enable you to select two heat ranges for different soldering jobs. The most common soldering gun provides both a 100 watt setting when the "trigger" switch is pressed to its first setting, and as the "trigger" switch is advanced to the next step, the soldering gun will provide 150 watts, when more heat is needed. A larger or heavy-duty soldering gun is also available, but a little harder to locate is the 200 to 250 watt solder gun. The first "trigger" switch position provides 200 watts, while the second switch position provides the 250 watt heat setting. When splicing wires together either using the "Western Union" or parallel splice or the end splice, a soldering gun should be used especially if the wire gauge is below size 22 ga. Otherwise the solder may not melt properly and the connections may reflect a "cold" solder joint and therefore a poor or noisy splice. Soldering wires to binding post connections should be performed with a soldering gun to ensure proper heating to the connection. Most larger connectors should be soldered or pre-tinned using a soldering gun for even solder flow.

Preparing the soldering iron

If your iron is new, read the instructions about preparing it for use. If there are no instructions, use the following procedure. It should be hot enough to melt solder applied to its tip quickly (half a second when dry, instantly when wet with solder). Apply a little solder directly to the tip so that the surface is shiny. This process is called "tinning" the tool. The solder coating helps conduct heat from the tip to the joint face, the tip is in contact with one side of the joint. If you can place the tip on the underside of the joint, do so. With the tool below the joint, convection helps transfer heat to the joint. Place the solder against the joint directly opposite the soldering tool. It should melt within a second for normal PC connections, within two seconds for most other connections. If it takes longer to melt, there is not enough heat for the job at hand. Keep the tool against the joint until the solder flows freely throughout the joint. When it flows freely, solder tends to form concave shapes between the conductors. With insufficient heat solder does not flow freely; it forms convex shapes-blobs. Once solder shape changes from convex to concave, remove the tool from the joint. Let the joint

cool without movement at room temperature. It usually takes no more than a few seconds. If the joint is moved before it is cool, it may take on a dull, satin look that is characteristic of a "cold" solder joint. Reheat cold joints until the solder flows freely and hold them still until cool. When the iron is set aside, or if it loses its shiny appearance, wipe away any dirt with a wet cloth or sponge. If it remains dull after cleaning, tin it again.

Overheating a transistor or diode while soldering can cause permanent damage. Use a small heatsink when you solder transistors, diodes or components with plastic parts that can melt. Grip the component lead with a pair of pliers up close to the unit so that the heat is conducted away. You will need to be careful, since it is easy to damage delicate component leads. A small alligator clip also makes a good heatsink to dissipate from the component.

Mechanical stress can damage components, too. Mount components so there is no appreciable mechanical strain on the leads.

Soldering to the pins of coil forms male cable plugs can be difficult. Use suitable small twist drill to clean the inside of the pin and then tin it with resin-core solder. While it is still liquid, clear the surplus solder from each pin with a whipping motion or by blowing through the pin from the inside of the form or plug. Watch out for flying hot solder, you can get severe burns. Next, file the nickel tip. Then insert the wire and solder it. After soldering, remove excess solder with a file, if necessary. When soldering to the pins of plastic coil-forms, hold the pin to be soldered with a pair of heavy pliers to form a heatsink. Do not allow the pin to overheat; it will loosen and become misaligned.

Preparing work for soldering

If you use old junk parts, be sure to completely clean all wires or surfaces before applying solder. Remove all enamel, dirt, scale, or oxidation by sanding or scraping the parts down to bare metal. Use fine sandpaper or emery paper to clean flat surfaces or wire. (Note, no amount of cleaning will allow you to solder to aluminum. When making a connection to a sheet of aluminum, you must connect the wire by a solder lug or a screw.)

When preparing wires, remove the insulation with wire strippers or a pocketknife. If using a knife, do not cut straight into the insulation; you might nick the wire and weaken it. Instead, hold the knife as if you were sharpening a pencil, taking care not to nick the wire as you remove the insulation. For enameled wire, use the back of the knife blade to scrape the wire until it is clean and bright. Next, tin the clean end of the wire. Now, hold the heated soldering-iron tip against the under surface of the wire and place the end of the rosin-core solder against the upper surface. As the solder melts, it flows on the clean end of the wire. Hold the hot tip of the soldering iron against the under surface of the tinned wire and remove the excess solder by letting it flow down on the tip. When properly tinned, the exposed surface of the wire should be covered with a thin, even coating of solder.

How to solder

The two key factors in quality soldering are time and temperature. Generally, rapid heating is desired, although most unsuccessful solder jobs fail because insufficient heat has been applied. Be careful; if heat is applied too long, the components or PC board can be damaged, the flux may be used up and surface oxidation can become a problem. The soldering-iron tips should be hot enough to readily melt the solder without burning, charring or discoloring components, PC boards or wires. Usually, a tip temperature about 100°F above the solder melting point is about right for mounting components on PC boards. Also, use solder that is sized appropriately for the job. As the cross-section of the solder decreases, so does the amount of heat required to melt it. Diameters from 0.025 to 0.040 inches are good for nearly all circuit wiring.

Always use a good quality multi-core solder. A standard 60% tin, 40% lead alloy solder with cores of non-corrosive flux will be found easiest to use. The flux contained in the longitudinal cores of multi-core solder is a chemical designed to clean the surfaces to be joined of deposited oxides, and to exclude air during the soldering process, which would otherwise prevent these metals coming together. Consequently, don't expect to be able to complete a joint by using the application of the tip of the iron loaded with molten solder alone, as this usually will not work. Having said that, there is a process called tinning where conductors are first coated in fresh, new solder prior to joining by a hot iron.

Solder comes in gauges like wire. The two most common types of solder, are 18 ga., used for general work, and the thinner 22 ga., used for fine work on printed circuit boards.

A well-soldered joint depends on

1. Soldering with a clean, well-tinned tip.
2. Cleaning the wires or parts to be soldered.
3. Making a good mechanical joint before soldering.
4. Allowing the joint to get hot enough before applying solder.
5. Allowing the solder to set before handling or moving soldered parts.

Making a good mechanical joint

Unless you are creating a temporary joint, the next step is to make a good mechanical connection between the parts to be soldered. For instance, wrap the wire carefully and tightly around a soldering terminal or soldering lug, as shown in Figure 3-7. Bend wire and make connections with long-nosed pliers. When connecting two wires together, make a tight splice before soldering. Once you have made a good mechanical contact, you are ready for the actual soldering.

The next step is to apply the soldering iron to the connection, soldering the connection as shown.

Figure 3-7   *Wire to lug soldering joint*

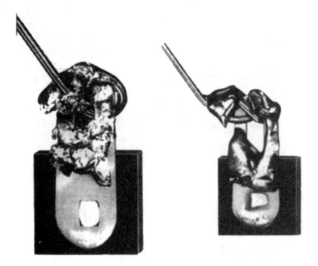

Figure 3-8 *Inferior wire to lug solder joint*

In soldering a wire splice, hold the iron below the splice and apply solder to the top of the splice. If the tip of the iron has a bit of melted solder on the side held against the splice, heat is transferred more readily to the splice and the soldering is done more easily. Don`t try to solder by applying solder to the joint and then pressing down on it with the iron. Be sure not to disturb the soldered joint until the solder has set. It may take a few seconds for the solder to set, depending upon the amount of solder used in making the joint. Now take a good look at the joint. It should have a shiny, smooth appearance—not pitted or grainy. If it does have a

pitted, granular appearance as seen in Figure 3-8. reheat the joint, scrape off the solder, and clean the connection. Then start over again. After the solder is well set, pull on the wire to see if it is a good, tight connection. If you find that you made a poor soldering job don't get upset, be thankful you found it and do it over again. A quick reference solder check list is shown in the listing in Table 3-3.

Soldering printed circuit boards

Most electronic devices use one or more printed circuit (PC) boards. A PC board is a thin sheet of fiberglass or phenolic resin that has a pattern of foil conductors "printed" on it. You insert component leads into holes in the board and solder the leads to the foil pattern. This method of assembly is widely used and you will probably encounter it if you choose to build from a kit. Printed circuit boards make assembly easy. First, insert component leads through the correct holes in the circuit board. Mount parts tightly against the circuit board unless otherwise directed. After inserting a lead into the board, bend it slightly outward to hold the part in place.

When the iron is hot, apply some solder to the flattened working end at the end of the bit, and wipe it on a piece of damp cloth or sponge so that the solder forms a thin film on the bit. This is tinning the bit.

Table 3-3
Soldering check list

1. Prepare the joint. Clean all surfaces and conductors thoroughly with fine steel wool. First, clean the circuit traces, then clean the component leads.

2. Prepare the soldering iron or gun. The soldering device should be hot enough to melt solder applied to the tip. Apply a small amount of solder directly to the tip, so that the surface is shiny.

3. Place the tip in contact with one side of the joint, if possible place the tip below the joint.

4. Place the solder against the joint directly opposite the soldering tool. The solder should melt within two seconds; if it takes longer use a larger iron.

5. Keep the soldering tool against the joint until the solder flows freely throughout the joint. When it flows freely the joint should form a concave shape; insufficient heat will form a convex shape.

6. Let the joint cool without any movement, the joint should cool and set-up with in a few seconds. If the joint is moved before it cools the joint will look dull instead of shiny and you will likely have a cold solder joint. Re-heat the joint and begin anew.

7. Once the iron is set aside, or if it loses its shiny appearance, wipe away any dirt with a wet cloth or wet sponge. When the iron is clean the tip should look clean and shiny. After cleaning the tip apply some solder.

Melt a little more solder on to the tip of the soldering iron, and put the tip so it contacts both parts of the joint. It is the molten solder on the tip of the iron that allows the heat to flow quickly from the iron into both parts of the joint. If the iron has the right amount of solder on it and is positioned correctly, then the two parts to be joined will reach the solder's melting temperature in a couple of seconds. Now apply the end of the solder to the point where both parts of the joint and the soldering iron are all touching one another. The solder will melt immediately and flow around all the parts that are at, or over, the melting part temperature. After a few seconds remove the iron from the joint. Make sure that no parts of the joint move after the soldering iron is removed until the solder is completely hard. This can take quite a few seconds with large joints. If the joint is disturbed during this cooling period it may become seriously weakened.

The most important point in soldering is that both parts of the joint to be made must be at the same temperature. The solder will flow evenly and make a good electrical and mechanical joint only if both parts of the joint are at an equal high temperature. Even though it appears that there is a metal-to-metal contact in a joint to be made, very often there exists a film of oxide on the surface that insulates the two parts. For this reason it is no good applying the soldering iron tip to one half of the joint only and expecting this to heat the other half of the joint as well.

It is important to use the right amount of solder, both on the iron and on the joint. Too little solder on the iron will result in poor heat transfer to the joint, too much and you will suffer from the solder forming strings as the iron is removed, causing splashes and bridges to other contacts. Too little solder applied to the joint will give the joint a half finished appearance: a good bond where the soldering iron has been, and no solder at all on the other part of the joint.

The hard cold solder on a properly made joint should have a smooth shiny appearance and if the wire is pulled it should not pull out of the joint. In a properly made joint the solder will bond the components very strongly indeed, since the process of soldering is similar to brazing, and to a lesser degree welding, in that the solder actually forms a molecular bond with the surfaces of the joint. Remember it is much more difficult to correct a poorly made joint than it is to make

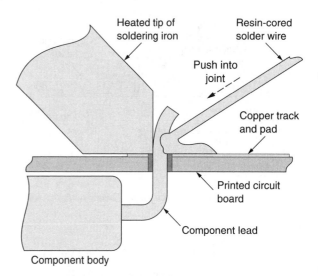

Figure 3-9 *Proper technique for soldering components on PC board*

the joint properly in the first place. Anyone can learn to solder, it just takes practice.

The diagram in Figure 3-9 shows how to solder a component lead to a PC board pad. The tip of the soldering iron heats both the lead and the copper pad, so the end of the solder wire melts when it's pushed into the contact. The diagram illustrated in Figure 3-10 shows how a good solder joint is obtained. Notice that it has a smooth and shiny "fillet" of solder metal, bonding all around to both the component lead and the copper pad of the PC board. This joint provides a reliable electrical connection.

Try to make the solder joint as quickly as possible because the longer you take, the higher the risk that the component itself and the printed circuit board pad and track will overheat and be damaged. But don't work so quickly that you cannot make a good solder joint.

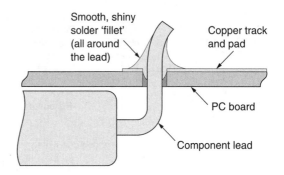

Figure 3-10 *Excellent solder joint*

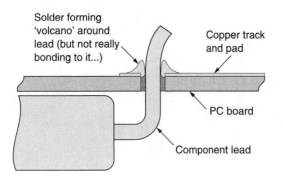

Figure 3-11   *Inferior solder joint: Example 1*

Having to solder the joint over again always increases the risk of applying too much heat to the PCB.

As the solder solidifies, take a careful look at the joint you have made, to make sure there's a smooth and fairly shiny metal "fillet" around it. This should be broadly concave in shape, showing that the solder has formed a good bond to both metal surfaces. If it has a rough and dull surface or just forms a "ball" on the component lead, or a "volcano" on the PCB pad with the lead emerging from the crater, you have a "dry joint." If your solder joint looks like the picture shown in Figure 3-11, you will have to re-solder the joint over again. Figure 3-12 shows another type of dry solder joint which would have to be re-soldered. These types of "dry" solder joints if now redone will cause the circuit to be unreliable and intermittent.

For projects that use one or more integrated circuits, with their leads closely-spaced pins, you may find it easier to use a finer gauge solder, i.e. less than 1 mm in diameter. This reduces the risk of applying too much solder to each joint, and accidentaly forming "bridges" between pads to PC "tracks."

The finished connection should be smooth and bright. Reheat any cloudy or grainy-looking connections. Finally, clip off the excess wire length, as shown in Figure 3-13.

Occasionally a solder "bridge" will form between two adjacent foil conductors. You must remove these bridges; otherwise a short circuit will exist between the two conductors. Remove a solder bridge by heating the bridge and quickly wiping away the melted solder with a soft cloth. Often you will find a hole on the board plugged by solder from a previous connection. Clear the hole by heating the solder while pushing a component lead through the hole from the other side of the board. Good soldering is a skill that is learnt by practice.

How to un-solder

In order to remove components, you need to learn the art of de-soldering. You might accidentally make a wrong connection or have to move a component that you put in an incorrect location. Take great care while un-soldering to avoid breaking or destroying good parts. The leads on components such as resistors or transistors and the lugs on other parts may sometimes break off when you are un-soldering a good, tight joint. To avoid heat damage, you must use as much care in un-soldering delicate parts as you do in soldering them. There are three basic ways of un-soldering. The first method is to heat the joint and "flick" the wet solder off. The second method is to use a metal wick or braid to remove the melted solder. This braid is available at most electronics

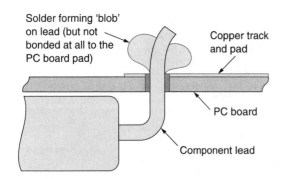

Figure 3-12   *Inferior solder joint: Example 2*

Figure 3-13   *Trimming excess component leads*

parts stores; use commercially made wicking material (braid). Place the braid against the joint that you want to un-solder. Use the heated soldering iron to gently press the braid against the joint. As the solder melts, it is pulled into the braid. By repeating this process, you can remove virtually all the solder from the joint. Then re-heat the joint and lift off the component leads. Another useful tool is an air-suction solder remover.

Most electronics parts stores have these devices. Before using a de-solder squeeze bulb, use your soldering iron to heat the joint you want to un-solder until the solder melts. Then squeeze the bulb to create a vacuum inside. Touch the tip of the bulb against the melted solder. Release the bulb to suck up the molten solder. Repeat the process until you've removed most of the solder from the joint. Then re-heat the joint and gently pry off the wires. This third method is easy, and is the preferred method, since it is fast and clean. You can use a vacuum device to suck up molten solder. There are many new styles of solder vacuum devices on the market that are much better than the older squeeze bulb types. The new vacuum de-soldering tools are about 8 to 12 inches long with a hollow Teflon tip. You draw the vacuum with a push handle and set it. As you re-heat the solder around the component to be removed, you push a button on the device to suck the solder into the chamber of the de-soldering tool.

De-soldering station

A de-soldering station is a very useful addition to an electronics workshop or workbench, but in many cases they just cost too much for most hobbyists. De-soldering stations are often used in production environments or as re-work stations, when production changes warrant changes to many circuit boards in production. Some repair shops used de-soldering stations to quickly and efficiently remove components.

Another useful de-soldering tool is one made specifically for removing integrated circuits. The specially designed de-soldering tip is made the same size as the integrated circuit, so that all IC pins can be de-soldered at once. This tool is often combined with a vacuum suction device to remove the solder as all the IC pins are heated. The IC de-soldering tips are made in various sizes; there are 8 pin, 14 pin and 16 pin versions,

which are used to uniformly de-solder all IC pins quickly and evenly, so as not to destroy the circuit board. The specialized soldering tips are often used in conjunction with vacuum systems to remove the solder at the same time.

Remember these things when un-soldering:

1. Be sure there is a little melted solder on the tip of your iron so that the joint will heat quickly.

2. Work quickly and carefully to avoid heat damage to parts. Use long-nosed pliers to hold the leads of components just as you did while soldering.

3. When loosening a wire lead, be careful not to bend the lug or tie point. Use pieces of wire or some old radio parts and wire. Practice until you can solder joints that are smooth, shiny, and tight. Then practice un-soldering connections until you are satisfied that you can do them quickly and without breaking wires or lugs.

Caring for your soldering iron

To get the best service from your soldering iron, keep it cleaned and well tinned. Keep a damp cloth on the bench as you work. Before soldering a connection, wipe the tip of the iron across the cloth, then touch some fresh solder to the tip. The tip will eventually become worn or pitted. You can repair minor wear by filing the tip back into shape. Be sure to tin the tip immediately after filing it. If the tip is badly worn or pitted, replace it. Replacement tips can be found at most electronics parts stores. Remember that oxidation develops more rapidly when the iron is hot. Therefore, do not keep the iron heated for long periods unless you are using it. Do not try to cool an iron rapidly with ice or water. If you do, the heating element may be damaged and need to be replaced or water might get into the barrel and cause rust. Take care of your soldering iron and it will give you many years of useful service.

Remember, soldering equipment gets hot! Be careful. Treat a soldering burn as you would any other. Handling lead or breathing soldering fumes is also hazardous. Observe these precautions to protect yourself and others!

Ventilation

Properly ventilate the work area where you will be soldering. If you can smell fumes, you are breathing them. Often when building a new circuit or repairing a "vintage" circuit you may be soldering continuously for a few hours at a time. This can mean you will be breathing solder fumes for many hours and the fumes can cause you to get dizzy or lightheaded. This is dangerous because you could fall down and possibly hurt yourself in the process. Many people highly allergic are also allergic to the smell of solder fumes. Solder fumes can cause sensitive people to get sinus infections. So ventilating solder fumes is an important subject. There are a few different ways to handle this problem. One method is to purchase a small fan unit housed with a carbon filter which sucks the solder fumes into the carbon filter to eliminate them. This is the most simple method of reducing or eliminating solder fumes from the immediate area. If there is a window near your soldering area, be sure to open the window to reduce the exposure to solder fumes. Another method of reducing or eliminating solder fumes is to buy or build a solder smoke removal system. You can purchase one of these systems but they tend to be quite expensive. You can create your own solder smoke removal system by locating or purchasing an 8 to 10 foot piece of 2 inch diameter flexible hose, similar to your vacuum cleaner hose. At the solder station end of the hose, you can affix the hose to a wooden stand in front of your work area. The other end of the hose is funneled into a small square "muffin" type fan placed near a window. Be sure to wash your hands after soldering, especially before handling food, since solder contains lead; also try to minimize direct contact with flux and flux solvents.

Chapter 4

AM, FM, and Shortwave Crystal Radio Projects

Parts list

Parts Bin

Simple Crystal Radio

L1 150 turns of 24 ga. enameled wire on 4" × 2" dia tube

C1 365 pF tuning capacitor (old radio)

D1 1N34 germanium diode (not a silicon diode)

C2 .005 μF, 35 volt disk capacitor

H1 high impedance headphones (crystal or 2000k type)

ANT 80-100 foot long wire antenna

GND cold water pipe or ground rod

Misc wood block, coil tube, hardware, screws, clips, washers, Krylon spray, wire, etc.

FM Crystal Radio Set

R1 47k ohm ¼ watt, 5% resistor

C1 82 pF capacitor

C2 80 pF air variable capacitor

C3 18 pF capacitor

D1 1N34 or rock crystal diode

L1 5 turns AWG #18 bare copper or silver wire, 12 mm inside diameter, tapped at 2.5 turns

H1 high impedance headphones (crystal type of 2k ohm type)

Ant antenna- 7" of #18 bare copper wire

Misc wood block, wire, hardware, clips, etc.

Supersensitive AM/Shortwave Crystal Radio

L1 11 turns, 22 ga. enameled wire on 2⅞" cardboard form

L2 (AM) 54 turns, 22 ga. enameled wire on 3" air-core form

Taps are brought out at 40T, 27T, 15T and 6T

L2 (SW) 15 turns, 22 ga. enameled wire on 3" air-core form

Taps are brought out at 12T, 9T, 6T and 3T

D1 germanium diode or
    galena crystal
    w/"cat's whisker"

C1 500 pF turning
    capacitor

C2 .001 μF, 35 volt
    disk capacitor

S1 single-pole –
    5-position rotary
    switch

S2 single-pole –
    6-position rotary
    switch

H1 crystal headphone
    or 2k headphones
    (Baldwin, Brandes)

ANT 80-100 foot long
    wire antenna

GND cold water pipe or
    ground rod

Misc wood block,
    hardware, wire, clips,
    screws, Krylon, etc.

A crystal radio is the simplest form of AM (amplitude modulation) receiver ever invented. It has great potential for experimentation and usually requires no source of power for its operation other than the radio signal itself, and costs little. Most people do not realize that crystal radios can be built to pickup shortwave as well as FM radio signals. Did you know that you can build very sensitive and selective crystal radios far better than most commercial AM radios! Ever thought of building a crystal radio? Building a crystal radio will give you immense satisfaction, and the results are sure to please.

Crystal sets date back to the earliest days of wireless (pre-W.W.I) and an enormous variety of circuit designs have been produced over the years. Their popularity has been variable as developments in other more elaborate forms of reception have taken place. However, fascination with this crystal radio design, building and experimentation, is still very strong today with national and international organizations and clubs offering competitions.

So how does a crystal radio work? An AM transmitter sends out its broadcast in the form of an electromagnetic

wave that radiates from its transmitting antenna. The AM transmitter sends out a fixed frequency carrier wave. When sound is present in the program material, the strength, or amplitude, of these waves is made to vary in response to the audio content of the program. The resulting wave is called an amplitude modulated wave. See Figure 4-1.

The purpose of a simple AM or crystal radio receiver is to pick up these AM waves and extract the audio signal so that it can be heard by the listener. It does this by a process called "detection." The detection process utilizes a device called a detector which effectively strips off either the upper half or lower half of the AM wave. It only remains to filter out the carrier wave to leave the audio signal and the job is done! The following diagram shows how the detector and filter work together to "recover" the original audio signal. The strength of the recovered audio signal is small but, given the right conditions, it can be large enough to drive a pair of headphones; all without any power source other than the signal itself.

The diagram shown in Figure 4-2 illustrates a basic crystal radio. In its simplest form, a crystal radio is composed of just our main components. A good antenna, a detector, a filter capacitor and pair of headphones. The basic diagram shows the input section at (A), the rectifier section at (B), the filter section at (C) and finally the headphones at section (D).

The antenna consists of a length of wire suspended above the ground, while the ground or earth connection could be a metal spike driven into the ground. When a

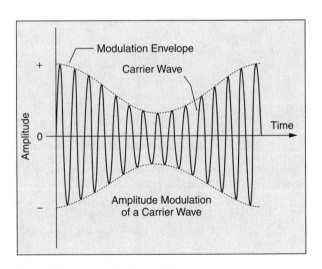

Figure 4-1 *Amplitude modulation*

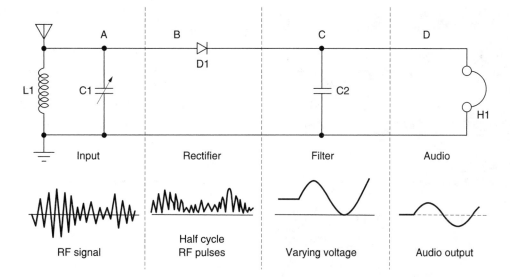

Figure 4-2 *Basic crystal radio*

capacitor is connected in parallel with an inductor and an alternating voltage applied across the combination, alternating current will flow. The amount of current that flows depends upon the frequency of the applied voltage. At a particular frequency, called the resonant frequency, almost no current flows. For frequencies above or below the resonant frequency, significant current will flow. In Figure 4-2, capacitor C1 plus the aerial capacitance forms the capacitor of our tuned circuit, while L1 forms the inductor. At the resonant frequency of the tuned circuit, almost no current flows to earth through L1 or C1, leaving virtually all of it free to flow through the detector. Alternating currents at broadcast frequencies either above or below the resonant frequency will tend to flow through C1 or L1 to earth. Thus we have added selectivity to the receiver. Notice that C1 is adjustable, as signified by the arrow. By varying C1 we can tune the receiver to select specific broadcast frequencies.

The main requirement for a detector is that it should act as a non-return "valve" or one-way "switch" to the alternating currents of the AM wave. If you could look at the alternating current in a circuit you would see it flowing back and forth, first one way and then the other. A detector placed in such a circuit allows the alternating current to flow easily in one direction but not in the other. Certain naturally occurring minerals were found to have this property and these became some of the earliest forms of detectors. One such mineral, the crystalline form of galena (lead sulphide), was found

to be particularly good at detection, and became very popular for building crystal radios. The detector consists of galena in a small cup or tin. A small coil of wire, with one free end was called the "cat's whisker" which was used to make contact with the crystal.

The remaining uni-directional current is filtered by the combined effect of the headphone impedance and the capacitor at C2 and only the desired audio signal current passes on through to the headphones at H1. One drawback with the very simple circuit shown is that it has little selectivity. That is, it picks up all AM broadcast signals with similar efficiency and often close together. So as time passed more advanced, more sensitive and more selective receivers were born.

Let's construct a simple crystal radio. In order to build the crystal radio shown in Figure 4-3, you will need to secure some tools and supplies. First you will need to find a clean well lit work bench or table to spread out all your tools, charts, diagrams and components. You will also need to secure a soldering iron, some 60/40 rosin core solder and a small jar of "Tip Tinner," a soldering iron tip cleaner/dresser, obtainable from your local Radio Shack store. You will also want to locate some small tools such a pair of end-cutters, a pair of needle-nose pliers, a magnifying glass and a set of Phillips and flat-blade screwdrivers for this project. Grab the crystal radio schematic, see Figure 4-2, and the resistor and capacitor identification charts, and place them in front of you, so we can get started building the project. The resistor identification chart in Table 4-1 will help you

Figure 4-3   *AM crystal radio*

the resistor's second digit of the value. The third colored band is the resistor's multiplier value. For example, a resistor whose first color band is yellow represents a digit four (4), while a second color band which is violet is seven (7). If the third color band is orange, then the multiplier would be (000) or one thousand, so the value would be 47,000 or 47k ohms. Resistors usually have a fourth colored band which represents the resistor's tolerance value. A silver band denotes a 10% tolerance value, while a gold band denotes a 5% tolerance resistor. No color band represents a tolerance value of 20%.

Our crystal radio projects in this chapter all utilize capacitors of some sort. Capacitors usually are available in two major classifications, polarized and non-polarized. The project in this chapter will utilize only non-polarized capacitors. Non-polarized capacitors can often be physically very small and not have their actual value printed on the body of the capacitor, so a chart was developed to help identify a capacitor using a three-digit code. See Table 4-2.

As this project will not require a circuit board, we will build the crystal radio on a 6″ × 6″ × ¾″ wood block base. You can elect to use an additional piece of wood at a right angle to the base as a front panel to secure the tuning capacitor as shown in Figure 4-3.

identify resistors needed for the some of the projects in this chapter. Each resistor will have three or four colored bands on the body of the resistor. These colors identify the resistor's value. The color bands start at one end of the resistor, the first colored band is the first digit of the resistor's value, while the second resistor represents

Table 4-1
Resistor color code chart

Color Band	1st Digit	2nd Digit	Multiplier	Tolerance
Black	0	0	1	
Brown	1	1	10	1%
Red	2	2	100	2%
Orange	3	3	1,000 (K)	3%
Yellow	4	4	10,000	4%
Green	5	5	100,000	
Blue	6	6	1,000,000 (M)	
Violet	7	7	10,000,000	
Gray	8	8	100,000,000	
White	9	9	1,000,000,000	
Gold			0.1	5%
Silver			0.01	10%
No color				20%

Table 4-2

Capacitance code information

This table provides the value of alphanumeric coded ceramic, mylar and mica capacitors in general. They come in many sizes, shapes, values and ratings; many different manufacturers worldwide produce them and not all play by the same rules. Some capacitors actually have the numeric values stamped on them; however, many are color coded and some have alphanumeric codes. The capacitor's first and second significant number IDs are the first and second values, followed by the multiplier number code, followed by the percentage tolerance letter code. Usually the first two digits of the code represent the significant part of the value, while the third digit, called the multiplier, corresponds to the number of zeros to be added to the first two digits.

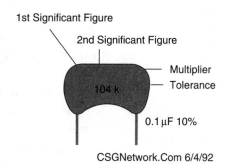

Value	Type	Code	Value	Type	Code
1.5 pF	Ceramic		1,000 pF /.001 µF	Ceramic / Mylar	102
3.3 pF	Ceramic		1,500 pF /.001 5µF	Ceramic / Mylar	152
10 pF	Ceramic		2,000 pF /.002 µF	Ceramic / Mylar	202
15 pF	Ceramic		2,200 pF /.0022 µF	Ceramic / Mylar	222
20 pF	Ceramic		4,700 pF /.0047 µF	Ceramic / Mylar	472
30 pF	Ceramic		5,000 pF /.005 µF	Ceramic / Mylar	502
33 pF	Ceramic		5,600 pF /.0056 µF	Ceramic / Mylar	562
47 pF	Ceramic		6,800 pF /.0068 µF	Ceramic / Mylar	682
56 pF	Ceramic		.01	Ceramic / Mylar	103
68 pF	Ceramic		.015	Mylar	
75 pF	Ceramic		.02	Mylar	203
82 pF	Ceramic		.022	Mylar	223
91 pF	Ceramic		.033	Mylar	333
100 pF	Ceramic	101	.047	Mylar	473
120 pF	Ceramic	121	.05	Mylar	503
130 pF	Ceramic	131	.056	Mylar	563
150 pF	Ceramic	151	.068	Mylar	683
180 pF	Ceramic	181	.1	Mylar	104
220 pF	Ceramic	221	.2	Mylar	204
330 pF	Ceramic	331	.22	Mylar	224
470 pF	Ceramic	471	.33	Mylar	334
560 pF	Ceramic	561	.47	Mylar	474
680 pF	Ceramic	681	.56	Mylar	564
750 pF	Ceramic	751	1	Mylar	105
820 pF	Ceramic	821	2	Mylar	205

Next, you will need to locate a plastic or cardboard tube approximately 3½ to 4″ long by 2″ in diameter on which to wind your coil. Punch a hole at both ends of the tube about ½″ from the actual end of the tube. Start winding the coil at one end of the tube using #24 ga. enameled wire, leave about 6″ of wire free at the first hole to allow connection to the circuit. Wind the coil with each winding touching the previous winding but not overlapping the prior winding. Wind about 150 turns of wire across the length of the tube. After winding about 150 turns take the free end of the remaining wire and put it through the last punched hole, leave about 6″ of wire so you can connect the coil to the circuit. When you are finished winding the coil, you can varnish the coil winding or just spray Krylon around the coil to keep the winding close to the coil, so they won't spread out or come loose. Fasten the coil to a 6″ × 6″ wood block; you can mount the coil upright as shown in Figure 4-3, or you can lay the coil down near one edge of the wood block and secure it to the block with some brackets or clips. You will have to use a knife and scrape off the enameled insulation of the two coil wires before you can fasten or solder them to the tuning capacitor. Now fasten the tuning capacitor to the wood block near the coil assembly, and wire the coil in parallel with the tuning capacitor. Next locate a germanium diode; make sure you use only a germanium diode and not the more common silicon diode. Remember to observe the correct polarity when connecting the diode in the circuit, the black to white colored band is the cathode lead. The cathode end of the diode points to the headphone in the circuit diagram. The anode side of the diode connects to one end of the "tank" or coil/capacitor and the cathode side of the diode connects to capacitor C2 and one end of the high impedance headphones. The free end of the headphones connects to the remaining end of the capacitor at C2 which connects to the free end of the "tank" circuit. You can connect the headphones via a terminal strip, a Fahnestock clip or headphone jack mounted on the wood block. Crystal radios generally require high impedance headphones, so you must use either a small crystal headphone or a pair of older high impedance (2000) ohm headphones; these are still available through EBAY or from Antique Electronics.

You could elect to build the radio using a galena crystal, with a "cat's whisker" instead of the germanium

diode to give the radio an "old-time" look if desired: see the Appendix for Antique Radio Supply.

Connect an antenna and a ground wire to rod in the ground or an electrical ground in your house and your crystal radio will be ready to test out. The ground wire can be attached to a metal cold water pipe, or to a metal rod stuck a couple of feet into the earth. Do not attach it to a line carrying gas or electricity. The crystal radio will operate forever without a power supply, but you do need a good antenna to make the radio work well. The simplest outdoor antenna for receiving signals is a horizontal wire with the lead-in wire to the set connected on one end. An insulated wire 50 to 100 feet long placed as high and clear of surrounding objects as possible works best: see Figure 4-4. Both ends of the antenna should have insulators and rigid support. Tree branches are not good because they sway in the wind. If you use a tree trunk, have slack in the wire so it can move without pulling the wire. Keep the antenna wire far away from chimneys, metal roofs, gutters, drain pipes, telephone, and especially power lines. The antenna and lead-in should be one continuous piece of wire. Use stranded copper wire, with only the lead-in extending beyond the insulator. A lightning arrestor and ground wire should also be used to avoid a direct hit by lightning.

A crystal radio will generally be able to detect radio signals within 25 to 35 miles, during the day and often much further at night. A simple crystal radio is not very selective, if there are many local AM radio stations that are close in frequency, then you may have some trouble tuning them in, and you may have to build a more selective radio, such as the All-Wave model below. The crystal radio project above is a great project for demonstrating the principles of AM radio and the fun and excitement of building a crystal radio or "free-power" radio.

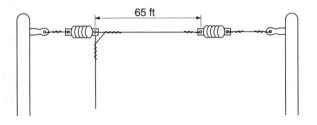

Figure 4-4 *Basic long-wire antenna*

FM radio crystal radio

Believe it or not, you can actually create a simple FM radio crystal radio. An FM crystal receiver is depicted in Figure 4-5. The key to understanding how this simple circuit can decode FM signals is to understand what happens when an FM signal is coupled to a tuned circuit. A good tuned circuit (one with a high "Q factor") will attenuate signals that are not near its resonant frequency. If you apply an FM signal to a good tuned circuit, as the frequency of the FM signal moves away from the resonant frequency of the tuned circuit, the tuned circuit will attenuate that signal. As the frequency of the FM signal moves toward the resonant frequency of the tuned circuit, the tuned circuit will let more of the signal pass through. Thus, a tuned circuit will impose amplitude changes on an FM radio signal that match the frequency changes. A crystal diode is sensitive to amplitude changes, so it converts the amplitude changes into a signal that the listener can hear through headphones.

Using a tuned circuit to induce amplitude changes onto an FM signal, then converting the amplitude changes to sound frequencies, is called slope detection. "Slope" refers to the slope of the tuned circuits' attenuation curve. If you have an AM receiver you can listen to an FM signal by using the *slope* of the tuned circuits in the receiver. This *slope tuning* method gives you none of the real advantages of FM. For instance, FM is noted for its immunity to static. An AM receiver

used to slope detect will hear static. The only real advantage to slope detection is that is can be cheap and convenient. When tuning in an FM signal with an AM receiver, the clearest sound occurs when the FM center frequency is offset from the center frequency of the receiver's tuned circuits. If the FM signal is at the receiver's center frequency, then frequency deviations both up and down produce downward amplitude changes. When the FM signal is off center frequency, up and down frequency changes create up and down amplitude changes in the receiver.

The FM crystal receiver circuit, shown in Figure 4-6, looks almost identical to a classic AM crystal circuit; but look closely, there are a few major changes. The FM crystal radio is relatively easy to build and can be built on a small block of wood as shown. The components values for the "tank" or resonant circuit were reduced to resonate at higher frequencies in the FM band. This was done by experimenting with smaller and smaller coils and capacitors values. The antenna is also much reduced in size (from that of AM crystal radio) to resonate at higher frequencies. The antenna is actually a 7″ long bare copper wire. The coil at L1 is just four turns #18 copper or silver wire, 12 mm inside diameter, tapped at 2.5 turns, The coil was wrapped around a "magic marker," then slipped off and expanded. As you can see, this is far different than the 150 turn coil for the AM crystal radio. The detector diode, once again, is a 1N34 germanium diode and not a silicon diode, or you could use a galena crystal as a detector if you desire. Capacitor C3 is 18 pF, but is not critical and can be 10 to 50 pF. Resistor R1 is a 47k ohm, ¼ watt, 5% carbon composition type. The air variable capacitor, used in the prototype, was an 80 pF variable type and it had two trimmers in it which should be adjusted for best reception.

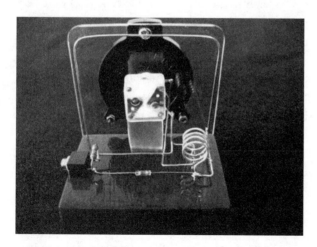

Figure 4-5 *FM crystal radio*

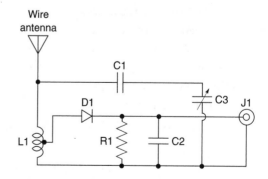

Figure 4-6 *Simple FM crystal radio*

Commonly available vernier dial and knob will fit the tuning capacitor and will work nicely. When building the FM crystal radio it is important to keep the components physically close together and the wiring short as possible. You can build the FM crystal radio in an open style format on a wood or Lucite plastic sheet. You can make your receiver a work of art, and impress your friends with an FM crystal radio built from scratch. Note that you will have to use high impedance headphones for this receiver. A detected FM signal is converted to AM due to an effect called slope detection that modulates amplitude.

This FM crystal radio works best near the transmitter, reception seems fine up to 15 miles away. Secondly, the sound level is not very loud; a quiet room is best for listening. One must be willing to move the set around to find a location for the best reception of signals. If desired, you could connect the audio output of the FM set directly to an audio amplifier's fed to a small speaker.

Supersensitive AM/Shortwave crystal radio

There are several characteristics that make the difference between a good and a great receiver. One of these characteristics is sensitivity. Great receivers can pick up weak signals and amplify them into a useful output. A second characteristic is selectivity. Great receivers can separate signals so that only one is heard at a time. The third and final thing that separates good receivers from great receivers is fidelity. Great receivers not only pick up weak signals but can reproduce a signal without distortion. The last project in this chapter is the All-Wave, AM/Shortwave crystal radio. This All-Wave design takes the crystal radio concept a little further and provides a more sensitive and more selective radio for you to build and enjoy.

The All-Wave Crystal Radio is a supersensitive and selective AM and shortwave receiver; see Figure 4-7. The design of the All-Wave radio has been around since the early 1930s and is still a very efficient receiver design that is fun to build and use. The receiver will

Figure 4-7 *Shortwave crystal radio*

tune from below the AM broadcast band to around the 40 meter band, so you will be able to pickup international shortwave reception quite well. You will be surprised how well it performs with a good pair of high impedance headphones.

The key to this efficient receiver's design lies with the two coils, L1 and L2, which are shown in the schematic at Figure 4-8. Coil L2 is a multi-tapped plug-in coil based on the band selected. For the AM radio band, L2 is the 54 turn multi-tapped coil, with tapped from the ground end at 6T, 15T, 27T and at 40T. The shortwave version of the L2 coil is a 15 turn multi-tapped coil with tapped at 3T, 6T, 9T, and 12 turns. Depending upon the band chosen, you simply plug in the AM or shortwave coil. The AM and shortwave coils (L2) were wound on a 3″ cardboard tube, but you could use a plastic form instead. The untuned primary coil (L1) was wound a slightly smaller 2 ⅞ form, so that it can be slipped inside of the L2 coil at the ground side of the coil assembly. The interchangeable coils (L2) were mounted on a non-metallic carrier, a piece of phenolic board with five pin-plugs secured to the carrier board. The coil form for L2 was glued to the phenolic carrier board which forms the plug-in coil assembly. The wire tapes are brought out to the five pin-plugs on the carrier board, so that the coil (L2) assembly can be plugged into sockets on the radio's wood base plate, see Figure 4-9. This will allow either the AM coil or the shortwave coil to be plugged into the crystal receiver. You could elect to build a fixed coil version of the crystal radio for either AM or shortwave reception if you desire, making the receiver a bit easier to build.

The detector for the All-Wave Crystal Radio can be either a 1N34 germanium diode or you could elect to

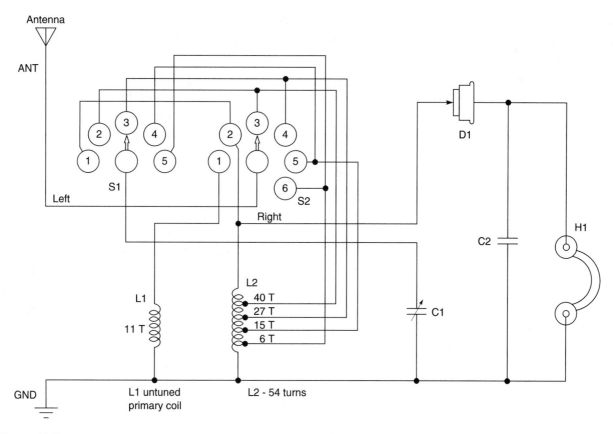

Figure 4-8 *Sensitive/selective AM/shortwave crystal radio. Courtesy of Midnight Science*

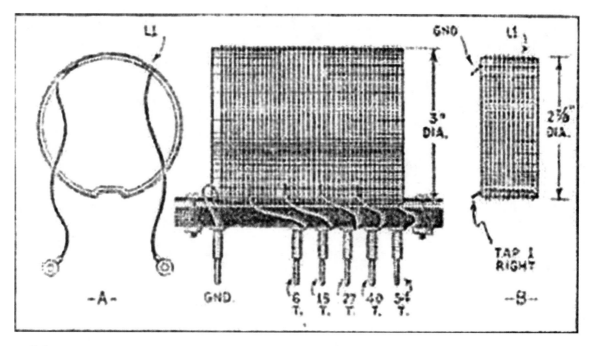

Figure 4-9 *Shortwave coil assembly*

build the radio more like the "original" version by using a galena crystal with a "cat's whisker." The galena crystal and "cat's whisker" are still available, see Appendix. Be careful to install the diode or galena crystal with the proper orientation.

The original All-Wave receiver used a 500 pF tuning capacitor at C1, but you could also use a 365 pF type which are easier to locate from sources such as Antique Radio Supply. Mount the tuning capacitor to the wood block base using standoff and screws. The ceramic disk capacitor at C2 acts as an audio filter, ahead of the high impedance headphones.

The All-Wave receiver needs to be used with either a crystal earphone or a pair of dual high impedance (2K) headphones such as the old Baldwin or Brandes headphones which are still available via EBAY or from Antique Radio Supply, see Appendix.

Switches S1 and S2 are used to select the taps between the antenna and the detector. The two switches are labeled **Right** and **Left** as shown in the diagram. The same station may come in on several taps, but use the one which places the station lowest on the turning capacitor setting for the loudest signals. Move the **Right** (S2) forward or clock-wise, one to three taps before advancing the **Left** switch (S1). The efficiency of this radio seems to lie partly in the low loss coils, but mostly in the one basic change in the circuit, in which the detector is connected permanently to the last turn of L2 towards ground.

The All-Wave Crystal radio is just about completed and you now have only to connect a good antenna and ground for the receiver to spring-to-life! The All-Wave Crystal Radio should be connected to a good ground such as a cold water pipe or 6 to 8 foot copper ground rod pounded into the ground. The All-Wave receiver will perform best with an 80 to 100 foot insulated long-wire antenna placed up-in-the-air outdoors away from metal objects. An alternate antenna might be a 50 to 60 foot insulated wire strung around your attic, away from any metal objects. You will be amazed at how well this receiver will work with a good antenna and a good pair of high impedance headphones.

In the event that your new All-Wave Crystal Radio fails to work, you will have to re-check a few details. First check the diode and how it was oriented in the circuit, you may have placed it backward and this will prevent the receiver from functioning. You will also need to check the coil connections and placement of L1 with respect to coil L2. There is no polarity regarding coil L1, but its placement at the ground end of L2 is important in order for the receiver to work properly. Also check the connections to L2, you may have reversed two end wires of the coil, and this will prevent the radio from working as well. Try the receiver once more, hopefully all of the connections are right now and the receiver should work fine. Reconnect up your ground and antenna, put on your high impedance headphones and you'll be ready to listen to your new All-Wave receiver. You may want to have a "log" book close to your new radio to log down the far away stations you will receive with this radio. During nighttime hours, you will be able to receive stations much further away, and when the shortwave coil is plugged in you will hear international shortwave stations from all around the world on your new All-Wave receiver. Have fun with your new All-Wave crystal radio.

Chapter 5

TRF AM Radio Receiver

Parts list

TRF AM Radio Project

R1   120k ohm, ¼ watt,
     5% resistor

R2   2.7k ohm, ¼ watt,
     5% resistor

R3   10k ohm
     potentiometer (volume)

R4   10k ohm
     potentiometer (trim)

R5   820 ohm, ¼ watt,
     5% resistor

R6,R9,R13   10k ohm,
     ¼ watt, 5% resistor

R7,R8   22k ohm, ¼ watt,
     5% resistor

R10   150k ohm, ¼ watt,
     5% resistor

R11,R12   4.7k ohm,
     ¼ watt, 5% resistor

C1   10 nF, 35 volt
     polyester capacitor

C2   500 pF tuning
     capacitor (variable)

C3   100 nF, 35 volt
     polyester capacitor

C4,C5   2.2 µF, 35 volt
     electrolytic capacitor

C6   33 µF, 35 volt
     electrolytic capacitor

C7   220 µF, 35 volt
     electrolytic capacitor

C8   2200 µF, 35 volt
     electrolytic capacitor

L1   220 µH coil; see text
     (Ocean State-LA-540)

D1,D2,D3,D4,D5,D6
     1N4148 silicon diodes

Q1   BC108B-NTE123A
     transistor

Q2   BC109C-NTE123AP
     transistor

Q3   BC179C-NTE234
     transistor

U1   ZN414 or MK484 -
     TRF Radio IC

U2   LM741 Op-amp IC

S1   SPST power switch

B1   9-volt transistor
     radio battery

SPK   8-ohm speaker or
     headphone

Misc   PC board, chassis,
     battery holder,
     battery clip, wire,
     hardware

AM radio is the name given for the type of radio broadcasting used in the medium wave band in all parts of the world and in the long wave band in certain areas, notably Europe. Medium wave frequencies in the US range from 540 to 1710 kHz with 10 kHz spacing. The letters AM stand for amplitude modulation, the technique by which speech and music are impressed on the radio waves. The sound waves from the studio are converted into electrical signals by the microphones, and these electrical signals are used to vary the amplitude, or the distance from 'peak' to 'trough' of the main constant frequency carrier wave generated at the transmitting station. As a result, the signal radiated by the station varies in amplitude in step with the variations in the loudness and pitch of the sound in the studio. In the receiver, the amplitude variations of the radio signal are separated from the carrier wave by a process known as demodulation, and after further amplification are fed to the loudspeaker.

Amplitude modulation is also used by international broadcasting stations operating in the shortwave band, and for some professional communications such as air traffic control, ship-to-shore traffic, and some mobile radio services.

AM radio stations in the medium wave or AM radio band can provide a service over an area the size of a town or of a whole continent, depending upon the power of the transmitter which can vary from less than a hundred watts to hundreds of kilowatts. Long wave AM radio stations cover even larger areas. Normally, all radio waves travel only in straight lines, which would limit a radio station to an area of only about 40 miles (65 km) in diameter—as far as the horizon. Medium and long wave stations can actually cover much larger areas because their signals are bent round the curvature of the Earth by the ionosphere (an electrically charged layer in the upper atmosphere which can reflect radio waves). In the US, where AM radio stations are operated on a commercial basis, coverage is generally and deliberately limited to one city and its environs. Major cities such as New York are served by large numbers of stations, each radiating a different program. AM radio stations in the US generally operate with omni-direction patterns during the day. In the evening many AM broadcasters switch to directional antenna beam patterns, and their broadcasts go much further due to ionospheric bending. This is why you can hear AM broadcast stations more distant in the evening.

AM radio first began with the experimental broadcast in 1906 by Reginald Fessenden, and was used for small-scale voice and music broadcasts up until World War I. The great increase in the use of AM radio came the following decade. The first licensed commercial radio services began on AM in the 1920s (the first licensed American radio station was started by Frank Conrad: KDKA in Pittsburgh, Pennsylvania). Radio programming boomed during the "Golden Age of Radio" (1920s–1950s). Dramas, comedy and all other forms of entertainment were produced, as well as broadcasts of news and music.

The quality of reproduction on AM radio is not as good as FM radio because of limitations in the amplitude modulation technique itself and because of overcrowding and subsequent interference in the medium wave band. Interference is particularly serious at night because the nature of the ionosphere alters after sunset, so that it can then reflect medium wave signals back to Earth a long way beyond their normal service area. Here, they interfere with other stations, producing whistles and other unwanted noises.

The AM method is still used for most popular broadcasting. Although the alternative FM radio system gives better quality, it is more difficult to receive, and is less suitable than AM for use in inexpensive transistor portables and car radios.

The most important concept in any communications system is the ability to send information from one place to another via a transmitter to a receiver, see Figure 5-1. This means you have to find a way to impress the audio signal information on the radio wave in such a way that it can be recovered at the other end. This process is known as modulation. In order to modulate a radio wave, you have to change either or both of the two basic characteristics of the wave: the amplitude or the frequency.

If you change the amplitude, or strength, of the signal in a way corresponding to the information you are trying to send, you are using a modulation system called amplitude modulation, or AM. The earliest means of radio communications was by Morse code, and the code key would turn the transmitter on and off. The amplitude went from nothing to full power whenever the key was pressed, a basic form of AM.

Modern AM transmitters vary the signal level smoothly in direct proportion to the sound they are transmitting. Positive peaks of the sound produce maximum radio energy, and negative peaks of the sound produce minimum energy.

The main disadvantage of AM is that most natural and man made radio noise is AM in nature, and

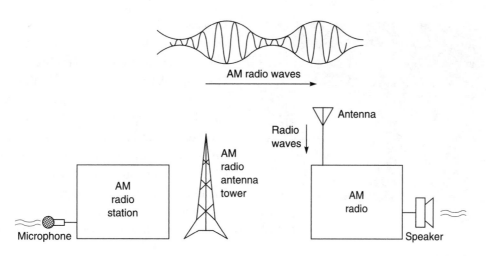

AM radio waves

Radio waves

Antenna

AM radio antenna tower

AM radio station

Microphone

AM radio

Speaker

Figure 5-1 *AM broadcast system block diagram*

AM receivers have no means of rejecting that noise. Also, weak signals are (because of their lower amplitude) quieter than strong ones, which requires the receiver to have circuits to compensate for the signal level differences.

An AM radio station takes a low-frequency audio signal such as people talking or music (2500 Hz) and mixes it or superimposes on a high-frequency carrier wave (540 kHz), see Figure 5-2. It sends this radio signal out into the air as electromagnetic waves using a tall antenna. The electromagnetic wave travels through the air, going in all directions. At home, your AM radio takes this incoming signal, in a process called detection. With amplitude modulation it is simple. All you need to

do is to rectify the signal, through a rectifier which removes the bottom half of every cycle. Essentially you are throwing away the high-frequency or carrier part, but keeping the low-frequency or audio part. At the output of the detector the original audio information is sent to the speaker in your radio. You then "hear" whatever was broadcast at the radio station.

The AM radio project that we are going to build in this chapter is called a TRF or tuned radio frequency type AM radio, see Figure 5-3. The TRF AM Radio design was at the forefront of radio design in the early 1920s. It utilized a tuned front end which allowed only RF energy of interest to be amplified by the RF amplifier, the signal was then passed on to the detector

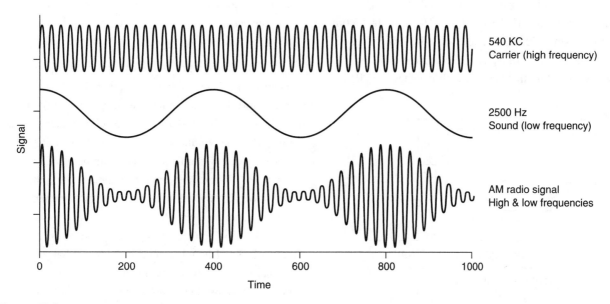

540 KC
Carrier (high frequency)

2500 Hz
Sound (low frequency)

AM radio signal
High & low frequencies

Signal

Time

Figure 5-2 *AM carrier and modulation signals*

Figure 5-3 *AM TRF receiver*

followed by an audio amplifier to drive headphones. Early TRF radios had up to three controls for tuning the receiver, but later all three controls were consolidated into "tuning" control. Our modern day TRF receiver shown in the block diagram in Figure 5-4 illustrates the three major components of the AM TRF receiver. The first block consists of the tuned RF front-end resonant coil/capacitor, the second block consists of the MK484 IC, an integrated circuit RF amplifier and detector, and finally the third block, which is an audio amplifier.

The TRF AM radio project was designed around the popular MK484 TRF integrated circuit module, shown in Figure 5-5. This receiver covers the medium wave band or (AM) radio band from approximately 540 to 1700 kHz. The TRF front end consists of two components, the ferrite aerial coil and the tuning capacitor and the MK484 TRF IC. The important word is tuned, since this AM radio is called a TRF, or tuned RF radio All the AM signals reaching the radio are very weak. Only that signal which matches the TRF

frequency is magnified by resonance so that it stands out at a very much higher level of signal strength. The original ZN414 IC, has now been replaced by the MK484, which is considerably easier to locate. The integrated circuit is a 3-pin, tuned radio frequency circuit, and incorporates several RF stages, automatic gain control and an AM detector. The receiver front-end is easily overloaded and the operating voltage of the IC is somewhat critical to achieve good results, but all-in-all the receiver is a great little radio for AM reception. The TRF radios were some of the first radios on the scene when the dawn of radio appeared in the early 1920s.

The front-end of the TRF receiver as mentioned consists of a tuned circuit composed of the ferrite inductor, a tuning capacitor and the TRF radio chip. A small square plastic 0-200 pF tuning capacitor was used for this radio project, but you could locate any suitable tuning capacitor within the range and it will work equally well. Often these small tuning capacitors will have additional on-board trimmer capacitors, but we used only the main tuning capacitor in the center of the device.

The second critical component of the TRF receiver is the coil at L1, which is wound on a ferrite bar. First, let's begin winding the coil at L1. The coil needs approximately 65 turns of #28 gauge enameled Litz wire on a 2½" long ferrite rod. The windings should measure approximately 25 mm wide. Start winding the coil, close to one end, by securing the enameled wire by a piece of insulating tape. Leave approximately 100 mm for connection. The quality of reception is dependent upon the care taken in winding the antenna coil.

The signal reception will vary, dependent upon location, and is better away from large buildings that can distort the signal. Wind the wire tightly around the ferrite rod, taking care to keep the coil together and not to overlap the windings. To help hold the windings in place a piece of double-sided tape can be placed on the ferrite rod. Before soldering the antenna coil to the circuit board, you will need to scrape off the enamel coating from the ends of the wire. The two components L1 and C2 are connected in parallel to form an LC network. The Litz wire coil and the ferrite bar give the LC network a high "Q" or quality factor. This is important to the selectivity of the receiver, that is the ability to tune into one radio station at a time.

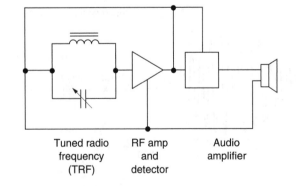

Tuned radio   RF amp   Audio
frequency   and   amplifier
(TRF)   detector

Figure 5-4 *TRF AM radio block diagram*

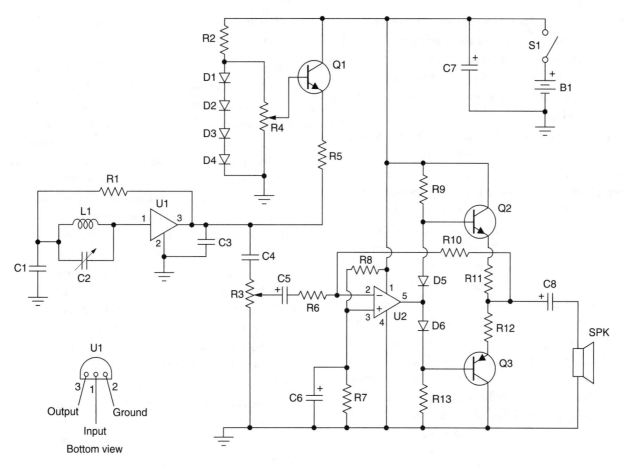

Figure 5-5 *TRF AM receiver broadcast band receiver*

The coil winding and the ferrite bar act as an efficient wire antenna, as no additional antenna is required. The only disadvantage is that the tuned circuit is directional and depends on the orientation of the bar with respect to the originating broadcast station. The coil winding and the ferrite bar act as an efficient wire antenna. No additional external antenna is needed. The only disadvantage is that the tuned circuit is directional. Signal strength depends on the orientation of the bar with respect to the origin of the signal. The coil and tuning capacitor may be taken from an old AM radio to save time, if desired.

The third main component of the TRF radio front-end is the TRF integrated circuit. The MK484 is a Japanese copy of the original ZN414. It contains an RF amplifier, active detector and automatic gain control (AGC to improve sensitivity) all in a 3-pin package. The input impedance is typically 4 M ohm. It operates over a range of 150 kHz to 3 MHz. DC supply of 1.1 V to 1.8 V and 0.3 mA current drain makes it ideal for

battery operation. The output is typically 40–60 mV of audio signal. It is important to realize that the radio IC does not create any sounds by itself. It can only take the RF signal provided from the TRF circuit, amplify it, separate the audio signal from the RF (called detection) and pass the audio signal on to be amplified. It has no selective or rejection components contained in it. (This is in contrast to superhetrodyne receivers.)

In this circuit, a small voltage regulator is built around the BC108B transistor, four 1N4148 diodes, the 2.7k and 10k preset resistor and the 820 ohm resistor. The 10k pot acts as a selectivity control for the whole receiver, controlling the operating voltage for the ZN414 (or MK484). If you live in an area that is permeated with strong radio signals, then the voltage may need to be decreased. I found optimum performance with a supply of around 1.2 volts.

The audio amplifier is built around an inverting 741 op-amp amplifying circuit. Extra current boost is provided using the BC109C/BC179 complementary transistor pair to drive an 8 ohm loudspeaker. The voltage gain of the complete audio amplifier is around 15. The audio output of the complete receiver is really quite good and free from distortion. I used a small wooden enclosure and the complete tuning assembly from an old radio.

Before we begin the AM radio project, you will want to locate a well lit and well ventilated work space. You will want a large work table or work bench to spread out all you components, diagrams and tools. Locate a small pencil tipped 27 to 33 watt soldering iron and a roll of 60/40, 22 ga. tin/lead rosin core solder. If you don't have a jar of "Tip Tinner," a soldering iron tip conditioner/dresser, then you will want to go to a local Radio Shack store and purchase one. Place all the component parts in front of you along with the necessary project diagrams, so it will be easy to work with everything in front of you. Grab a few small tools, such as a pair of end-cutters, a pair of needle-nose pliers, a pair of tweezers, a magnifying glass, a small flat-blade, and small Phillips screwdriver. Once everything is right in front of you we can begin building the AM radio project. Warm up your soldering iron and we will begin. It is a good idea to use a circuit board for this project, since circuit boards provide a more reliable method used to build circuits: you could elect to use prototype construction or point-to-point wiring on a perf-board as an alternative.

Refer to the chart in resistor color chart in Table 5-1 and the capacitor code chart in Table 5-2. Resistors used in this project will be mostly carbon composition type resistors and potentiometers. Each resistor will have three or four color bands on the body of the resistor. The first band closest to one edge will be the first color which represents the first value. For example, a brown color band will signify the first digit as a one (1). The second color band represents the second digit value. So if the second color is red, then the value of the second digit would be two (2). The third color band is the multiplier value. So, if the third color is orange then the multiplier value would be (000), so the resistor value would be 12,000 or 12k ohms. Often there is a fourth color band which represents the tolerance value. No fourth band denotes a resistor with a 20% tolerance, a silver band denotes a 10% tolerance value, and a gold band represents a 5% tolerance value. Now refer the capacitor code chart. Many small capacitors are too small to have their full value marked on the body of the capacitor, so a three-digit code is used to represent the value.

Let's begin installing the project resistors now; find and identify a couple of resistors at a time. Once you are

Table 5-1

Resistor color code chart

Color Band	1st Digit	2nd Digit	Multiplier	Tolerance
Black	0	0	1	
Brown	1	1	10	1%
Red	2	2	100	2%
Orange	3	3	1,000 (K)	3%
Yellow	4	4	10,000	4%
Green	5	5	100,000	
Blue	6	6	1,000,000 (M)	
Violet	7	7	10,000,000	
Gray	8	8	100,000,000	
White	9	9	1,000,000,000	
Gold			0.1	5%
Silver			0.01	10%
No color				20%

Table 5-2

Capacitance code information

This table provides the value of alphanumeric coded ceramic, mylar and mica capacitors in general. They come in many sizes, shapes, values and ratings; many different manufacturers worldwide produce them and not all play by the same rules. Some capacitors actually have the numeric values stamped on them; however, many are color coded and some have alphanumeric codes. The capacitor's first and second significant number IDs are the first and second values, followed by the multiplier number code, followed by the percentage tolerance letter code. Usually the first two digits of the code represent the significant part of the value, while the third digit, called the multiplier, corresponds to the number of zeros to be added to the first two digits.

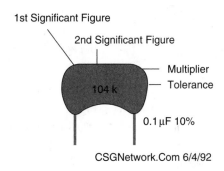

Value	Type	Code	Value	Type	Code
1.5 pF	Ceramic		1,000 pF /.001 µF	Ceramic / Mylar	102
3.3 pF	Ceramic		1,500 pF /.0015 µF	Ceramic / Mylar	152
10 pF	Ceramic		2,000 pF /.002 µF	Ceramic / Mylar	202
15 pF	Ceramic		2,200 pF /.0022 µF	Ceramic / Mylar	222
20 pF	Ceramic		4,700 pF /.0047 µF	Ceramic / Mylar	472
30 pF	Ceramic		5,000 pF /.005 µF	Ceramic / Mylar	502
33 pF	Ceramic		5,600 pF /.0056 µF	Ceramic / Mylar	562
47 pF	Ceramic		6,800 pF /.0068 µF	Ceramic / Mylar	682
56 pF	Ceramic		.01	Ceramic / Mylar	103
68 pF	Ceramic		.015	Mylar	
75 pF	Ceramic		.02	Mylar	203
82 pF	Ceramic		.022	Mylar	223
91 pF	Ceramic		.033	Mylar	333
100 pF	Ceramic	101	.047	Mylar	473
120 pF	Ceramic	121	.05	Mylar	503
130 pF	Ceramic	131	.056	Mylar	563
150 pF	Ceramic	151	.068	Mylar	683
180 pF	Ceramic	181	.1	Mylar	104
220 pF	Ceramic	221	.2	Mylar	204
330 pF	Ceramic	331	.22	Mylar	224
470 pF	Ceramic	471	.33	Mylar	334
560 pF	Ceramic	561	.47	Mylar	474
680 pF	Ceramic	681	.56	Mylar	564
750 pF	Ceramic	751	1	Mylar	105
820 pF	Ceramic	821	2	Mylar	205

sure you have correctly identified each resistor, and you know where each resistor is placed on the PC board, you can go ahead and install them on the circuit board. Solder the resistors in place and then trim the excess component leads with your end-cutters. Cut the excess component leads flush to the edge of the circuit board. Install the remaining resistors, solder them in place and remember to trim the excess leads from the board.

Next we will move on to installing the capacitors. This project uses two types of capacitors, small mica or polyester types and electrolytic types. The smaller capacitors may not have their actual value printed on them, but might have a three-digit code on them. Refer to the capacitor code chart to help identify each capacitor before installing them on the PC board. Slowly identify each capacitor and place it in the correct location on the circuit board. Solder these small capacitors in their respective locations after referring to the schematic and parts layout diagrams. This project also utilizes larger electrolytic capacitors. Electrolytic capacitors have polarity and this must be observed if the circuit is to work correctly. Take a look at one of the larger electrolytic capacitors and you will notice that it may have a black or white color band or a plus (+) or minus (−) marking on one side of the capacitor body. When placing these types of capacitors on the circuit board, be sure you have oriented the capacitor correctly.

Install the electrolytic capacitors on the circuit board and solder them in place; remember to trim the excess component lead with your end-cutter.

Let's move on to installing the silicon diodes. The AM radio project uses six silicon diodes, and they too have polarity which must be observed. Each diode will have either a white or black band at one end of the diode's body. This band denotes the cathode lead of the particular diode. Refer to the parts layout and schematic diagram when installing the six resistors. Place the diodes on the circuit board and solder them in place on the board. Don't forget to cut the extra lead lengths.

Locate the semiconductor pin-out diagram shown in Figure 5-6. Now, go ahead and identify all three of the transistors for the project and install the transistors on the circuit board. There are two NPN types and one PNP type transistors in the AM radio project. Transistors usually have three leads, a Base lead, a Collector lead and an Emitter lead. The symbol for a transistor is a vertical line which is the Base lead and two lines running at 45°angles with respect to the Base lead. The angled lead with the arrow pointing towards or away from the Base lead is the Emitter lead and the remaining lead is the Collector lead. Be sure you can identify each of the transistor leads before placing them on the circuit board. Solder the transistors in place on

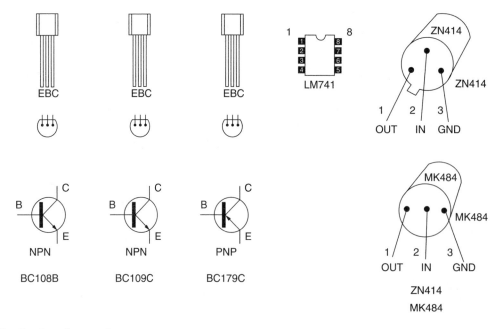

Figure 5-6 *Semiconductor pin-outs*

the circuit board; when finished remember to trim the excess component leads.

The ZN414 or MK484 TRF module looks very much like a common three-lead transistor. Carefully refer to the pin-out diagram for the MK484 before installing, in order to prevent damaging it upon power-up. One lead is the input lead from the tuned circuit. One lead is the ground lead and the remaining lead is the output lead which is directed to the actual circuit.

The AM radio project uses a single dual in-line integrated circuit, an LM471. We elected to use an IC socket as insurance against a possible circuit failure at some point in time. Integrated circuit sockets are inexpensive and make life easier in the event of a circuit problem, since it is much easier to unplug an IC rather than attempting to unsolder a multi-pin integrated circuit. Go ahead and solder-in a good quality IC socket for U2. Take a look at U2 and you will notice that at one end of the plastic package you will find a cut-out, notch or indented circle. Pin 1 of the IC will be just to the left of the notch or cut-out. Be sure to align pin of the IC to pin 1 of the IC socket, pin 1 of U2 is the null offset pin which is not used in this circuit. Note that pin 2 of U2 is the inverting input which is connected to the junction of R6 and R8.

Next locate the ferrite inductor coil or L1; you will notice that it is in parallel with the tuning capacitor C2. The parallel tuned circuit is placed between capacitor C1 and the input of U1, the MK484 TRF IC. These components are best placed at one end of the PC board if possible, or can be placed off the PC board on the front of your enclosure. Now, locate the volume control, the 10k ohm potentiometer; it is a chassis mounted control which can be mounted on the front panel of your enclosure.

Let's wrap up the project. First you will want to solder the battery clip; note that the black wire of the battery clip will go to the circuit board minus or ground bus and the red battery clip wire will go in series with one post on the on-off switch at S1, the remaining post of the SPST toggle switch will connect to the Collector of Q1. Finally locate two 4″ pieces of #22 ga. insulated hookup wire. Bare both ends of the two wires, solder one end of each wire to the speaker leads. Now take one free wire end and solder it to the common ground of the circuit and then solder

the remaining free wire end to the minus end of capacitor C8.

This completes the construction of the AM radio project. Let's take a well deserved rest and when we return we will check the circuit board for possible "cold" solder joints and possible "short" circuits. Pick up the circuit board with the foil side facing upwards toward you. Look closely at the circuit board, first we will inspect the circuit board for possible "cold" solder joints. The solder joints should all look clean, shiny and bright. If any of the solder joints look dull, dark or "blobby" then you should remove the solder from that joint and re-solder the joint over again. Now pick up the circuit board once again and we will look for possible "short" circuits. "Short" circuits are usually caused from one of two reasons. The first cause of "short" circuits are loose solder "blobs" or solder balls which can sometimes stick to the circuit board. The second cause for "short" circuits is to have "stray" component leads stick to the foil side of the board. Rosin core solder residue can often be sticky and often solder "blobs" or "stray" component leads can bridge across circuit traces causing a "short" circuit. Look over the board and remove anything that looks like it might be bridging any circuit traces. Once the circuit inspection has occurred, you can prepare the circuit for its first test.

Connect up a fresh 9 volt battery to the AM radio circuit. Once the battery is connected to the circuit, you can "flip" the on-off switch to the on position and you should begin to hear some "hiss" from the speaker. If you do not hear anything, turn up the volume control to mid-position. At this point you should begin to hear some sounds from the speaker. Move the tuning capacitor control across the AM radio band and if all goes well you should begin to hear some local stations. Move the ferrite core coil or the circuit board around in a 360° pattern to see where the station is loudest. Hopefully the radio works fine without any further adjustments.

In the event that the AM radio circuit does not work at all, you will have to remove the battery from the circuit and inspect the circuit board one more time, but this time you will be looking for components placed incorrectly. Remember earlier, we talked about certain components have polarity such as diodes, electrolytic capacitors, transistors, etc. Now, you will need to inspect the PC board closely to see if you might have placed a component incorrectly. Its easy to make a mistake and insert a capacitor or diode incorrectly; it happens to professionals too! Another possible cause for failure is that you may have placed an IC incorrectly in the socket! Finally, it's easy to incorrectly install one of the transistors. Look at each transistor carefully, you may want to refer to the manufacturer's

specification and/or pin-out information sheets along with the circuit layout and schematic diagram to confirm transistor pin-outs. Once you have inspected the circuit board for misplaced components and have corrected any mistakes, you can reapply power to the AM radio circuit and check out the circuit once again. Have fun listening to your new AM radio, you will have the great satisfaction of listening to the radio that you just built.

Chapter 6

Solid-State FM Broadcast Receiver

Parts list

FM Radio Receiver

R1,R15   4.7k ohm, ¼w, 5% resistor (yellow-violet-red)

R2,R18   1K resistor, ¼w, 5% resistor (brown-black-red)

R3,R16,R17   270 ohm, ¼w, 5% resistor (red-violet-brown)

R4,R11   47k ohm, ¼w, 5% resistor (yellow-violet-orange)

R5,R7,R8,R9,R14   10K, ¼w, 5% resistors (brown-black-orange)

R6   470 ohm, ¼w, 5% resistor (yellow-violet-brown)

R10   2.2k ohm, ¼w, 5% resistor (red-red-red)

R12,R13   10k ohm, potentiometers

R19   2 ohm, ¼w, 5% resistor (red-black-gold)

C1   4.7 or 5pF, 35v disk capacitor (marked 5 or 5K)

C2,C3,C7,C8,C16,C18, .01µF, 35v disk capacitor (marked .01 or 103 or 10 nF)

C4,5,15,27   .001µF, 35v disk capacitor (marked .001 or 102 or 1 nF)

C6,C14   100 pF, 35v disk capacitor (marked 100, 101, or 101K)

C9,C17,C22,C26   4.7 or 10 µF, 35v electrolytic capacitor

C10,C13,C19   100 to 220 µF, 35v electrolytic capacitor

C11,C12   22 pF, 35v disk capacitor (marked 22)

C20,C21,C23,C24,C28,C29 .01 µF, 35v disk capacitor (marked .01 or 103 or 10 nF)

C25   .1 µF, 35v disk capacitor (marked .1 or 104)

FL1   10.7 MHz ceramic filter; Digi-Key Electronics

L1   slug tuned plastic molded coil

L2   10.7 MHz shielded can style IF transformer (421F122)

D1 varactor diode,
  33 pF

D2 Zener diode,
  6.2 volt

U1 ULN2111 or MC1357 IC

U2 LM386 IC

U3 SA602 IC

Q1,Q3 2N3904 NPN
  transistor

Q2 2SC2498 or 2SC2570A
  NPN transistor

B1 9 volt transistor
  radio battery

Misc PC board, battery
  holder, battery clip,
  RCA jacks

⅛″ earphone jack, wire,
  antenna

Figure 6-1 *FM receiver*

The FM broadcast receiver will permit you to listen to your favorite radio shows as well as allow you to tune outside of the standard FM broadcast band to "listen-in" to FM room bugs, FM mikes and other interesting hidden transmitters. The FM receiver could also be connected to an SCA adapter, so that you may listen to 'MUSAK' radio show used in stores and restaurants. The FM radio project is just plain fun and will teach you about FM radio theory.

The FM broadcast receiver demonstrates the basics of FM radio theory in one compact, simple circuit! Separate circuits are "sections" of the receiver, making it easy to visualize every function. The latest in IC circuitry is used in the FM receiver, which allows high performance.

The FM receiver project is designed to receive wide-band FM transmissions typical of FM broadcast stations, wireless mikes, and telephone transmitters or "bugs," rather than Narrow Band FM transmissions (NBFM) as used on police or public service radios.

Why not build your own FM Broadcast receiver like the one shown in Figure 6-1. Not only will you learn how FM receiver works but you will be able to build and utilize your new FM radio. The FM receiver is great for listening to FM broadcasts and news as well as listening-in to commercial SCA—free music, stock quotes, sports, etc. The receiver has an SCA or sub-carrier

output which can be fed into an SCA converter to allow this reception of these signals.

The receiver tunes from 70 to 110 MHz, ideal for tuning in out-of-band FM wireless mikes, "bugs" and other "hidden" transmitters. The FM receiver also features plenty of audio output, and operates from a standard 9 volt battery for truly portable operation.

FM modulation

Frequency modulation (FM) is a form of modulation which represents information as variations in the instantaneous frequency of a carrier wave. Contrast this with amplitude modulation, in which the amplitude of the carrier is varied while its frequency remains constant. In analog applications, the carrier frequency is varied in direct proportion to changes in the amplitude of an input signal. Digital data can be represented by shifting the carrier frequency among a set of discrete values, a technique known as frequency-shift keying. The diagram in Figure 6-2 illustrates the FM modulation scheme, the RF frequency is varied with the sound input rather than the amplitude.

FM is commonly used at VHF radio frequencies for high-fidelity broadcasts of music and speech, as in FM broadcasting. Normal (analog) TV sound is also broadcast using FM. A narrowband form is used for voice communications in commercial and amateur radio settings. The type of FM used in broadcast is generally

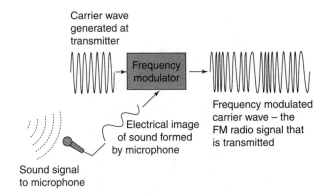

Figure 6-2   *FM modulation*

called wide-FM, or W-FM. In two-way radio, narrowband narrow-fm (N-FM) is used to conserve bandwidth. In addition, it is used to send signals into space.

Wideband FM (W-FM) requires a wider bandwidth than amplitude modulation by an equivalent modulating signal, but this also makes the signal more robust against noise and interference. Frequency modulation is also more robust against simple signal amplitude fading phenomena. As a result, FM was chosen as the modulation standard for high frequency, high fidelity radio transmission: hence the term "FM radio." FM broadcasting uses a well-known part of the VHF band between 88 and 108 MHz in the USA.

FM receivers inherently exhibit a phenomenon called capture, where the tuner is able to clearly receive the stronger of two stations being broadcast on the same frequency. Problematically, however, frequency drift or lack of selectivity may cause one station or signal to be suddenly overtaken by another on an adjacent channel.

Frequency drift typically constituted a problem on very old or inexpensive receivers, while inadequate selectivity may plague any tuner. Frequency modulation is used on the FM broadcast band between 88 and 108 MHz as well as in the VHF and UHF bands for both public service and amateur radio operators.

FM receiver circuit description

The FM receiver block diagram in Figure 6-3 illustrates the fundamentals of FM reception. The schematic diagram of the FM broadcast receiver is depicted in Figure 6-4. FM broadcast signals captured by the whip antenna or from the input jack are applied to the RF amplifier Q2, a high gain, low noise microwave style transistor. Input signals are amplified about 100 times or equivalent to 20 dB. After being boosted, these input signals are routed to the SA602 mixer-oscillator chip for conversion down to the 10.7 MHz Intermediate Frequency (IF). The SA602 is a very popular chip used in a great variety of receivers from cellular telephones to satellite receivers. The reason for its popularity is that it contains most of a receiver's "front end" circuitry on a single, easy-to-use chip. Internal to the SA602 is an oscillator tuned by an external tank circuit; C11, 12, L1 and varactor diode D1. A varactor diode acts like a voltage variable capacitor and the voltage applied across it comes from tuning pot R12. This IF signal is an exact reproduction of the desired signal, but at 10.7 MHz. We then band-pass filter it with ceramic filter FL-1.

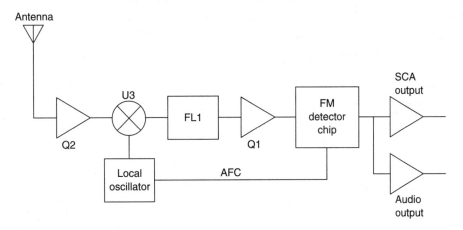

Figure 6-3   *FM broadcast receiver block diagram*

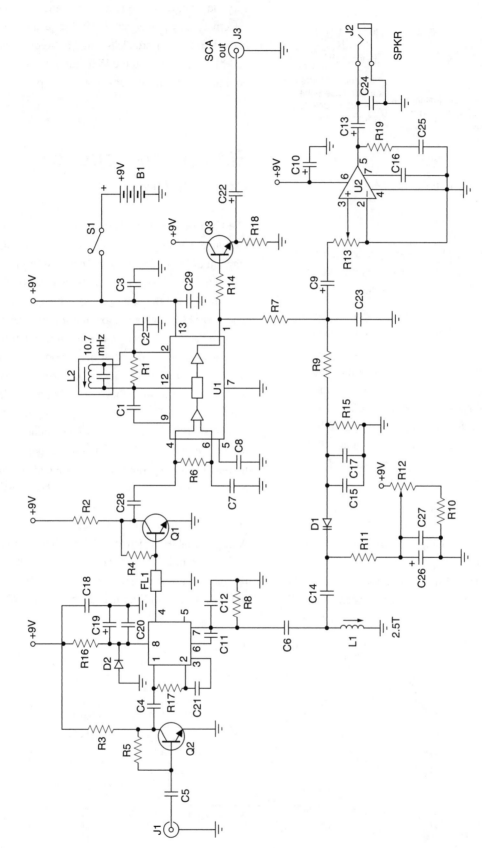

Figure 6-4 *FM broadcast receiver: Courtesy of Ramsey Electronics*

The ceramic filter allows only the desired 10.7 MHz signal to pass, rejecting any adjacent channel signals. Transistor Q1 amplifies the IF signal before it is applied to the detector chip U1.

One of the primary advantages of FM is that the program information (modulation) is contained in carrier frequency variations and not in amplitude changes like AM, which would be susceptible to noise and interference. This means that we can limit or clip the signal, thus removing any noise and not affecting the desired information. The first section of U1 performs this limiting function, providing high gain and clipping any amplitude (noise) variations. FM is detected in the second section using a process known as quadrature detection. We'll not get into this too deeply because it requires a knowledge of trigonometry identities—let's just say it extracts the frequency modulated program information. Capacitor C1 and coil assembly L2 comprise a phase shift network that is required by the quadrature detector. Demodulated signals are amplified and sent out through pin 1 to the volume control, AFC circuitry and the SCA buffer amp. The AFC circuitry is used to properly keep the receiver locked on to the tuned-in radio station. Resistor R9 and capacitor C17 filter out any audio voltage variations so that only major frequency changes will shift the tuning voltage to lock the frequency.

Audio signals from the volume control are amplified to speaker levels by U2, an LM-386 audio amplifier chip. Transistor Q3 provides a buffer stage to external SCA devices you may connect up to the receiver. SCA units decode (demodulate, actually) "hidden" audio and data services transmitted over standard FM broadcast channels. These services include stock quotes, sports scores, talk shows and even background music—commercial free!

Circuit assembly

Look for a clean well lit area in which to assembly your shortwave receiver. You will need a 27 to 30 watt pencil tip soldering iron, some 60/40 tin/lead rosin core solder. You will also need a few hand tools including a pair of end-cutters, needle-nose pliers, a few screwdrivers and a pair of tweezers to pick up small parts. You will want

to locate a small jar of "Tip Tinner" from your local Radio Shack store. "Tip Tinner" is a soldering iron tip cleaner/dresser which is invaluable when soldering circuit board. Locate the schematic and layout diagrams and we will begin to construct your new shortwave receiver. Place the project circuit board in front of you and we will begin building the receiver.

Locate all of the resistors for the project and place them in front of you. Each resistor has a color code which describes its value. Check the color code against the parts list and schematic before installing them onto the circuit board. Refer to the resistor color chart in Table 6-1. Place about four or five resistors on the PC board and place these resistors in their respective locations carefully making sure each on is in the correct location. Let your soldering iron tip heat both the component lead wire and PC board trace enough so that the wire itself and the foil trace both become hot enough together to melt a bit of solder so that it flows smoothly from the component leads to the PC board trace. Now, go ahead and solder each of the resistors onto the circuit board. Trim off the excess component leads with and small pair of end-cutters. Pick up and identify the next grouping of resistors and install them on the circuit board, follow up by soldering this next batch of resistors to the circuit board, remember to trim the extra lead lengths form the PC board.

Once all the resistors have been installed on to the circuit board, we can move on to identifying all of the capacitors for the project. Note that there will be different looking capacitors for this project, some will be small disks, while others will be larger body devices with minus and plus marking on them. These larger capacitors with polarity marking are electrolytic capacitors and they have polarity which must be observed for the circuit to work correctly. Look through the parts pile and locate the capacitors and place them in front of you. You will notice a number of small disk capacitors, we will install these first. Look closely at each of the disk capacitors at their markings. Sometimes the capacitors are very small and they may not have their actual value printed on them but will instead have some sort of code. Refer to the parts list for these codes or the capacitor code chart Table 6-2. Locate four or five of these small disk capacitors, identify them and install them on the circuit board, while referring to the schematic and layout diagrams.

Table 6-1

Resistor color code chart

Color Band	1st Digit	2nd Digit	Multiplier	Tolerance
Black	0	0	1	
Brown	1	1	10	1%
Red	2	2	100	2%
Orange	3	3	1,000 (K)	3%
Yellow	4	4	10,000	4%
Green	5	5	100,000	
Blue	6	6	1,000,000 (M)	
Violet	7	7	10,000,000	
Gray	8	8	100,000,000	
White	9	9	1,000,000,000	
Gold			0.1	5%
Silver			0.01	10%
No color				20%

After inserting the disk capacitors on to the circuit board, you can go ahead and solder them to the circuit board. Remember to cut the excess component leads. Next, move on and install another grouping of small disk capacitors, then solder them in place on the printed circuit board. Cut the extra component leads after soldering in the capacitors. When you are finished installing the small capacitors we will move on to the larger electrolytic types.

Next, we will identify and install the electrolytic capacitors. Electrolytic capacitors each have polarity markings on them in the form of a black band with either a plus or minus marking, which denotes the polarity of the device. You must pay particular attention to these markings in order for the FM receiver to work properly. Refer to the schematic and parts layout diagrams when installing C9, C17, C22 and C26, which are all 4.7 to 10 µF electrolytic capacitors. Install these four capacitors and solder them to the circuit board. Remember to trim the excess component leads. Cut the excess leads flush to the edge of the PC board with a pair of small end-cutters. Now locate and install electrolytic capacitors C10, C13 and C19, these are all 100 to 200 µF electrolytic types. After placing them on the PC board in their proper location and with respect to polarity marking you can go ahead and solder them to the PC board.

At this time, locate diodes D1, the varactor "tuning" diode and D2, a Zener diode from the parts pile. Install D1, varactor diode. It has a black body with green and red bands. Observe the correct positioning of the banded end by following the silkscreen. The green band is positioned as shown on the silkscreen. Now, install D2, the Zener diode. It has a gray painted glass body with a black band. Orient the band as shown on the layout diagram. A Zener diode acts as a voltage regulator, keeping the voltage across it constant. This keeps frequency drift to a minimum. Once both of the diodes have been placed on the PC board you can solder them in place and then trim the extra lead lengths flush to the edge of the circuit board.

Next, we are going to locate and install the two inductors; see Table 6-3. Locate the slug tuned plastic molded coil, which is coil L1. Once you have located it, you can place it on the PC; remember to check the schematic and parts layout diagram for placement details. Now, locate L2, the 10.7 MHz shielded metal quadrature coil. Notice that it will only go into the PC board one way. Place L2 on the board and now solder both L1 and L2 to the PC board. Trim the excess component leads from the board.

At this time, you can locate FL1, the 10.7 MHz ceramic IF filter. This part looks like a disk capacitor

Table 6-2

Capacitance code information

This table provides the value of alphanumeric coded ceramic, mylar and mica capacitors in general. They come in many sizes, shapes, values and ratings; many different manufacturers worldwide produce them and not all play by the same rules. Some capacitors actually have the numeric values stamped on them; however, many are color coded and some have alphanumeric codes. The capacitor's first and second significant number IDs are the first and second values, followed by the multiplier number code, followed by the percentage tolerance letter code. Usually the first two digits of the code represent the significant part of the value, while the third digit, called the multiplier, corresponds to the number of zeros to be added to the first two digits.

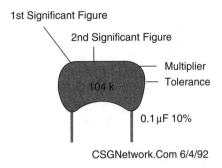

CSGNetwork.Com 6/4/92

Value	Type	Code	Value	Type	Code
1.5 pF	Ceramic		1,000 pF /.001 μF	Ceramic / Mylar	102
3.3 pF	Ceramic		1,500 pF /.0015 μF	Ceramic / Mylar	152
10 pF	Ceramic		2,000 pF /.002 μF	Ceramic / Mylar	202
15 pF	Ceramic		2,200 pF /.0022 μF	Ceramic / Mylar	222
20 pF	Ceramic		4,700 pF /.0047 μF	Ceramic / Mylar	472
30 pF	Ceramic		5,000 pF /.005 μF	Ceramic / Mylar	502
33 pF	Ceramic		5,600 pF /.0056 μF	Ceramic / Mylar	562
47 pF	Ceramic		6,800 pF /.0068 μF	Ceramic / Mylar	682
56 pF	Ceramic		.01	Ceramic / Mylar	103
68 pF	Ceramic		.015	Mylar	
75 pF	Ceramic		.02	Mylar	203
82 pF	Ceramic		.022	Mylar	223
91 pF	Ceramic		.033	Mylar	333
100 pF	Ceramic	101	.047	Mylar	473
120 pF	Ceramic	121	.05	Mylar	503
130 pF	Ceramic	131	.056	Mylar	563
150 pF	Ceramic	151	.068	Mylar	683
180 pF	Ceramic	181	.1	Mylar	104
220 pF	Ceramic	221	.2	Mylar	204
330 pF	Ceramic	331	.22	Mylar	224
470 pF	Ceramic	471	.33	Mylar	334
560 pF	Ceramic	561	.47	Mylar	474
680 pF	Ceramic	681	.56	Mylar	564
750 pF	Ceramic	751	1	Mylar	105
820 pF	Ceramic	821	2	Mylar	205

Table 6-3
Inductor coil winding information

L1   2½ turn coil #22 ga. insulated magnet wire on slug
tuned coil form. Unwind original coil and rewind.
2½ turns onto Series 48 adjustable coil form
48A Series Coil (48A117MPC) Circuit Specialists.

L2   10.7 MHz shielded can style IF transformer (red)
Part # (421F122) Mouser Electronics.

but has three leads. Amazingly, its function is equal to a small handful of inductors and capacitors hooked up in a bandpass filter configuration. For those techno-types, its bandwidth is 280 kHz centered at 10.7 MHz with an insertion loss of only 6 dB. Go ahead and solder FL1 in now.

Let's locate the transistors for the receiver now! Next, refer to the semiconductor pin-out diagram shown in Figure 6-5. Transistors are three semiconductors which generally have three leads: Base lead, a Collector lead and an Emitter lead. The transistor symbol shows that the Base lead is a vertical line. Both the Collector and Emitter leads are diagonal lines which point to the Base lead. The Emitter lead has a small arrow which points toward or away from the Base lead. Try and locate transistor Q1, a 2N3904 type. Watch placement of its flat side, refer to the parts layout diagram and the

schematic for proper placement. Install Q2, 2SC2498 or 2SC2570A NPN transistor. Transistors are VERY critical of correct orientation of their leads. Be sure to pay close attention to the position of the flat side of the transistor. Finally, install Q3, 2N3904 transistor. Be sure to observe flat side orientation of the transistors when mounting them. Now go ahead and solder the transistors in place on the PC board.

Let's take a short break and when we return we will install the integrated circuits and the final components. Before installing the integrated circuits, please consider using integrated circuit sockets as an insurance policy against having to try and un-solder a 14 pin IC from the circuit board which is a difficult task even for the seasoned electronics technician. Integrated circuits DO fail and if one ever does, you can simply unplug it and replace it with a new one. Refer to the schematic and parts layout diagrams when installing the integrated circuits. Solder in the IC sockets in their respective locations and then we will install the actual ICs.

Integrated circuit packages will generally have a cut-out, notch or indented circle at the top end of the IC package. The cut-out or notch will usually be at the top center while if there is an indented circle it will be at the top left of the IC package. Pin 1 of the IC will always be left of the cut-out or notch, so you can easily identify the pin-outs. Using your schematic and parts

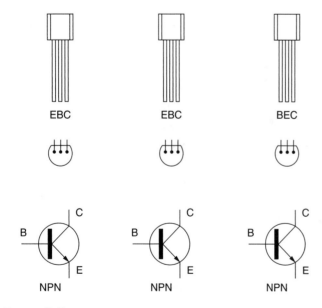

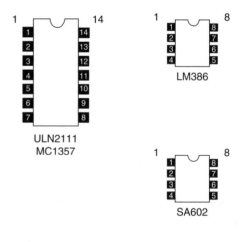

Figure 6-5 *Semiconductor pin-outs*

layout diagrams observe the correct orientation of the integrated circuits before inserting them into the sockets. First locate and install U1, the ULN2111 FM demodulator chip. Since this chip has 14 pins, be especially careful not to bend any pins under as you insert it into the PC board. When you solder the chip, verify that you are soldering all 14 pins. Next, let's install U2, the LM386 audio amplifier IC chip. The LM386 is a very popular part, and with good reason. Tucked inside that little plastic case are about 20 transistors that take a feeble audio signal and amplify it up to room-filling volume! Finally, go ahead and install U3, the SA602 IC mixer-oscillator. The marked end of the IC (band or dot) must face the edge of the PC board. If you wish, install an 8-pin DIP socket, still remembering to orient and install the SA602 correctly. Remember to firmly press the integrated circuits into their sockets.

The receiver project has two potentiometers, R12 is the "tuning" control, while R13 is the volume control. Locate the two 10k ohm potentiometers and install them, being sure to refer to the schematic and parts layout diagrams. Note that both potentiometers are on the front panel of the PC board. Install potentiometer R12, the tuning control. Insert it into the PC board so it seats firmly and is straight and even. Solder the three pins and the two mounting tabs. Next, install potentiometer R13, the volume control in the same manner.

Now locate S1, the on-off push button switch. Solder all six pins firmly. We've installed some "landmark" parts so far just to get us used to soldering our PC board. Finally, we are going to place the panel jacks on the rear portion of the circuit board. Locate and install J1, the RCA antenna jack, solder all four points. Now locate and install J2, the speaker jack, solder all three pins. At last, install J3, the SCA output jack, solder all four points.

There are only a few more components to go before we finish our FM receiver project. Install the 9 volt battery clamp. Use a scrap component lead wire looped through the holes in the clamp and into the holes in the PC board. Solder the wire to the clamp but don't use too much solder as this will prevent you from seating the battery fully into the clamp. Now, go ahead and install the 9 volt battery snap connector. Make sure you put the right color wire into the proper hole on the PC board. At long last, you are all done and ready to listen to tunes from your radio.

Before we apply power to the FM receiver for the first time we should make one last inspection of our work. First you will want to make sure that you have installed the diodes, electrolytic capacitors and the transistors with the correct orientation. Failure to do so may result in a damaged circuit, when power is first applied. Don't forget to check the orientation of the integrated circuits for proper location and orientation. Now we will inspect the PC board for "cold" solder joints. Pick up the PC board with the foil side facing upwards towards you. Take a look at all of the solder connections and make sure that they all look clean, smooth and shiny. If you find a solder joint that looks dirty, dark, dull or "blobby" then un-solder the connection, remove the solder and re-solder the connection once again until it looks good. Finally, we are going to inspect the PC board for possible "short" circuits. Often "cut" components leads can form "bridges" which can "short" out the circuit board traces, thus causing power to flow into the wrong places, and possibly damage the circuit. Also look carefully for any solder blobs that can form "bridges" between circuit traces. Now that you have inspected the PC circuit board, we can move on to testing and aligning your new receiver.

Testing and alignment

A hex head alignment tool is required to tune up your FM receiver. If you do not already have a set of plastic or nylon coil alignment tools and expect to build other radio or transmitter type kits, such tools are well worth having and can be found at any electronics store, including Radio Shack. While a hex style Allen wrench will fit the coil slug, the metal itself will drastically change the coil's inductance while it is in the slug. For this reason you should not use an Allen wrench unless you don't mind the trial and error process of turning the slug, withdrawing the tool, checking tuning, reinserting the tool, re-tuning, checking and so on.

You can fabricate your own non-metallic tuning tool with some patience, a bit of sandpaper and a piece of plastic rod. Try a large wooden matchstick, kebab skewer, or small plastic crochet needle. Carefully sand down the end to a hex shaped form and check for proper fit within the coil slug. It is better to have the tool fit loosely rather than too tight, since a tight tool will cause stress on the delicate slug.

Connect up a fresh 9 volt battery (preferably heavy duty alkaline type), a speaker or earphone, a short length of wire (about two feet) to serve as antenna, or the telescopic whip if the optional matching case and knob set was purchased. Now, set the volume and the tuning controls to the center of their range, then press the power switch "ON," you should hear some rushing noise from your speaker. Next, rotate the volume control and see if it does its job, then slowly turn L2's slug back and forth for maximum noise. This noise should be similar to what you would hear when a regular FM radio is tuned between stations. L2 may be tuned with a regular blade style screwdriver.

Try turning the tuning control, and see if you can receive any stations, but don't be alarmed if nothing is received since we haven't adjusted the oscillator coil yet. The oscillator coil, L1, is adjustable to receive from 70 to 110 MHz. We have to adjust it to cover the frequency range of interest. For alignment purposes we'll work with the understanding that the standard FM 88-108 MHz band is what we wish to receive.

Using an insulated alignment tool, slowly rotate L1's slug gently so that the slug is even with the top of the coil form. (Do not force the slug; if it binds, rotate the slug back and gently "rock it" to-and-fro as you would a thread forming tap. Slugs are by nature very brittle and will crack easily.) Now carefully rotate the slug into the coil form seven turns. The slug will be in about ¼″. Rotate the tune control until you hear an FM broadcast signal. Adjust L2 for best sound. Install R9, 10k ohm (brown, black, orange). R9 adds in AFC voltage to stabilize against drift. Without it your radio may "wander." After a signal is received, rotate the tune control to receive other stations. If the stations seem to "jump away" as you try to tune one in, the coil L1 is adjusted incorrectly and should be re-tuned. The reason for this odd phenomenon is that the AFC circuit is pushing the signal away rather than pulling it in. This is caused by having the oscillator 10.7 MHz below the desired signal rather than 10.7 MHz above it. A similar example of this is closing your eyes, hearing a sound, but not being able to determine if it is in front of, or behind you. Readjusting L1 will correct this situation.

Make sure the slug is about three turns from the top of the coil form.

Calibrating the receiver is easy. First tune in a station known to operate at the low end (88 MHz) of the FM broadcast band. Rotate the tune control almost fully counter-clockwise to where that station should logically be located on the dial. Turn L1's slug until that same station is received. The tune control should now tune in the entire FM broadcast band. An easy way of making your own dial is to paste a small rectangle of paper behind the tuning dial knob. Then as you tune in different frequencies, pencil in on the paper the received frequency.

Additional receiver considerations

Your FM receiver was designed for clean, undistorted audio from average FM broadcast stations, while keeping battery current drain low. You may significantly increase the FM radio's audio output by adding a single 10 uF electrolytic capacitor across pins 1 and 8 on the LM-386 audio amplifier IC. The "+" side should connect to pin 1. An easy way of adding this capacitor is to simply "tack" solder it right on the IC pins on the solder side of the PC board. A 50-cent of speaker, lying naked on your workbench, will not give you a fair test of your FM receiver audio quality or volume. Speaker quality and well-designed enclosures have their clear purposes! A reasonable speaker in a box delivers a pleasant listening volume as well as significantly improved fidelity.

Out-of-band operation

Although many people use their receiver for comfortable FM broadcast reception, some James Bond types will want to use it to receive clandestine transmissions from phone bugs, room transmitters, etc. It's a simple matter to re-tune coil L1 to receive the frequency band of interest. First make sure you verify that your receiver receives the FM broadcast band correctly before re-tuning.

SCA output

Many FM broadcast stations now transmit sub-carriers on their signals. Sub-carriers are signals that are modulated on the carrier, just like the normal audio signals, except that they are too high in frequency to be heard. Normal audio signals range in frequency from 20 to 20,000 hertz. Most sub-carriers start at 56,000 hertz (56 kHz). These sub-carriers are themselves modulated, sometimes with audio signals, such as background music, but more and more these days with various forms of data. Some of the information carried on these sub-carrier data services include stock quotes, weather reports, news, sports and even paging signals. There is no limit to the variety of data that can be sent on a sub-carrier signal, and broadcasters are finding new things to send all the time!

The SCA output is a baseband demodulated sub-carrier signal output before any low pass filtering or de-emphasis. It provides a handy connection point for add-on SCA adapters such as the Ramsey SCA-1 kit. You can also connect this output to a receiver capable of receiving 30 to 75 kHz FM signals, or a PLL demodulator you've built yourself. Check out the data sheet from Signetics concerning the NE-565 PLL chip; an SCA demodulator is featured.

Chapter 7

Doerle Single Tube Super-Regenerative Radio Receiver

Parts list

Parts Bin

Doerle Receiver – Original #19 Dual Triode

R1 2.2 megohm $\frac{1}{2}$w, 10% resistor

R2 330k ohm $\frac{1}{2}$w, 10% resistor

R3 47k ohm, $\frac{1}{2}$w, 10% resistor

R4 100k ohm, potentiometer panel mount

R5 25 ohm reostat

C1 5-18 pF antenna trimmer capacitor

C2 140 pF main tuning capacitor

C3 .001 mF, 200 volt disk capacitor

C4 100 pF, 200 volt disk capacitor

C5 .1 mF, 200 volt disk capacitor

C6 .005 mF, 200 volt disk capacitor

L1 2.5 mH choke coil – Digi-Key Electronics

L2 "tickler" coil (see text)

L3 main tuning coil (see text)

T1 #19 triode tube – 2.5 volt filaments

B1 Type A filament batteries (6 volts)

B2 Type B battery 90 volt battery pack

HP 200 ohm high impedance headphones

Misc chassis, tuning dial, wire, sockets, standoffs, screws, etc.

Doerle Receiver – 6SN7 Version

R1 2.2 megohm $\frac{1}{2}$w, 10% resistor

R2 1000 ohm, $\frac{1}{2}$w, 10% resistor

R3 470k ohm potentiometer panel mount

R4 100k ohm, $\frac{1}{2}$w 10% resistor

R5 100k ohm potentiometer

C1 8-18 pF trimmer capacitor

C2 100 pF main tuning
    capacitor

C3 50 pF variable tuning
    control (fine tune)

C4,C5 470 pF, 200 volt
    disk capacitor

C6 100 pF, 200 volt
    disk capacitor

C7 .01 mF, 200 volt
    disk capacitor

C8 .22 mF, 200 volt
    capacitor

L1 2.5 mH choke coil
    Digi-Key Electronics

L2 "tickler" coil
    (see text)

L3 main tuning coil
    (see text)

T1 6SN7 triode tube

B1 type A battery -
    6.3 volt power supply
    or batteries

B2 type B battery - 150
    volt power supply or
    batteries

HP 200 ohm high
    impedance headphones

Misc chassis, tuning
    dial, wire, standoffs,
    screws, etc.

Figure 7-1 *Doerle receiver*

In this chapter we will build a copy of the original Doerle radio and we will also show you how to build the Doerle radio receiver using a more modern tube such as the 6SL7, which may be a lot easier to locate these days. If you are a "purist" you may want to try and locate the original #19 triode tube, which is still actually available. You can build either version, which you can construct to look just like the original radio looked back in the 1930s.

This radio is a study in simplicity, the low parts count kept the original cost down. A single sheet of metal, probably steel, painted in black wrinkle finish became a chassis. The single #19 tube powered by batteries eliminated unnecessary tubes and power supplies. The result is a simple, high-performance shortwave radio that beginners could afford to build and operate during the depths of the Great Depression.

First let's examine the circuit diagram, shown in Figure 7-3 for the original Doerle radio. The antenna is

Old Time Radio lovers will enjoy building this single tube regenerative AM/Shortwave receiver called the Doerle Radio, shown in Figures 7-1 and 7-2. It is a classic radio from around 1934, which was offered as a radio kit that a radio enthusiast could build at home. This very sensitive receiver is a fun-to-build radio which can be powered from batteries as they did in the "old days" or you could construct an AC power supply for it. The original Doerle radio receiver was offered as a kit for $3.00. The early Doerle receiver used a #19 dual triode receiving tube, one section of the tube acted as a regenerative detector and the second section of the tube acted as an audio amplifier. Doerle perfected this receiver design in 1933 when the new #19 tube became readily available.

Figure 7-2 *Doerle receiver – inside view*

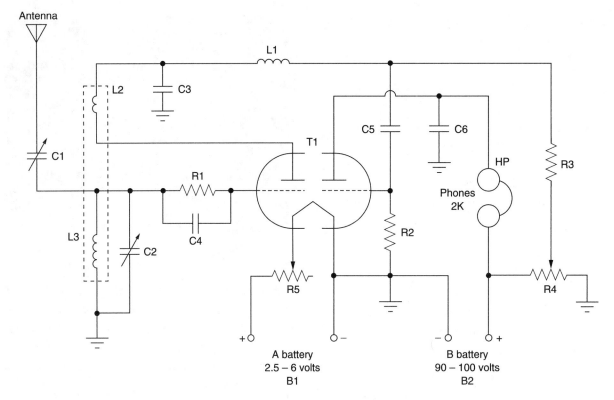

Figure 7-3 *Original #19 dual triode receiver*

coupled through a small 15 pfd screwdriver adjusted variable capacitor to the main tuning coil wound on a four pin form. A 10 pfd variable capacitor across it gives a fairly good trade-off between slow tuning rate and wide tuning range. The signal is fed through a grid leak configuration to one of the grids of the #19 tube. From the plate, the signal goes to the tickler wound on the same four pin coil form. The top side of the tickler coil connects to a 500 pfd capacitor and a 2.1 mH radio frequency choke (RF). The choke allows DC plate voltage into the tickler but prevents the radio frequency (RF) alternating current from getting into the rest of the circuit. The 500 pfd capacitor takes the RF to ground, thereby allowing RF voltage to build up on the tickler winding. Without this 500 pfd capacitor the receiver would probably not regenerate properly.

Variations in the DC voltage delivered to the detector plate feed through a .1 mfd capacitor to the second grid. A 250K, ½ watt resistor pulls the grid to DC ground, to provide the necessary zero bias for this class B amplifier tube, yet the resistance is high enough so that the AC avoids the ground and readily flows into the second grid.

Voltage on the detector plate is changed by a 50K resistor attached to the wiper of a 50K regeneration potentiometer that is connected from the B+ battery to

ground. Plate current in the detector flows through these two resistors. Variations in current creates a varying voltage drop across this pair. The varying voltage is the audio signal that is applied to grid of the audio triode.

The B+ voltage source is fed through headphones to the second plate to complete the audio amplifier part of the circuit. The sound that is heard in the headphones comes from the variations in the audio plate current. A .006 mfd capacitor on the plate is large enough to take any stray RF to ground, but is not so large to take audio frequency (AF) voltage to ground. This capacitor aids receiver stability and audio quality.

A 20 ohm rheostat in series with one of the filament legs allows adjustment of the voltage. A rheostat is a low resistance, high wattage potentiometer. In the 1930s, you would have used two No. 6 1.5 volt dry cells (they were about 6″ tall and 2 ¾″ diameter) as your filament power supply. The rheostat would cut the voltage down to two volts needed by the tube's filament. As the voltage provided by the batteries dropped with use, you would adjust the rheostat to maintain the proper voltage on the filament. In other words, the rheostat allowed you to continue using the filament batteries as their voltage dropped to the point of total failure. Since the filament current for a #19 triode tube is just over 250 mA, so a 20 ohm resistor would allow you to use filament batteries up to 6 volts. And quite frequently that 6 volts was the battery borrowed from the family car.

Building the original Doerle receiver

The first step in building this or any project is getting the parts. In fact, this may be the most difficult step. When it comes to building a replica of an old radio like this, you just cannot go out and buy everything you need. You may have to adapt and use whatever you have on hand or can scavenge. For a project of this type you may want to build the circuit using point-to-point wiring between terminal and components rather than a circuit board to give the radio an "old-time" look. You are probably going to need both a soldering iron and a soldering gun to build this radio. You will also want to consult with the resistor color code chart in Table 7-1 and the capacitor code chart in Table 7-2 in order to identify the correct components.

Building this old radio can be lots of fun, but locating all of these old components can be difficult, and in fact you won't be able to locate the exact same items. You'll get something close. You'll have to mix and match whatever you can find. Most of these components are available at flea markets if you visit enough of them. Many of the components are available new from dealers, such as Antique Radio Supply, see Appendix. But often you'll have to pay substantially higher yet fair prices. The used #19 tube was purchased from a flea market dealer for $5. Yet, that same tube can be obtained brand new from Antique Electronic Supply for about $6, see Appendix. Again, you may have to scrounge, make do, adapt, experiment, and make a few mistakes. But the results are well worth it.

One idea is to build a prototype receiver with whatever components you can find, no matter how new or old, and get it working. Then, over the next few years, visit flea markets, antique radio meets, check out dealers' catalogs, and accumulate the older, more authentic parts. After you accumulate some of those rare, old, beautiful parts from the 20% and 30%, rebuild the receiver. And then a few years later rebuild it again. And every time you rebuild it, add older and more authentic parts, until you get an exact copy of the #19 Doerle receiver. The radio will become more and more of an authentic replica of a receiver from the early days of radio. With patience, practice, and plenty of enjoyment, you can build a fine radio.

Looking at the illustration of the Doerle-1 version, we see in the center a large, slow-motion dial drive

Table 7-1

Resistor color code chart

Color Band	1st Digit	2nd Digit	Multiplier	Tolerance
Black	0	0	1	
Brown	1	1	10	1%
Red	2	2	100	2%
Orange	3	3	1,000 (K)	3%
Yellow	4	4	10,000	4%
Green	5	5	100,000	
Blue	6	6	1,000,000 (M)	
Violet	7	7	10,000,000	
Gray	8	8	100,000,000	
White	9	9	1,000,000,000	
Gold			0.1	5%
Silver			0.01	10%
No color				20%

Table 7-2

Capacitance code information

This table provides the value of alphanumeric coded ceramic, mylar and mica capacitors in general. They come in many sizes, shapes, values and ratings; many different manufacturers worldwide produce them and not all play by the same rules. Some capacitors actually have the numeric values stamped on them; however, many are color coded and some have alphanumeric codes. The capacitor's first and second significant number IDs are the first and second values, followed by the multiplier number code, followed by the percentage tolerance letter code. Usually the first two digits of the code represent the significant part of the value, while the third digit, called the multiplier, corresponds to the number of zeros to be added to the first two digits.

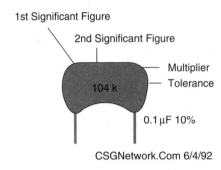

Value	Type	Code	Value	Type	Code
1.5 pF	Ceramic		1,000 pF /.001 µF	Ceramic / Mylar	102
3.3 pF	Ceramic		1,500 pF /.0015 µF	Ceramic / Mylar	152
10 pF	Ceramic		2,000 pF /.002 µF	Ceramic / Mylar	202
15 pF	Ceramic		2,200 pF /.0022 µF	Ceramic / Mylar	222
20 pF	Ceramic		4,700 pF /.0047 µF	Ceramic / Mylar	472
30 pF	Ceramic		5,000 pF /.005 µF	Ceramic / Mylar	502
33 pF	Ceramic		5,600 pF /.0056 µF	Ceramic / Mylar	562
47 pF	Ceramic		6,800 pF /.0068 µF	Ceramic / Mylar	682
56 pF	Ceramic		.01	Ceramic / Mylar	103
68 pF	Ceramic		.015	Mylar	
75 pF	Ceramic		.02	Mylar	203
82 pF	Ceramic		.022	Mylar	223
91 pF	Ceramic		.033	Mylar	333
100 pF	Ceramic	101	.047	Mylar	473
120 pF	Ceramic	121	.05	Mylar	503
130 pF	Ceramic	131	.056	Mylar	563
150 pF	Ceramic	151	.068	Mylar	683
180 pF	Ceramic	181	.1	Mylar	104
220 pF	Ceramic	221	.2	Mylar	204
330 pF	Ceramic	331	.22	Mylar	224
470 pF	Ceramic	471	.33	Mylar	334
560 pF	Ceramic	561	.47	Mylar	474
680 pF	Ceramic	681	.56	Mylar	564
750 pF	Ceramic	751	1	Mylar	105
820 pF	Ceramic	821	2	Mylar	205

Table 7-3

Short-wave band listening frequencies

Shortwave Band	Shortwave Band
120 meters = 2300–2495 kHz	25 meters = 11.500–12.160 MHz
90 meters = 3200–3400 kHz	22 meters = 13.570–13.870 MHz
75 meters = 3900–4000 kHz	19 meters = 15.030–15.800 MHz
60 meters = 4750–5060 kHz	17 meters = 17.480–17.900 MHz
49 meters = 5730–6295 kHz	16 meters = 18.900–19.020 MHz
41 meters = 6890–6990 kHz	13 meters = 21.450–21.750 MHz
41 meters = 7100–7600 kHz	11 meters = 25.670–26.100 MHz
31 meters = 9250–9990 kHz	

friction-driven by a small knob at the bottom of the set. This allows fine tuning of stations. To the upper left are the antenna and ground terminals. Just to the right in a hole in the front panel is exposed the screw which adjusts the antenna coupling capacitor. Below left is the filament rheostat knob. At the upper right are the headphone terminals with the regeneration control located at the lower right. Although the receiver as shown in the original drawings may appear complex, it really is not. This is one of simplest receivers that you could build.

The original chassis for the Doerle receiver was an L-shaped piece of sheet metal, most likely steel, and painted with a black wrinkle finish. While the front panel was perfectly rectangular, the base was trapezoidal. Dimples were embossed into the base plate which served as feet.

For our replica Doerle receiver, you can start with an 11″ × 9″ sheet of .063″ (¹⁄₁₆″) aluminum to construct the chassis. The most complicated mechanical problems that you may encounter in building the receiver is the mounting #19 dual triode tube up on standoffs, mounting the variable capacitors and building the coils on the coil-form, see Figure 7-4.

The original Doerle model used a simple single bearing variable capacitor. We elected to use a heavy duty Hammarlund tuning capacitor; it is of significantly higher quality and is much heavier, so you will have to support the rear end of the capacitor with a bracket of some type. Once a Z-bracket as fabricated and attached to the breadboard, the prototype dial-drive-capacitor assembly was solid, yet turned smoothly.

Since a solid, rigid mount for the tuning capacitor is essential if you are to achieve frequency stability and eliminate microphonics, you must get this part of the assembly correct. And since your components will need their own custom mounting hardware, you'll need to experiment until you get it right. That's why the breadboard prototype is so useful. Take your time.

In continuing the capacitor assembly, I used a pair of ½″ long 6-32 machine screws in the top and bottom pair of holes on the drive mechanism. In the left-right pair of holes, I used a pair of 1½″ long 6-32 machine screws. The extra length became mounting studs to which an aluminum cross-arm was attached. In the cross-arm was drilled a ⅜″ hole to accommodate the mounting flange of the capacitor.

After the capacitor holes were drilled, the burrs removed, and edges of the panel were rounded and smoothed, I roughed up the surface with fine sandpaper.

Figure 7-4 *Doerle coil assembly*

Next, the panel was bent into an L shape. You can use an old sheet metal brake, or you can could do the same by clamping the metal between a couple of 1 by 2's or lengths of angle iron held tightly together in a vise or with C-clamps. Since only a single, simple bend is required, you might get by clamping the sheet metal to a table top and by bending the sheet over the edge.

In order to obtain smooth tuning with a large range, you should locate some sort of planetary drive mechanism in front of the main tuning capacitor, such as a used National Velvet Vernier dial drive, which can be removed from an old radio. The first step was to dismantle it into the drive mechanism, the knob, and the dial plate. To mount the actual planetary drive mechanism, I had to cut a ⅝″ hole in the plate with a socket punch. Next, the drive collar was slipped through the ⅝″ hole. With a marker I marked the location of the four mounting holes.

In moving the capacitor and dial drive to the finished receiver, I chose to position the angle mounting bracket vertically as shown in the photos. This made fabrication and installation of a new base bracket much easier, and allowed more clearance for the tube and coil socket as well as for the terminal strip below the dial drive.

When the components are "test" mounted, you can move from the prototype stage to the finished L-shaped panel, if you are happy with the final layout. Then you can spray on a thick coat of fast drying metal primer. Once dry, I wet-sanded it under a stream of water until smooth. Then came the black wrinkle paint. If you wish to try and duplicate the original, you will want to try and create the wrinkle finish. If you are not that concerned about having the radio look exactly like the original radio, you can just paint the chassis with a single coat of black paint. Black Krylon paint is available in a spray can in your local home improvement store. For best results in creating a wrinkle finish, you must put down a heavy layer of paint. It must be heavy so that it can wrinkle up as it dries. If the coat is too light, or if you try to apply a second light coat over the existing layer, you won't get the best results. It might even be shiny where it is supposed to be flat black.

It is recommended that you paint one surface at a time and let it dry. Then reposition the chassis so that another surface is horizontal and then paint that. Of course, it will take four times as long to paint the radio than if you get in a hurry and paint the whole thing at once, but the results will be more predictable. You can bake it in an electric kitchen oven at the lowest possible heat. Nasty smelling solvents are baked out of the paint, so your wife or mother may not be too happy with you!

After mounting the tuning capacitor, the only other assembly of any complexity is the coil socket and antenna trimmer mount. Mounting the other components was fairly easily done. You can use machine screws, spacers of wood, plastic, porcelain, and metal. You can slice wood spacers from hardwood planks, but you'll need a table saw to do a good job. Plastic spacers can be cut from plastic tubing, but just be sure to get the rigid stuff used by plumbers for water supply lines. Porcelain spacers are something, which can be scavenged at a local hamfest or radio flea market. A small trimmer capacitor is mounted on porcelain standoffs in order to insulate both terminals from the chassis. Between the standoffs is a drilled ¼″ hole in the front panel to allow screwdriver adjustment. That sounds simple enough. But behind the trimmer is the socket that accepts the four-pin lube socket coil form. What I used here were a couple of 2″ 6-32 machine screws cut to the appropriate length with heavy clippers. They were then fed through the coil socket mounting holes and through brake line spacers into the porcelain standoffs. Two soldering lugs were sandwiched between each metal spacer and porcelain standoff. One lug on each side was soldered to a capacitor lead. The other lug on each side became connector for the trimmer. Since the metal spacer pressed tightly against the capacitor lead, it was imperative that the four pin socket be made of insulating material. Here I used an old phenolic wafer socket. A metal socket would short out the capacitor. It sounds complicated but it should be fairly easily understood by studying the photographs.

Wiring was done with some modern plastic covered hookup wire, and with some genuinely old cloth-covered hookup wire salvaged from a piece of old tube gear. Watch for old tube gear with long lengths of cloth-covered wiring at flea markets. New old stock cloth covered wire is around but it is not always easy to find. Sometimes you can look for old wire while scavenging at hamfests.

The coils used in the original circuit were 1½″ dia. by 3″ tall phenolic coil forms. You can use phenolic bases from four prong tubes for the coil forms and they can be purchased from Antique Electronic Supply. The low frequency or main "short-wave" coil is close-wound with about 30 turns of #26 wire. You wind the "tickler" coil, which is 9 or10 turns of #24 ga. wire below the tuning coil, and both coils were covered with clear fingernail polish. The coil covers approximately 2.5 to 4.5 MHz. The exact number of turns you'll need is going to depend on the tuning capacitor you use and the stray capacitance that results from your particular component layout, among other things. You'll have to experiment to get the right number of turns. The original Dorele receiver had plug-in coils for different radio bands. In order to construct an AM radio coil, you will have to wind another main coil along with another "tickler" coil on a second coil form. The main coil will need about 40 turns of #26 ga. wire, with a 9–10 turn "tickler" winding.

Battery terminals can also be purchased at a radio flea market. Using a table saw you could elect to slice a ¼″ strip from an oak 1×6. That gave me a mounting strip of ¾″ × 5 ½″ × ¼₁″ thick, just right for this radio.

Firing up your original Doerle receiver

Now your #19 Doerle radio receiver is bolted together and wired up. It's time to try it out. First, check the wiring, then check it again. It's easy to make stupid wiring errors, sometimes after checking the wiring once or twice! Have a friend look over the circuit before applying power to the circuit for the first time. The coils used in the original drawing look like 1½″ dia. by 3″ tall coil forms. You can use phenolic bases from four prong tubes. They're not from old tubes but from Antique Electronic Supply.

The low frequency coil is close-wound with about 30 turns of #26 wire. You can wind about 10 turns of tickler below the tuning coil, and cover both coils with clear fingernail polish. The coil covers approximately 2.5 to 4.5 MHz. The exact number of turns you'll need is going to depend on the tuning capacitor you use and the stray capacitance that results from your particular component layout among other things. You'll have to experiment to get the right number of turns.

The original ten turns of tickler turned out to be too many, so you may have to improvise to get it right. You'll discover this problem very quickly if, when you advance the regeneration control, the receiver goes into a deafening howl. You may need to remove one turn of tickler coil at a time until regeneration starts up smoothly near the end of the control clockwise rotation. You should adjust the tickler with the tuning capacitor open, that is, at the receiver's highest frequency. Here, more regeneration is necessary than at lower frequencies. And you must adjust the antenna trimmer control for the minimum capacity that gives excellent sensitivity. Adjusting the antenna trimmer "couples" or "de-couples" the connection between the receiver and the antenna, and can have a pronounced effect on regeneration adjustment and overall sensitivity.

The #19 tube filaments can be powered by alkaline flashlight batteries. Alkaline batteries should be used rather than cheaper zinc-carbon types since alkaline can deliver higher currents, and because they'll last much longer. Alkaline batteries will give many more hours of use for your money. You can hook the filament batteries up to the set with the rheostat set to maximum resistance. You can then rotate the rheostat, with a voltmeter connected to the filament leads, in order to allow the meter to show two volts. Then put a reference dot on the panel with a marker pen. You can do something similar to the finished machine. As the batteries age, their voltage will drop making it necessary to adjust the control to maintain the two volts on the filaments.

If you do not want to use flashlight batteries, you may consider nickel-cadmium batteries. They're inexpensive, each cell delivers about 1.2 volts, so you will need two, three or even four in series, which will deliver as much as 4.8 volts. And when they run down, you can recharge them. But before you go to that trouble: try flashlight batteries. You may be surprised by how long they last. As for a 90 volt B+ battery, you will most likely not be able to find them and if you do they will be expensive. Newark Electronics still had them, the last time I checked their catalog. It is easy enough to build a 90 volt "stack" of batteries. You can build a battery "pack" by buying ten 9 volt transistor radio batteries, and taping them together with electrical tape. When you do this, alternate their terminals so that you can easily wire the batteries in series.

The tube manuals lists the filament current at 260 mA. A voltmeter will show that the filament battery voltage drops about .2 volt over 15 to 20 minutes of operation. The batteries will last quite a long time. As for B+, a "stack" of 9 volt transistor radio batteries produced about 97 volts. With the receiver fired up and properly adjusted, the current draw was only 4 mA, significantly less than what a transistor radio battery is capable of delivering. After several minutes, the B+ battery dropped to 95.9 volts, so you can expect to get long life out of this battery pack. Several changes of filament batteries will be needed before the B+ battery dies.

So how does a Doerle radio receiver perform? For such a simple radio, it performs very well. Just a quick run through the band allowed me to copy amateur CW and marine Morse code. The foreign shortwave broadcasters came blasting-in.

Improvements? Yes, I can think of some. First, a volume control would be useful. I suspect that radio stations today run far more power than they did in the 30s. Therefore, the signals in the headphones are so loud as to be painful. A volume control would be useful to cut some signals down to size. I find the tuning a little too fast. Tuning a single sideband signal in is touchy and difficult in some cases. This could be fixed by going to five pin coils forms where the tuning capacitor could be attached to a tap in the tuning coil. Such a modification, however, changes the Doerle kit which was built for shortwave listeners, not amateurs. If you're a ham, you'll probably want to modify the design so that a ham band is spread out along the whole dial. And the last change that I would make is to improve the antenna trimmer adjustment. Adjusting the trimmer with a screwdriver can be painful after awhile. You don't do it very often, but when necessary it's not a convenient thing.

The 6SL7 Doerle receiver

In 1934, there was only one dual triode available for battery operation, the #19 dual triode! But the 1930s were years of explosive growth in the electronics industry. Just months after the #19 triode was released new metal tubes appeared, and in another ten years, seven and nine pin miniature tubes became available. These new series of tubes offered a variety of new dual triode tubes, many of which were better suited for the Doerle receiver circuit: see the 6SL7 Doerle receiver schematic shown in Figure 7-5.

If you elect to build this modified version of the Doerle, you will need to locate a 6SL7. The 6SL7 is a glass octal high-μ or (hi-gain) twin triode. The μ (mu) of a tube is the amplification factor, or voltage gain. Although the value of what is considered "high μ" has increased over the years, today "high" means a μ of from 60 to 100 or more. That means that the maximum voltage gain possible is 60 to 100, but usually it is significantly less.

High voltage gain tubes usually don't develop very much power. High power tubes usually have lower voltage gain. You can get both in the same tube. The #19 is an audio power tube with a relatively low μ. The 6SL7, on the other hand, has a μ of 70, but delivers little power. If you build both the #19 and 6SL7 versions, you'll immediately hear the increased headphone volume of the latter tube. Headphones don't need much power, so high voltage gain tubes works well, although more gain is usually better. A medium-μ tube like the #19 triode will give great performance. The 6SL7 was very common, and is still available. It is a higher-power, lower-voltage-gain tube with the same pin-out arrangement as the 6SL7.

The only major change introduced into this version of the Doerle is the mechanical layout. In this version, the prototype used an old National ACN dial drive, found at a local hamfest or flea-market, for a couple of dollars. I mounted it on an aluminum face place wrapped around a piece of 1×6 oak. The actual planetary drive is identical to that under the phenolic dial plates of the Velvet Vernier used in the model 19 Doerle. The ACN is much larger, $5'' \times 7''$ overall, and allows the user to mark a paper scale so that the dial becomes direct reading.

The antenna coupling trimmer was replaced by an 8-18 pF air variable stripped from an old radio. A custom bracket was snipped out of aluminum sheet and quickly bent up. By mounting the bracket on the wood base away from other meta components the capacitor electrically "floated" as needed. A knob allows the operator to reach around the side and easily adjust coupling when necessary. But you can take a similar small capacitor and cannibalize it by removing one or more plates to reduce the capacity. Just try get something less than 50 pfd for antenna coupling.

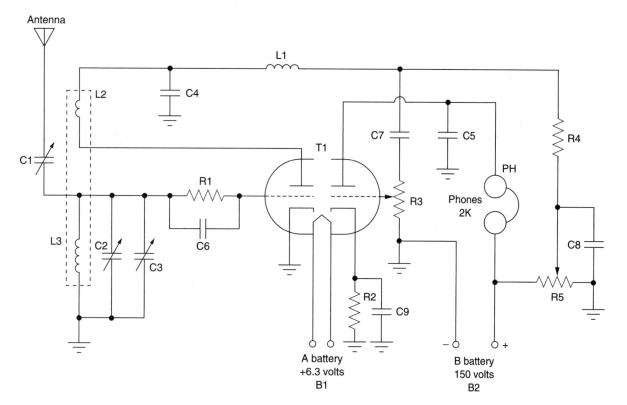

Figure 7-5 *New 6SL7 Doerle received*

A half megohm potentiometer replaced the 330 kilohm grid resistor used in the audio side of the original receiver, by connecting the panel-mounted pot to the rest of the circuit with miniature coaxial cable to reduce AC hum. In the original circuit, the audio grid was biased as before, but now has control over the volume. With a high gain tube like the 6SL7, it is important to reduce the audio output so that you don't destroy your hearing. A one kilohm resistor and a 10 mfd capacitor in parallel are inserted in the cathode lead of the audio triode to provide at least some negative bias to help the tube amplify with less distortion. You may want to increase or decrease the value of the resistor to achieve the necessary grid voltage specified in the tube manual. The voltage you measure across the cathode resistor is your grid bias.

If you compare the original circuit with this version, you'll find that almost all the same components can be used. The few differences are minor. And, again, don't be afraid to substitute. Just about anything within reason will work.

The biggest problem with this version of the Doerle receiver is that you will likely need a power supply,

since you have to supply 6.3 volts AC into the filaments. Remember the original Doerle radio had 2.5 volt filaments, and you'll need about 150 volts for the plates on the 6SL7.

Does the 6SL7 version of the Doerle radio work? It sure does! You can use the same coils used in the original version of the Doerle radio for the 6SL7 version of the radio. When powering this version of the Doerle receiver be sure to set the volume control all the way down before applying power, since the audio output is much greater than that of the original radio. Put on a pair of headphones and turn on the 150 volts B+. Remember, you must use high impedance type headphones for this type of radio, not 8 ohm types. The Internet is a good source of the old 2000 ohm high impedance headphones. Bringing up the regeneration and audio controls brought in stations immediately. Many stations appeared across the dial, and they were surprisingly easy to tune in.

One important change in this circuit is the use of two capacitors in the main tuning LC circuit. In the circuit shown, you'll see a 50 pF fixed capacitor is parallel with a 100 pF variable capacitor. Since capacitance in parallel adds, the combination of the two essentially creates a virtual variable capacitor with a range of 50 to 150 pF. The net effect is to slow down the rate of tuning. The amount of band that can be covered by a single coil is reduced, but makes tuning far less "touchy." This is important when tuning a sideband signal because very slight changes in tuning are needed to clear up the signal. You might want to replace the 50 pF with a 20–150 pF trimmer or something similar. By adjusting the trimmer, you can change the "speed" at which signals are tuned in. More capacitance reduces the speed, but it will also

decrease the band of frequencies covered by the coil being used. You might want to replace the 50 pF fixed capacitor with a 140 pF for use as a band-set capacitor and a 35 pF for use as a band-spread capacitor. Both should be equipped with slow-motion dial drives. You can use just about any dial drive that turns up at a flea market or in a radio parts catalog. You can mix and match components. Many times you'll need more mechanical skills than you will need electronic skills for building old radios, so dig through your junk box, visit a hamfest and have fun building an old radio!

Chapter 8

IC Shortwave Radio Receiver

Parts list

Shortwave Receiver Project

R1,R5,R20   100k ohm, $\frac{1}{4}$w
5% resistor [brown-black-yellow]

R2,R4   22k ohm, $\frac{1}{4}$w,
5% resistor

R3,R7,R9,R10   1k ohm,
$\frac{1}{4}$w 5% resistor
[brown-black-red]

R13,R17,R18,R24   1k ohm,
$\frac{1}{4}$w 5% resistor
[brown-black-red]

R6,R8,R11,R19,R23
10k ohm, $\frac{1}{4}$w
5% resistor
[brown-black-orange]

R12   1M ohm, $\frac{1}{4}$w
5% resistor
[brown-black-green]

R14,R21   10K
potentiometers, 103,
or 10nF

R15   2 ohm, $\frac{1}{4}$w
5% resistor
[red-black-gold]

R16,R22,R25   270 ohm,
$\frac{1}{4}$w 5% resistor
[red-violet-brown]

R26   100 ohm, $\frac{1}{4}$w
5% resistor
[yellow-violet-orange]

C1,C3,C20,C21,C25
.001 µF, 35 vdc disk
capacitor

C2   330 pF, 35 vdc
mylar capacitor

C4,C6,C8,C28   .01 µF
disk capacitors, 35
vdc (marked .01, 103,
or 10nF)

C5,C17   100 pF disk
capacitors, 35 vdc
(marked 100, 101,
or 100K)

C7,C10,C12,C18,C19,C29
1 µF ceramic disk
capacitor, 35 vdc
(marked .1 or 104)

C9,C24   1 µF, 35 vdc
electrolytic capacitor

C11,C22,C23,C27   10 µF,
35 vdc electrolytic
capacitor

C13,C14   220 µF, 35 vdc
electrolytic capacitor

C15   (see text)

C16   (see text)

C26   1000 µF, 35 vdc
electrolytic capacitor

C30   10 pF, 35 vdc
mylar capacitor

L1,L2   1 µH molded type
inductor
(brown-black-gold)-
Digi-Key

L3   12 µH molded type
inductor (brown-red-
black) -Digi-Key

T1,T2   shielded can
inductor
(42IF-103) Mouser
Electronics

T3 shielded can inductor (42IF-123) Mouser Electronics

D1 1N270 diode, glass bead style

D2 Red LED (RSSI)

D3 Varactor diode, transistor style with 2-leads (MVAM108)

Q1,Q2,Q3 2N3904 NPN transistors

U1 LM358 8-pin DIP IC

U2 LM386 8-pin DIP IC

S1 DPDT PC-mount pushbutton switch

J1 RCA-type PC-mount jack

J2 subminiature $\frac{1}{8}$ inch PC-mount jack

B1 9 volt transistor radio battery

Misc PC board, battery holder, battery clip, wire, hardware, speaker

Figure 8-1 *Shortwave radio*

Have you ever wanted to listen-in into the fascinating world of radio? The shortwave radio project will allow you to listen to international stations such as the BBC, Radio Canada, Radio Moscow, Radio Sweden. Using just a few feet of wire as you antenna you will be able to reach out and hear the world. Shortwave listening is fun for folks of all ages from 8 to 80 years of age.

The shortwave radio project shown in Figure 8-1 covers the shortwave bands from 4 to 10 MHz. The receiver uses varactor diode for smooth tuning across the band. The superheterodyne shortwave receiver provides both great sensitivity and selectivity. The tuning control as well as the volume and RF gain controls are all located on the front panel for easy operation. The shortwave receiver is powered from a common 9 volt transistor radio battery, so the receiver can be taken virtually anywhere. The receiver would make an ideal scout, school or club project.

You'll easily tune into broadcasts from many other countries as well. As you become more and more

familiar with the world of shortwave broadcasting, you'll be deciding on your own favorite band. You will hear a variety of other "interesting" sounds, but just remember that this receiver is designed for AM only. If a Morse code signal really sounds "good," it is because it is being transmitted in AM tone-modulated form, or perhaps the signal is so close to an AM broadcast carrier that the carrier acts as a "beat-frequency-oscillator" (BFO). Even though this receiver will permit you to tune through several different ham radio bands, the signals are not likely to be intelligible. Reception of CW and SSB signals on an AM receiver requires a BFO. This is not a complicated feature, but it is not a feature of this receiver. Our companion receivers designed for the ham bands will let you tune into these SSB and CW broadcasts.

Shortwave listening as a hobby in itself

Many people worldwide enjoy listening to shortwave broadcasts of all kinds, and they keep written records of what they hear. Almost every nation on Earth has some sort of shortwave broadcast service, though many are much more challenging to tune than the powerful signals of Radio Moscow and the BBC. In addition, these "SWLs" (Shortwave Listeners) listen to ham operators, government and commercial stations and even clandestine operations. Some shortwave listeners enjoy collecting QSL cards from stations which they have logged. Shortwave listening is, for some, a step toward getting a ham radio license. For others, it is a great hobby in itself.

The Shortwave Receiver is a good introductory receiver for this hobby. After you decide exactly what kinds of listening are of the most interest to you, you'll be in a better position to choose a more elaborate receiver.

To learn more about this SWL hobby, look for a copy of *Popular Communications* at newsstands. An inexpensive and interesting general introduction to all kinds of radio listening is the book, *Shortwave Listening Guide* by William Barden, Jr. (1987; Radio Shack Catalog Number 62-1084). To learn more about the hobby of ham radio, write to ARRL (American Radio Relay League), 225 Main Street, Newington, CT 06111.

Circuit description

Take a moment and examine the shortwave receiver block diagram in Figure 8-2. The simplified signal flow of the block diagram shows the basic sections of the receiver. The corresponding components are noted under each main block and can be cross-referenced to the schematic.

The main shortwave receiver diagram is illustrated in Figure 8-3. The receiver circuit begins at the antenna input at J1. RF signals (Fc = carrier frequency) from the

antenna are applied to the RF input and filtering allowing only the signals of interest to pass through. The high pass filter helps eliminate unwanted signals picked up by the antenna, so improving the overall reception quality of the radio. After the input signal is filtered, it moves to the mixer stage. Notice on the diagram that there are two inputs to the mixer. We have discussed one of these input signals coming into the mixer but not the other. Now, look down at the local oscillator block. The local oscillator (LO for short) acts as your tuning control for what frequencies you can receive by generating a signal on the board close in value to that which will be used by the mixer. There is a direct relationship between the generated frequency of the local oscillator (LO) and the exact receive frequency (Fc) you want to listen to. This will become clear when we finish discussing the block diagram. The LO section is a Colpitts oscillator that takes advantage of smooth varactor diode tuning. The varactor (D3) forms an L/C (inductor/capacitor) tank circuit with T3. Increasing the voltage on the varactor diode with R21 (tuning pot) increases the capacitance of D3, thus increasing the frequency output of the LO section.

Now that we know the two signals coming into the mixer stage, both the Fc (receive carrier frequency) and the LO (generated local oscillator), we can better cover its operation. The mixer takes these input signals and performs a few very basic operations. The technical explanation of how the mixer combines these

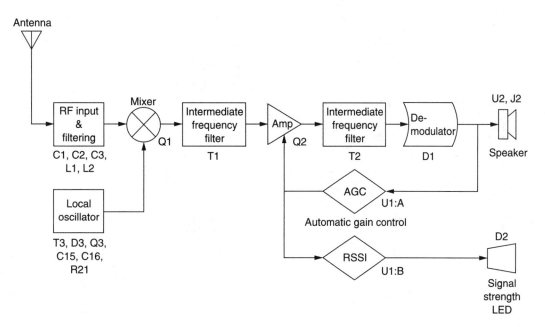

Figure 8-2 *Shortwave receiver block diagram*

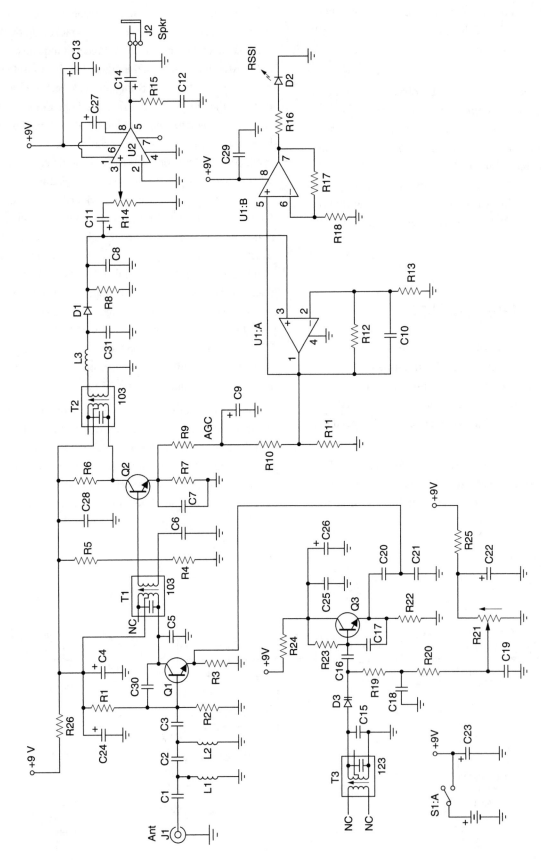

Figure 8-3   *Shortwave receiver schematic. Courtesy of Ramsey Electronics*

signals through Fourier Series is interesting but very drawn out. The point of using the block diagram, however, is to simplify matters. Therefore, the function of the mixer is to obtain the "product" and the "sum" of the input signals. This means you take the input signals and merely "add" and "subtract" their values to determine what you get on the output. The general formulas to use are quite simply, "Fc + LO = Mixer Output" and "Fc – LO = Mixer Output."

The output of the mixer stage unfortunately has other signals besides the ones we want. This brings us to the next stage, Intermediate Frequency Filtering. An Intermediate Frequency (IF) is a signal somewhere between the RF signal Fc we started with and the final audio message we are trying to get. The desired IF we are dealing with is a fixed number, such as 455 kHz. As long as the proper relationship between Fc and the LO exists, the IF value will stay constant. Due to the fact that the IF frequency stays the same all the time, the Intermediate Frequency Filter can be very narrow. The filter will remove any other signals coming from the mixer that are not in the proper pass band and yield a clean signal for further processing.

The next stage is an amplifier with an adjustable gain feedback loop. The gain control is dependent on the amount of signal being received (this Automatic Gain Control "AGC" will be covered in a moment). The amplifier boosts the signal level of the incoming IF and gives us a stronger signal to work with. After the amplifier stage is another Intermediate Frequency Filter. This helps remove any unwanted residual signals still present and cleans up the amplified IF for a high quality signal. At this point the audio signal we are trying to obtain is riding on the IF signal.

The demodulator circuit finally extracts the message from the IF section through a process called envelope detection. Now that we have our message back in the audio realm, it is directed through the audio circuitry to the speaker output. The demodulated audio branches off before the audio circuitry and is used to perform some useful functions. The RSSI LED, Received Signal Strength Indicator Light Emitting Diode, gives us a general signal level feature. The stronger the signal we receive, the brighter

the LED will glow. This is great for help pulling in those weak transglobal transmitter signals. The gain of the RSSI circuit is controlled by the value the feedback resistor R17. If you are commonly using a small whip antenna and listening to weak signals, you may need to increase the value of R17 to customize the response indication of D2, the signal strength LED.

The final stage, and the real secret to the sensitivity of the receiver, is the Automatic Gain Control (AGC). The gain control looks at the amount of signal level present at the output of the demodulator and varies the amount of gain the amplifier has accordingly. If there is a strong signal coming through the demodulator the AGC circuit lowers the gain of the amplifier. If the received signal coming through the demodulator is very weak, the AGC circuit increases the gain of the amplifier (Q2) allowing us to receive signals from around the world and listen to them with clarity.

Receiver assembly

Look for a clean well lit area in which to assemble your shortwave receiver. You will need a 27 to 30 watt pencil tip soldering iron, some 60/40 tin/lead rosin core solder, a few hand tools including a pair of end-cutters, a needle-nose pliers, a few screwdrivers and a pair of tweezers to pick up small parts. Locate your schematic and layout diagrams and we will begin to construct your new shortwave receiver. Let your soldering iron tip heat both the component lead wire and PC board trace enough so that the wire itself AND the foil trace BOTH become hot enough TOGETHER to melt a bit of solder so that it flows smoothly from the pin to the PC board trace. Locate the printed circuit board and all the components and we will start.

Before we begin constructing the shortwave receiver, refer to the resistor color code chart in Table 8-1 and the capacitor code chart in Table 8-2. Each resistor will have three or four color bands on them. The first color band will start close to one edge of the resistor body. The first color band will represent the first digit of the resistor value. The second color band will represent the second digit value, while the third color band illustrates

Table 8-1

Resistor color code chart

Color Band	1st Digit	2nd Digit	Multiplier	Tolerance
Black	0	0	1	
Brown	1	1	10	1%
Red	2	2	100	2%
Orange	3	3	1,000 (K)	3%
Yellow	4	4	10,000	4%
Green	5	5	100,000	
Blue	6	6	1,000,000 (M)	
Violet	7	7	10,000,000	
Gray	8	8	100,000,000	
White	9	9	1,000,000,000	
Gold			0.1	5%
Silver			0.01	10%
No color				20%

the resistor multiplier value. The fourth band, if there is one, will represent the resistor's tolerance value. A silver band is a 10% tolerance, while a gold band depicts a 5% tolerance value. No fourth color band notes a resistor with a 20% tolerance. Locate all of the resistors for the project and place them in front of you. Each resistor has a color code which describes its value. Check the color code against the parts list and schematic before installing them onto the circuit board. Place about four or five resistors on the PC board and place these resistors in their respective locations carefully making sure each one is in the correct location. Next solder each of the resistors onto the circuit board. Trim off the excess component leads with a small pair of end-cutters. Pick up and identify the next grouping of resistors and install them on the circuit board, follow up by soldering this next batch of resistors to the circuit board, remember to trim the extra lead lengths form the PC board.

Once all the resistors have been installed onto the circuit board we can move on to identifying all of the capacitors for the project. Note that there will be different-looking capacitors for this project, some will be small disks, while others will be larger body devices with minus and plus marking on them. These larger capacitors with polarity marking are electrolytic capacitors and they have polarity which must be observed for the circuit to work correctly. Look through the parts pile and locate the capacitors and place them in front of you. You will notice a number of small disk capacitors, we will install these first. Look closely at each of the disk capacitors at their markings. Sometimes the capacitors are very small and they may not have their actual value printed on them but will instead have some sort of code. Refer to the parts list for these codes. Locate four or five of these small disk capacitors, identify them and install them on the circuit board, while referring to the schematic and layout diagrams. After inserting the disk capacitors onto the circuit board, you can go ahead and solder them to the circuit board. Remember to cut the excess component leads. Next move on and install another grouping of small disk capacitors, then solder them in place on the printed circuit board. Cut the extra component leads after soldering in the capacitors. When you are finished installing the small capacitors we will move on to the larger electrolytic types.

Capacitors C9 and C24 are 1.0 µF electrolytic capacitors and capacitors, while C11, C22, C23, and C27 are 10 µF electrolytic capacitors. Capacitors C13 and C14 are 220 µF, while C26 is a 1000 µF electrolytic capacitor. When installing these capacitors you must orient them with respect to their polarity. Each electrolytic will have either a plus or minus marking

Table 8-2
Capacitance code information

This table provides the value of alphanumeric coded ceramic, mylar and mica capacitors in general. They come in many sizes, shapes, values and ratings; many different manufacturers worldwide produce them and not all play by the same rules. Some capacitors actually have the numeric values stamped on them; however, many are color coded and some have alphanumeric codes. The capacitor's first and second significant number IDs are the first and second values, followed by the multiplier number code, followed by the percentage tolerance letter code. Usually the first two digits of the code represent the significant part of the value, while the third digit, called the multiplier, corresponds to the number of zeros to be added to the first two digits.

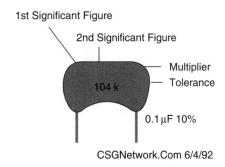

CSGNetwork.Com 6/4/92

Value	Type	Code	Value	Type	Code
1.5 pF	Ceramic		1,000 pF /.001 µF	Ceramic / Mylar	102
3.3 pF	Ceramic		1,500 pF /.0015 µF	Ceramic / Mylar	152
10 pF	Ceramic		2,000 pF /.002 µF	Ceramic / Mylar	202
15 pF	Ceramic		2,200 pF /.0022 µF	Ceramic / Mylar	222
20 pF	Ceramic		4,700 pF /.0047 µF	Ceramic / Mylar	472
30 pF	Ceramic		5,000 pF /.005 µF	Ceramic / Mylar	502
33 pF	Ceramic		5,600 pF /.0056 µF	Ceramic / Mylar	562
47 pF	Ceramic		6,800 pF /.0068 µF	Ceramic / Mylar	682
56 pF	Ceramic		.01	Ceramic / Mylar	103
68 pF	Ceramic		.015	Mylar	
75 pF	Ceramic		.02	Mylar	203
82 pF	Ceramic		.022	Mylar	223
91 pF	Ceramic		.033	Mylar	333
100 pF	Ceramic	101	.047	Mylar	473
120 pF	Ceramic	121	.05	Mylar	503
130 pF	Ceramic	131	.056	Mylar	563
150 pF	Ceramic	151	.068	Mylar	683
180 pF	Ceramic	181	.1	Mylar	104
220 pF	Ceramic	221	.2	Mylar	204
330 pF	Ceramic	331	.22	Mylar	224
470 pF	Ceramic	471	.33	Mylar	334
560 pF	Ceramic	561	.47	Mylar	474
680 pF	Ceramic	681	.56	Mylar	564
750 pF	Ceramic	751	1	Mylar	105
820 pF	Ceramic	821	2	Mylar	205

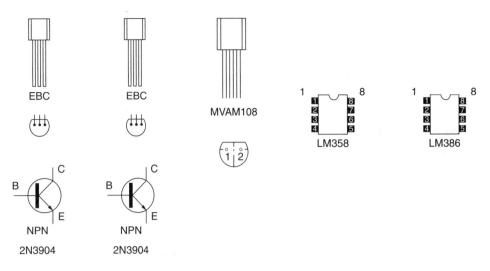

Figure 8-4 *Semiconductor pin-outs*

on them. You must look at the schematic and layout diagram and observe the polarity marking on the diagram match up with the markings on the capacitor when installing them to ensure that the circuit will work. Failure to observe polarity marking may result in damage to the circuit when it is first powered up. Identify these electrolytic capacitors and install on the PC board. Solder them in place and then cut the extra lead lengths flush to the edge of the circuit board.

Take a short break and locate the semiconductor pin-out diagram, shown in Figure 8-4, which illustrates the output pins of the transistors and integrated circuits in the receiver. Next, let's move on to identifying and installing the two diodes; note that diode D2 is the varactor or tuning diode and D1 is a glass bead style diode, while D3 is the red LED RSSI indicator. Diodes have marking on them that denote polarity and it must be observed for the circuit to work correctly. You will notice that a diode will have a small black or white colored band at one end of the diode body. This band denotes the diode's cathode. You will remember that a diode's symbol looks like a triangle pointing to a line. Well, the triangle is the anode and the line corresponds to the black or white band on the diode. Refer to the schematic when installing the two diodes and make sure you have installed them in the correct PC locations before soldering them in place. The RSSI LED is mounted on the front of the PC board near the potentiometers.

Now locate the three NPN transistors and we will install them on the PC board. Remember that transistors

each will most likely have three leads, a Base lead, a Collector lead and an Emitter lead. The symbol for a transistor is a vertical line, with two other leads, at 45° angles, pointing to this vertical line. One of these angled leads will have a small arrow pointing towards or away from the vertical line, this lead is the Emitter lead. The vertical lead is the Base lead and the remaining lead is the Collector. Locate and identify these four transistors and install them on the PC board. Be sure you can identify each lead and that each lead goes to the correct PC hole on the board; refer to the layout diagram and schematic before soldering the transistors to the printed circuit board. Don't forget to trim the excess lead from the transistors after the transistors have been soldered to the board.

Now, it's time for a little "destruction"! (If you jumped ahead and installed T3, we've got bad news for you.) Before this, two shielded transformers can be installed, and the internal capacitors need to be removed. Looking at the underside of these two transformers, you'll see a tubular part, probably white with a brown band, somewhat like a resistor. These are brittle and easily crushed with any sharp object that can be pressed against them with mild force (small nail, small screwdriver). You'll find these capacitors will easily disintegrate into particles. DO NOT crush the capacitors in the other two inductors. After crushing the capacitor, install T3. Next locate the remaining inductor cans, i.e. T1 and T2 and install them on the PC board; remember orientation of the inductor cans is important. Now locate the small "resistor like" inductor, i.e. L1,

L2 and L3. Solder the inductors to the circuit board and remember to trim the excess leads.

Finally we are going to install the integrated circuits to the circuit board. Consider installing IC sockets, instead of mounting the IC's directly to the PC board. Integrated circuit sockets are good insurance in the event of a possible circuit failure at some later date. It is much easier to unplug an IC rather than trying to un-solder one from a PC board. Mount three IC sockets to the PC boards at the respective locations of U1 and U2. Before installing the ICs we will need to make sure that you can identify the pin-outs on each of the ICs. Have a look at the eight pin integrated circuits, note that at the top of each of them you will see a notch in the center of the device or a small indented circle at the left side at the top of the package. The IC pin to the left of the notch or indented circle is pin one (1). Refer to the schematic and layout diagram to observe where pin one (1) is for each of the integrated circuits before installing them to the PC board. Once you are sure of the pin-outs for each of the ICs and where they go on the PC board, you can insert them into the proper IC socket. If there are any jumpers on the circuit board, you can go ahead and install them now.

Well, we have made quite a bit of progress on building your shortwave receiver. Take a short break and when we return we will mount the remaining final components. At this time look through the remaining parts and look for the two chassis mounted potentiometers R1 and R2. Place the potentiometes at the front of the PC board and solder them in place. Now, locate and install the power on-off switch at S1, and solder it in place on the printed circuit board.

Finally locate and install J1, an RCA chassis jack and J2, the ⅛″ phone jack onto the PC board. Locate the battery clip and solder it to the PC board, where you see the plus and minus markings. The red battery clip wire will go the plus marking on the board, while the black wire will go to the minus marking on the board. Install the battery clamp. Position battery and holder so as not to cover nearby PC board mounting holes. Use the method for securing the clamp that is most convenient for you, such as a (1) wire looped through clamp and PC board holes and soldered, (2) small screws, (3) double-faced adhesive strips or (4) hot-melt glue.

Your shortwave broadcast receiver is now almost finished. Now, you will need to decide what frequency

Table 8-3
Band selection components

C16 = .001	T3 Slug In	T3 Slug out
C15	6.5–13 MHz	9–18.3 MHz
C15	5.1–7.9 MHz	7.1–10.7 MHz
C15	4.3–5.3 MHz	6–8.3 MHz
C16 = 100 pF	**T3 Slug in**	**T3 Slug out**
C15	8.4–14.2 MHz	11.5–19.5 MHz
C15	6–8.1 MHz	8.5–11 MHz
C15	4.9–6.3 MHz	6.7–8.4 MHz

range you would like your receiver to tune, and then select the values for C15 and C16 from the chart, shown in Table 8-3. A suggested configuration would be C15 = 47 pF and C16 = 100 pF. This will give you complete coverage from 6 to 11 MHz by adjusting T3 in or out. This is a very active section of the shortwave band and will provide you with hours of listening enjoyment any time of the day. Select and install C15 and C16.

Finally, you will need to locate an enclosure for the receiver, and then install the circuit board into the enclosure. Place the knobs on the potentiometer shafts, locate a 9 volt transistor radio battery and connect it to the receiver battery clips. Connect up an antenna and you are ready to test out your new receiver.

Before turning on your receiver, double-check that you have correctly oriented all three of the ICs, that you have correctly oriented the transistors, that you have installed the diodes correctly, and that you have installed the electrolytic capacitors correctly.

Finally we will inspect the circuit board for "cold" solder joints and possible "short" circuits. Pickup the PC board with the foil side of the board facing you and inspect the solder connections. You will want to make sure that all of the solder connections look clean, smooth and shiny. If any of the solder joints look dull, dirty or "blobby" then you should un-solder that particular joint and re-solder it over again so that it looks clean, smooth and shiny. Next, let's inspect the CP board for any possible "short" circuits. Often "stray" cut component leads will "bridge" across between circuit traces and "short" out a PC trace. Another possible source of "short" circuits is a solder blob which can "bridge" across a circuit trace. Carefully inspect the PC board for any solder blobs or "stray" wires before connecting up a 9 volt battery.

Initial testing and adjustment

Now, connect up a speaker or headphones, connect up some sort of antenna and install a fresh 9-volt battery and we can begin testing the receiver. We have come to the "moment-of-truth," go ahead and turn the receiver power switch to the "on" position. After adjusting the volume to a pleasant level, you should hear some shortwave stations by turning the Tune Control, no matter how any of the adjustable coils happen to be set. While listening to any kind of station, whether broadcast or Teletype, etc., use a small screwdriver to adjust the slugs in transformers T1 and T2 for the best-sounding reception. The Tuning Control covers varying segments of the bands selected by adjustment of oscillator coil T3. Adjustment of T3 anywhere between the full In position to the full Out position will give the user full range between the minimum and maximum frequency coverage set by C15 and C16.

Both T1 and T2 must be adjusted with a non-metallic alignment tool such as is used in radio-TV service. If you do not have one, a suitable tool can be made by patiently sanding a screwdriver-like blade on the end of a wooden match stick, kebab skewer or small plastic crochet needle. Again, please be aware that a metal screwdriver blade will drastically increase the coil inductance and make adjustment quite difficult. T1 and T2 are simply adjusted for strongest reception of any signal range that is tuned in.

If you do not have any sort of testing or frequency reference equipment whatsoever, the easiest way to begin using the receiver, is to set the Tune Control at its midpoint, slowly tune T3 with your alignment tool as though it were a tuning dial. Stop when you come into the middle of a group or cluster of foreign broadcast stations. Try tuning around these stations with the Tune Control. If you like what you hear, readjust both T1 and T2 for best reception. Eventually, you will get a clue as to what general frequency band you are hearing, because many stations periodically announce

their frequencies, particularly at sign-on and sign-off times.

If you like precision, use a frequency counter or calibrated receiver to find the SR2's strong oscillator signal, remembering that there is a 455 kHz IF difference (above or below) between the local oscillator frequency and the broadcast signal you are hearing.

Shortwave antenna ideas

The type of antenna you'll want to use for your shortwave receiver depends on the degree of interest you have in shortwave listening, on whether you are limited to an indoor or balcony antenna. If you cannot use an outdoor antenna, you can make yourself a whip antenna from telescoping radio antenna from your local Radio Shack store. A better antenna would be to throw a 10 to 20 foot piece of insulated wire out the window, secured to a pole or tree away from your window. You could also place a length of wire in your attic to act as an antenna. An even better antenna would be a dipole antenna "cut" for the desired radio band of most interest. In order to construct your own dipole antenna, you can refer to the diagram depicted in Figure 8-5. Simply use the formula: 468 divided by the frequency in Megahertz for example. If you wanted to construct a dipole for 7.410 MHz, you would divide 468 by 7.410. The results would be 63.157 feet, so one-half of that number is 31.5 feet. Each leg of the dipole antenna would then be 31.5 feet. If you do construct your own dipole antenna, you should keep the antenna away from metal objects and use insulated wire and be sure to keep

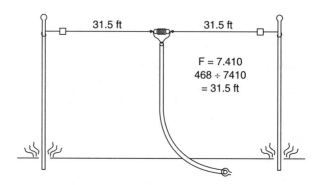

Figure 8-5 *Shortwave dipole antenna*

away from all power lines. The lead-in wire for your dipole antenna to your new receiver can be a length of RG8X min-coaxial cable.

Shortwave listening

Shortwave listening is a fun hobby for young and old alike. Once you have your antenna all set up, it is time to begin listening to your new radio. The chart in Table 8-4 illustrates the most common shortwave bands and their shortwave frequency in Megahertz vs. their wavelength in meters. Many times the shortwave bands are referred to in meters, such as the 40 or 80 meter band, so this chart will help you see the relationship between the two. As a shortwave radio listener, you will want to know what International broadcasters are broadcasting when you have time to listen to your new radio. Most people have spare time in the evening for

their hobby. So you will want to know which bands are best for night time use, and who is broadcasting at that time. Refer to the chart in Table 8-5, which illustrates the broadcaster and their frequencies, etc. Usually the best shortwave bands, for evening use are the lower frequency bands such as the 40, 41 and 49 meter bands. Use this chart or check on the Internet for the latest shortwave listings. Have fun!

Troubleshooting tips

If you experience difficulty, think of your receiver as consisting of several sections or stages: oscillator-mixer, IF audio and AGC, and final audio output (LM386). The first step in case of problems is to make sure that the tunable oscillator is working, which can be done by listening for its signal on another receiver. After the oscillator circuit is confirmed as working, standard signal tracing procedures should isolate any problem, which will be either an incorrectly installed part, a defective part or a bad solder connection. Correct orientation or polarity of all diodes, transistors, electrolytic capacitors and ICs is essential.

Problem: Strong shortwave broadcast audible throughout tuning range.

Solution: This can occur if your antenna is "too good" or if the RF Gain control is turned up too high. The high sensitivity of the front end is designed for simple antennas, with most reception quite satisfactory with only 25 feet or so of wire.

Problem: Local AM radio station audible throughout tuning range.

Solution: Whether this will even happen depends on how close you are to a local AM station. It is very important that all component leads be as short as possible, since just a bit of wire can help D1 and the several stages of audio amplification give you a free, unwanted "classic crystal radio." A grounded metal case for the SR2 is one possible solution.

Table 8-4
Popular shortwave bands

Frequency vs. Meters

Broadcast Frequency – kHz/MHz	Frequency in Meters
2300–2495 kHz	120 Meters
3200–3400 kHz	90 Meters
3900–4000 kHz	75 Meters
4750–5060 kHz	60 Meters
5850–6200 kHz	49 Meters
7100–7350 kHz	41 Meters
9400–9900 kHz	31 Meters
11600–12050 kHz	25 Meters
13570–13800 kHz	22 Meters
15100–15800 kHz	19 Meters
17480–17900 kHz	16 Meters
18900–19020 kHz	15 Meters
21450–21850 kHz	13 Meters
25600–26100 kHz	11 Meters

Table 8-5

Evening shortwave broadcast listings

English shortwave broadcasts to the Americas sorted by time:

1700–1800 NA	U.S.A.	WYFR	13695, 17555
1700–1800 NA, LA	Japan	R. Japan	9535
1700–1900 Am	U.S.A.	WHRI	9840, 15285, 15650
1700–2200 NA	U.S.A.	WBCQ	18910
1705–1905 Am	Canada	R. Canada Int'l	9610
1705–1905 NA	Canada	R. Canada Int'l	9800 drm
1800–1900 NA	U.S.A.	WYFR	13695, 17535, 17555
1800–1900 NA, Su-F	U.S.A.	WBCQ	7415
1830–1845 NA, Eu	Israel	Israel B/C Auth.	6985, 7545, 9345
1900–2000 NA	U.S.A.	WYFR	13695, 17535, 17555
1900–2000 LA	U.S.A.	WYFR	6085
1900–2000 Am	U.S.A.	WHRI	9840, 13760, 15285
1900–2100 NA	U.S.A.	KJES	15385
1900–2100 NA, Sa-Su	Netherlands	R. Netherlands	15315bo, 15525bo, 17725ca
1900–2200 NA	U.S.A.	WWCR4	9975
1900–2400 NA	U.S.A.	WBCQ	7415
2000–2025 NA, Eu, Af	Israel	Israel B/C Auth.	6280, 7545, 15640
2000–2100 Am	U.S.A.	WHRI	11765, 15285
2000–2100 LA	U.S.A.	WYFR	17575
2000–2200 NA	U.S.A.	WYFR	17535, 17555
2000–2400 NA	Costa Rica Univ.	Network	13750
2030–2130 NA, Eu	Cuba	R. Havana Cuba	9505, 11760
2045–2130 Am	Vatican City	Vatican R.	9800 drm
2100–2127 Eu, As, NA	Czech Republic	R. Prague	5930, 9430
2100–2130 NA, Sa	Canada	CBCNQ	9625
2100–2200 NA, Eu, Af	U.S.A.	WEWN	6890, 15785
2100–2200 NA	Japan	R. Japan	17825
2100–2200 LA	United Kingdom	BBC	9480sc, 11675gr
2100–2200 Am, Sa-Su	U.S.A.	WRMI	7385
2100–2200 Am	U.S.A.	WHRI	9660, 11765
2100–2300 NA	U.S.A.	WWCR1	7465
2100–2400 Am, M-F	U.S.A.	WRMI	7385
2130–2156 NA, Eu	Romania	R. Romania Int'l	6055, 6115, 7145, 9755
2130–2200 NA, M-Sa	Canada	CBCNQ	9625
2130–2200 NA	Netherlands	R. Netherlands	9800ca drm
2200–2257 NA	Netherlands	R. Netherlands	15425fg drm
2200–2300 NA, M-F	U.S.A.	WBCQ	5110
2200–2300 LA	United Kingdom	BBC	5975fg, 9480sc
2200–2300 NA	Canada	R. Canada Int'l	9800 drm
2200–2300 Am	U.S.A.	WHRI	7490, 9660

Table 8-5

Evening shortwave broadcast listings—cont'd

English shortwave broadcasts to the Americas sorted by time:

2200–2300 NA, M-F	U.S.A.	WBCQ	18910
2200–2300 NA, Sa-Su	U.S.A.	WWRB	3185
2200–2400 LA	Guyana	V. of Guyana	3290
2200–2400 NA, Su-F	Canada	CBCNQ	9625
2200–2400 Am	Anguilla	Wld Univ Network	6090
2200–2400 NA	U.S.A.	WWCR3	5070
2200–2400 NA	U.S.A.	WWCR4	9985
2200–2400 NA, Eu, Af	U.S.A.	WEWN	7560, 9975
2200–2400 NA	U.S.A.	WWRB Overcomer	6890
2230–2257 NA, Af	Czech Republic	R. Prague	5930, 9435
2300–2330 NA	Germany	Deutsche Welle	9800ca drm
2300–2350 NA	Turkey	V. of Turkey	5960
2300–2356 NA, Eu	Romania	R. Romania Int'l	6055, 6115, 7105, 9610
2300–2400 NA	Egypt	R. Cairo	11885
2300–2400 NA	China	China R. Int'l	6040, 11970
2300–2400 NA	U.S.A.	WBCQ	5110
2300–2400 Am	U.S.A.	WINB	9265
2300–2400 NA	Cuba	R. Havana Cuba	9550
2300–2400 NA, LA	China	China R. Int'l	5990ca
2300–2400 NA	U.S.A.	WHRA	5850
2300–2400 Am	U.S.A.	WHRI	7315, 7490
2300–2400 LA	U.S.A.	WYFR	15170, 15400
2300–2400 NA	U.S.A.	WWCR1	3215
2300–2400 NA	U.S.A.	WWRB	3185, 5050
2300–2400 LA, Su-F	U.S.A.	WWRB	5745
2330–2357 NA	Czech Republic	R. Prague	5930, 7345
2330–2359 NA	Sweden	R. Sweden	9800ca drm
2330–2400 NA	Lithuania	R. Vilnius	7325
2330–2400 Am, Su	U.S.A.	WRMI	9955
2335–2400 LA, Su-M	Austria	Osterreich 1	9870
2343–2358 LA, Tu-Sa	Austria	Osterreich 1	9870

Relay site codes: ae-United Arab Emirates, al-Albania, an-Antigua, ar-Armenia, as-Ascension, au-Austria, bo-Bonaire Neth. Antilles, bu-Bulgaria, ca-Sackville Canada, ch-China, cl-Chile, cu-Cuba, cy-Cyprus, de-Delano USA, fg-French Guiana, fr-France, ga-Gabon, ge-Germany, gr-Greenville USA, it-Italy, ja-Japan, ka-Kazakhstan, ko-South Korea, la-Latvia, li-Lithuania, ma-Madagascar, ml-Mali, mo-Moldova, ne-Netherlands, om-Oman, po-Sines Portugal, ru-Russia, rw-Rwanda, sc-Cypress Creek SC, se-Seychelles, si-Singapore, sl-Sri Lanka, so-Slovakia, sp-Spain, sw-Sweden, ta-Taiwan, th-Thailand, uk-United Kingdom, uz-Uzbekistan, wb-WBCQ USA, wr-WRMI USA, wy-WYFR USA, za-South Africa.

Notes – Days of week: Su-Sunday, M-Monday, Tu-Tuesday, W-Wednesday, Th-Thursday, F-Friday, Sa-Saturday, exW-except Wednesday.
Target areas: Af-Africa, Am-America, As-Asia, Eu-Europe, LA-Latin America, ME-Middle East, NA-North America, Oc-Oceania/Australia; Other: alt-alternate frequency, drm-Digital Radio Mondiale, occ-occasional use, se-Special English.

Courtesy of Daniel Sampson; www.primetimeshortwave.com

Chapter 9

80/40 Meter Code Practice Receiver

Parts list

80/40 Meter Direct Conversion Receiver Specifications

Sensitivity 3-uV or better

Modes CW, RTTY (FSK), SSB, AM

Audio output 50 mW typical

Frequency coverage Portions of the 80, 75, and 40 meter bands

Power Source 9 volt transistor battery, alkaline preferred

PC Board 4.000" × 4.700"

80/40 Meter Direct Conversion Receiver

R2 15 ohm, $\frac{1}{4}$w, 5% resistor (brown, green, black)

R3,R8   100 ohm, $\frac{1}{4}$w, 5% resistor (brown, black, brown)

R1 1000 ohm variable potentiometer

C1 8.2 pF ceramic disk (8.2)

C18 100 pF ceramic trimmer capacitor

C19 15-50 pF variable capacitor (main tuning) - see text

C2,C6,C7 .1 µF monolithic (.1 or 104)

C3 see chart below

C4 not used

C5 see chart below

C8,C9,C10 .1 µF monolithic (.1 or 104)

C20 .01 µF ceramic disk (.01 or 103)

C12,C13,C14 1 µF, 35 volt electrolytic capacitor

C16,C17 470 µF, 35 volt electrolytic capacitor

C11,C15 10 µF, 35 volt electrolytic capacitor

U1 SA602 RF Mixer IC, 8-pin DIP

U2 78L05 voltage regulator

U3 LM386 audio amplifier, 8-pin DIP

S1 Power On-off switch

J1 RCA antenna jack

J2 $\frac{1}{8}$" headphone jack

B1 9 volt transistor radio battery

Misc PC board, IC sockets, chassis, knobs, wire, battery

holder, battery clip,
hardware, etc.

80 Meter Band Tuning components

C3,C5 470 monolithic
    capacitor (471)

L1 6.8 µH variable
    inductor, shielded

L2 3.3 µH molded choke
    (orange-orange-gold)

L3 33 µH molded choke
    (orange-orange-black)

40 Meter Band Turning Components

C3 470 pF monolithic
    (471)

C5 680 pF monolithic
    (681)

L1 1.5 µH variable
    inductor, shielded

L2 1 µH molded choke
    (brown-black-gold)

L3 10 µH molded choke
    (brown-black-black)

* Inductors available
    from Digi-Key
    Electronics, see
    Appendix.

Whether you're a novice or an "old-timer," you'll be pleasantly surprised at how this dual band Direct Conversion Receiver performs! The receiver can be constructed to receive either the 80 meter or 40 meter amateur radio bands. You will be able to receive AM, CW and SSB, signals from hams around the world.

Far-away signals are easily copied; and the sensitivity of the 80/40 Meter Direct Conversion Receiver rivals more expensive receivers. This receiver is ideal for practicing and increasing your CW or code efficiency or just having fun listening to shortwave broadcasts or ham radio operators, at home or away. Build the receiver for your favorite band, tuning options allow you to customize the tuning range from full-band to 100 kHz or less. Powered by a common 9 volt transistor battery, it's always ready for action on vacation and camping trips!

Figure 9-1 *80/40 meter code receiver*

The 80/40 Meter Direct Conversion Receiver shown in Figure 9-1, can easily be constructed by a radio neophyte in a few hours with the easy-to-follow instructions.

Circuit description

The circuit diagram shown in Figure 9-2 illustrates the schematic for the 80/40 Meter Direct Conversion Receiver. A dipole or suitable antenna is fed directly to input jack J1, which is coupled to potentiometer R1, which serves as a continuously variable RF attenuator and also serves as the receiver's gain control. Using the RF gain control to set volume is advantageous as it also reduces strong in-band signals that could otherwise overload the receiver front end. Series inductors L2 and L3, and trimmer capacitor C18, provide front-end bandpass filtering, and an impedance match to 50 to 75-ohm antenna systems. IC U1 is an NE602/612 mixer and Local Oscillator (LO) in an 8-pin dip package. The mixer section is an active Gilbert cell design for good conversion gain and low noise figure. The LO section uses heavy capacitive loading to minimize frequency drift. Tuning is capacitive, using a modern Hi-Q miniature plastic variable. The local oscillator (LO) stability is enhanced by a 78L05 voltage regulator.

Since this is a direct conversion receiver, the LO is tuned to the carrier frequency of Single Sideband Signals, or to a difference of 300 to 800 Hz on CW

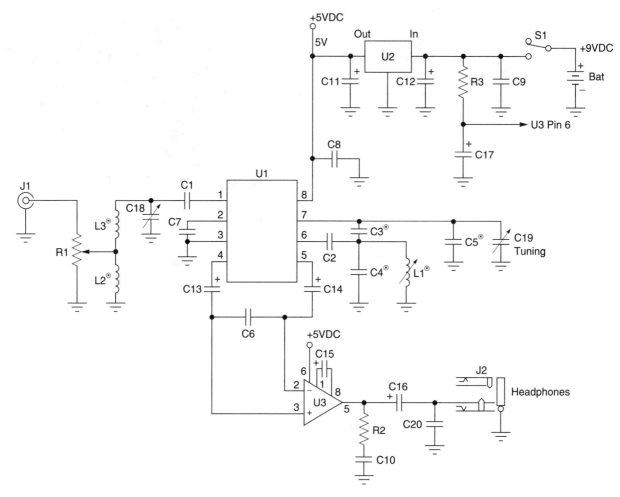

Figure 9-2 *80/40 meter direct conversion amateur radio receiver. Courtesy of Vectronics*

signals, to produce an aural output that is differentially coupled to a LM386 audio IC. The audio amplifier IC is coupled to the headphone jack at J2 by capacitor C16. The receiver circuit is power from a single 9 volt transistor radio battery.

Let's get started! Before we begin building the 80/40 meter receiver, you will need to locate a clean well lit and well ventilated work area. A large table or workbench would be a suitable work surface for your project. Next you will want to locate a small 27 to 33 watt pencil tipped soldering iron, as well as a roll of #22 gauge 60/40 tin/lead solder and a small jar of "Tip Tinner," a soldering iron tip cleaner/dresser, from your local Radio Shack store. You will also want to secure a few hand tools for the project, such as a pair of small end-cutters, a pair of tweezers and a pair of needle-nose pliers. Locate a small Phillips and a small flat-blade screwdriver, as well as a magnifying glass to round out your tool list. Grab the schematic, parts layout

diagram as well as the resistor and capacitor identifier charts and we will begin our project. Place all the project components on the table in front of you. The 80/40 direct conversion radio is an RF or radio frequency project and it is best constructed on a printed circuit board with large ground plane areas covering the board for the best RF grounding techniques. Once you have all the parts and PC board in front of you, heat up the soldering iron and we'll get started!

First, find your resistor identifier chart in Table 9-1, which will help you select the resistors from the parts pile. Resistors used in this project are mostly small ¼ watt carbon composition type resistors, which have colored bands along the resistor body. The first color band should be closest to one end of the resistor body. This first color band represents the first digit of the resistor value. For example, resistor R2 has four color bands, the first one is a brown band followed by a green band followed by a black band. The fourth band is gold.

Table 9-1

Resistor color code chart

Color Band	1st Digit	2nd Digit	Multiplier	Tolerance
Black	0	0	1	
Brown	1	1	10	1%
Red	2	2	100	2%
Orange	3	3	1,000 (K)	3%
Yellow	4	4	10,000	4%
Green	5	5	100,000	
Blue	6	6	1,000,000 (M)	
Violet	7	7	10,000,000	
Gray	8	8	100,000,000	
White	9	9	1,000,000,000	
Gold			0.1	5%
Silver			0.01	10%
No color				20%

The first band is brown, which denotes a digit one (1), the second band is green, which represents a digit five (5), and the third band is black, which represents a multiplier of zero (0), so the resistor value is 15 ohms with a tolerance value of 5%. Identify the remaining resistors for the project and we can begin "populating" the PC board. Place a few resistors on the board at one time, so as not to confuse the process. Make sure that you place the correct resistor into the correct PC location before soldering it in place. Once you solder a few resistors in place on the PC board you can use your end-cutter to trim the excess component leads. Cut the excess leads flush to the edge of the circuit board. Then place a few more resistors on the PC board and solder them on to the board. Remember to trim the component leads as necessary.

Next we will locate the capacitors for the 80/40 meter receiver. Capacitors are listed as electrolytic and non-electrolytic types. The non-electrolytic types are generally smaller in value and size as compared with the electrolytic types. Non-electrolytic capacitors, in fact, can be so small that their actual value cannot be printed on them, so a special chart was devised as shown in Table 9-2. The chart illustrates the three digit codes which are often used to represent capacitors. For example, a .001 μF capacitor will have (102), while a .01 μF capacitor will have (103) marked on it to

represent its true value. Use the chart to identify the small capacitors in the project. Non-electrolytics have no polarity markings on them so they can be installed in either direction on the PC board.

Electrolytic capacitors are usually larger in size and value and they will have a white or black band on the side of the capacitor body or a plus (+) or minus (−) marking on the body of the capacitor near the leads. These markings are polarity markings and that indicate the direction in which the capacitor must be mounted on the PC board. Failure to observe polarity when installing the capacitor may result in damage to the capacitor or to the circuit itself, so pay particular attention to capacitor polarity when placing the capacitors on the board. Note that the receiver can be built for either the 80 meter ham band or the 40 meter ham band. You will need to look at the parts list when deciding which band you want to receive. The inductors L1, L2 and L3 as well as capacitors C3, C4 and C5 determine the band selection.

Let's go ahead and place some of the non-electrolytic capacitors on the PC board. Identify a few small capacitors at a time and place them on the PC board and solder them in place. Trim the component leads after. Go ahead and install the remaining non-electrolytic capacitors after choosing your desired band. Install the capacitors and solder them in place, remember to cut

Table 9-2
Capacitance code information

This table provides the value of alphanumeric coded ceramic, mylar and mica capacitors in general. They come in many sizes, shapes, values and ratings; many different manufacturers worldwide produce them and not all play by the same rules. Some capacitors actually have the numeric values stamped on them; however, many are color coded and some have alphanumeric codes. The capacitor's first and second significant number IDs are the first and second values, followed by the multiplier number code, followed by the percentage tolerance letter code. Usually the first two digits of the code represent the significant part of the value, while the third digit, called the multiplier, corresponds to the number of zeros to be added to the first two digits.

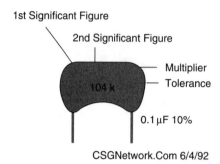

Value	Type	Code	Value	Type	Code
1.5 pF	Ceramic		1,000 pF /.001 µF	Ceramic / Mylar	102
3.3 pF	Ceramic		1,500 pF /.0015 µF	Ceramic / Mylar	152
10 pF	Ceramic		2,000 pF /.002 µF	Ceramic / Mylar	202
15 pF	Ceramic		2,200 pF /.0022 µF	Ceramic / Mylar	222
20 pF	Ceramic		4,700 pF /.0047 µF	Ceramic / Mylar	472
30 pF	Ceramic		5,000 pF /.005 µF	Ceramic / Mylar	502
33 pF	Ceramic		5,600 pF /.0056 µF	Ceramic / Mylar	562
47 pF	Ceramic		6,800 pF /.0068 µF	Ceramic / Mylar	682
56 pF	Ceramic		.01	Ceramic / Mylar	103
68 pF	Ceramic		.015	Mylar	
75 pF	Ceramic		.02	Mylar	203
82 pF	Ceramic		.022	Mylar	223
91 pF	Ceramic		.033	Mylar	333
100 pF	Ceramic	101	.047	Mylar	473
120 pF	Ceramic	121	.05	Mylar	503
130 pF	Ceramic	131	.056	Mylar	563
150 pF	Ceramic	151	.068	Mylar	683
180 pF	Ceramic	181	.1	Mylar	104
220 pF	Ceramic	221	.2	Mylar	204
330 pF	Ceramic	331	.22	Mylar	224
470 pF	Ceramic	471	.33	Mylar	334
560 pF	Ceramic	561	.47	Mylar	474
680 pF	Ceramic	681	.56	Mylar	564
750 pF	Ceramic	751	1	Mylar	105
820 pF	Ceramic	821	2	Mylar	205

the extra leads flush to the edge of the circuit board. Now, you can go ahead and install the electrolytic capacitors onto the circuit board. Go ahead and solder them in place and trim the leads as necessary.

This receiver project uses a number of small inductors. These small inductors will generally have color bands on them to help identify them. Molded chokes appear, at first glance, to be similar to resistors in both shape and band marking. However, a closer look will enable you to differentiate between the two—chokes are generally larger in diameter and fatter at the ends than resistors. When doing your inventory, separate out any chokes and consult the parts list for specific color-code information. Note, that inductor L1 is an adjustable slug tuned type. Remember that specific chokes are used for each band: see parts list before mounting the chokes. Chokes do not have polarity so they can be mounted in either direction on the PC board.

The 80/40 meter receiver utilizes two integrated circuits and a regulator IC. Take a look at the diagram shown in Figure 9-3, which illustrates the semiconductor pin-outs. When constructing the project it is best to use IC sockets as an insurance against a possible circuit failure down-the-road. Its much easier to unplug an IC rather than trying to un-solder it from the PC board. IC sockets will have a notch or cut-out at one end of the plastic socket. Pin one (1) of the IC socket will be just to the left of the notch or cut-out. Note that pin 1 of U1 connects to C1, while pin one (1) of U2 connects to C15. When inserting the IC into its respective socket make sure you align pin one (1) of the IC with pin one (1) the socket. Failure to install the IC properly could result in damage to the IC as well as to the circuit when power is first applied.

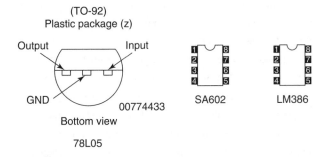

Figure 9-3 *Semiconductor pin-outs*

Let's finish the circuit board by mounting the volume control and the adjustable capacitors. The volume control potentiometer at R1 is a right angle PC board mounted type, which is placed at the edge of the PC board as is the main tuning capacitor. Capacitor C18, which is connected at the junction of L3 and C1, is an 8.2 pF trimmer type; go ahead and solder it to the circuit board using two pieces of bare #22 ga. stiff single conductor wire.

Locate main-tuning capacitor C19, now locate the mounting location for main tuning capacitor C19. The tuning capacitor was mounted on its side using double-sided sticky tape. Remove the protective cover from the double-sided tape, and firmly press the body of capacitor C19 to mount it to the PC board. Firmly press the double-sided tape over the silk-screened outline for the body of C19. You may decide to mount the tuning capacitor in a different way. At the rear of the capacitor: locate the four internal trimmer capacitors, and using a small jeweler's screwdriver or alignment tool, fully open (unmesh) all four trimmers. Note: tuning shaft faces front of board. Bend the two rotor lugs so they are parallel to the front face of the capacitor as shown above. Connect the two rotor lugs to the PC board ground points as shown using scraps of lead wire trimmed from resistors as jumper wires. Cut a 6″ length of 24-AWG insulated hook-up wire in half. Remove about ¼″ of the insulation from each of the cut ends. Solder the jumpers to the capacitor rotor lugs, and to the ground foil run on the bottom of the PC board. Since the tuning capacitor has four sections, you can increase the tuning ranges by paralleling different sections to give a greater tuning range: see Table 9-3. If you used just the 140 pF section, your tuning range would be 190 kHz, but if you combined the 140 pF section with the 40 pF section your tuning range would become 180 pF and so forth. Capacitor jumpers: 180 pF = use 140-pF and 40-pF sections paralleled; 222 pF = use 140-pF and 82-pF sections paralleled; 262 pF = use 140-pF, 82-pF and a 40-pF sections paralleled; 302 pF = use all four capacitor sections in parallel.

The 80/40 meter code practice receiver can be built for either of your favorite ham bands. First you will need to decide which band you are interested in. Then you will have to select the proper components from the chart in Table 9-4, for the band of interest. Note that you will have to select three capacitors and three coils for each band.

Table 9-3

Band tuning range

80 Meter Tuning

302 pF	350 kHz
262 pF	310 kHz
222 pF	270 kHz
180 pF	230 kHz
140 pF	190 kHz
82 pF	110 kHz
40 pF	55 kHz

40 Meter Tuning Range

302 pF	N/A
262 pF	N/A
222 pF	N/A
180 pF	310 kHz
140 pF	250 kHz
82 pF	150 kHz
40 pF	75 kHz

Let's take a few minutes for a short well-deserved break, and when we return we will look over the circuit board for any possible "cold" solder joints or "short" circuits. Pick up the circuit board with the foil side facing upwards toward you. First, we will examine the circuit board for "cold" solder joints. Take a look at all of the solder joints on the board, they should look clean, bright and shiny. If you see any solder joints which look dark, dull or "blobby," then you should remove the solder with a solder-sucker and then reapply more solder and re-solder the joint all over again so that it

Table 9-4

80/40 Band change components

40-Meter Components	*80-Meter Components*
C3 - 470 pF capacitor	C3 - 470 pF capacitor
C4 - not used	C4 - not used
C5 - 680 pF capacitor	C5 - 470 pF capacitor
L1 - 1.5 μH molded coil	L1 - 6.8 μH molded coil
L2 - 1 μH molded coil	L2 - 3.3 μH molded coil
L3 - 10 μH molded coil	L3 - 33 μH molded coil

looks good. Next, we will look the board over for any possible "short" circuits which are often caused by "stray" component leads cut from the board or from small solder blobs which may have bridged between circuit traces. Rosin core solder can often leave a sticky residue which will collect component leads and solder blobs, so look the PC board over carefully for any foreign objects on the foil side of the board.

The 80/40 meter receiver radio prototype was mounted in a metal chassis box on plastic standoffs. The PC board was aligned in the enclosure so that the volume control, and main tuning control and power switch were all mounted at one side of the case. Alight the PC board along the edge of the chassis box, so that you drill holes in the chassis for the volume and tuning controls. The ⅛″ switched headphone jack at J2 was mounted on the front of the chassis, while the antenna input jack at J1 was mounted on the rear of the case. The 9 volt transistor radio battery was mounted in a battery holder which was fastened to the bottom panel of the chassis box.

Your new 80/40 meter direct conversion receiver project is almost completed. Apply the battery clip to the 9 volt battery. Connect up an antenna to the antenna jack, a dipole antenna cut for the band of interest would be the best type of antenna. For test purposes, you could connect a long piece of wire to the antenna jack for a temporary antenna. Plug in an 8 ohm headphone and then, turn on the power switch to the "On" position. Now, adjust the volume control to the mid-position and then turn the main tuning control. If all-goes-well, you should begin to hear some radio stations. You may want to test the radio in the early morning or evening when these two bands are most active for best results. Hopefully you will hear some stations while you tune across the band.

If for some reason, your receiver is DEAD, then you will want to turn off the receiver and remove the battery clip and do another inspection of the circuit board. This inspection is best done by another pair of eyes, someone who is skilled at electronics might be your best choice for inspection. You will want to make sure that you installed the electrolytic capacitors correctly with respect to their polarity. You will also want to make sure that the integrated circuits are aligned correctly in their respective sockets: it's easy to make a mistake. Check to make sure that the regulator IC at U2 was correctly installed with the input of the regulator going to capacitor C12, and the output of the regulator going to capacitor C11. Finally, make sure that you have the battery clip polarity wired correctly. Once you have examined the circuit board once more, you can go ahead and reconnect the battery and antenna and try the receiver again to see if it works now.

If your receiver is now working correctly, we can move on to the alignment of your new receiver. There are two primary methods of aligning the receiver's local oscillator.

Receiver alignment method 1: using an additional receiver

Place your new 80/40 meter receiver next to an amateur or general coverage receiver that covers the band of interest. Connect a short 3 or 4 foot wire antenna to the receiver, route the antenna wire so it passes near the direct conversion receiver PC board. Next, select a tuning tool that properly fits coil L1. The 20 meter kit requires a hex type alignment tool. The other kits require a small blade type tool that matches the slot in the coil core, or use a small jeweler's screwdriver. Now, set coil L1 tuning core to the top of the coil form. At this time, tune the receiver to the lowest operating frequency by turning the tuning capacitor shaft fully CCW (counter clockwise). Next, go ahead and set the monitor receiver to the lowest frequency you wish to receive on the direct conversion receiver. The receiver should be set to a wide filter bandwidth, and if it has a BFO, it should be on. Install a fresh 9 volt alkaline battery to the receiver, and turn it on by depressing the push-action power switch.

Finally, carefully adjust the core in coil L1 until it is heard sweeping across the monitor receiver frequency. Set the coil for zero beat, or strongest S-meter reading on the receiver. Note that metal alignment tools will cause some frequency shift, and interact with the tuning when removed or inserted into the coil. You will have to adjust the coil in small increments, and observe the point where the proper frequency is reached when the tool is removed.

Receiver alignment method 2: using a signal generator

First, set the signal generator to the direct conversion receiver's lowest operating frequency. Set the generator output level for a 30-μV CW signal. Next, connect the generator RF output to the VEC direct conversion antenna jack (J1) using coax cable and the appropriate mating connectors or adapters. Now, connect a speaker or headphones to jack J2 on the receiver. At this point, set the receiver gain fully CW (clockwise) and the tuning knob fully CCW (counter clockwise). Next, select a tuning tool that properly fits coil L1. The 20 meter kit requires a hex type alignment tool. The other kits require a small blade type tool that matches the slot in the coil core. Or, a jeweler's screwdriver may be used. Go ahead and set coil L1 tuning core to the top of the coil form. Now, connect a fresh 9 volt alkaline battery to the receiver, and turn it on using the push-action power switch. Finally, carefully adjust the core in coil L1 until the generator signal is heard. Set the coil for zero beat. Note that metal alignment tools will cause some frequency shift, and interact with the tuning when removed or inserted into the coil. You will have to adjust the coil in small increments, and observe the point where the proper frequency is reached when the tool is removed.

Final alignment

Once the tuning range is set, the RF input stage must be peaked for best sensitivity. This can be done using a signal generator, or on-the-air signals. Find an alignment tool or jeweler's screwdriver that fits trimmer capacitor C18. Use one of the following three methods to peak C18.

1. Using a signal generator, peak C18 for best sensitivity. Reduce the signal generator output, or reduce the R1 Gain control setting, to prevent overload as C18 is brought into resonance.

2. Using off-air signals, adjust C18 until signals are loudest. Reduce the R1 Gain control as C18 is

brought into resonance to prevent overload. Verify that C18 is properly peaked while monitoring a known on-air signal.

3. With an outside antenna connected, peak C18 for maximum atmospheric noise in the headphones or speaker. Verify that C18 is properly peaked while monitoring a known on-air signal.

Antenna considerations

With your new receiver completed and aligned, you can attach a dipole antenna for the chosen band of operation. Direct conversion receivers need good antennas. Using a random length of wire will give poor results. Half-wave dipoles are very effective antennas and are simple to construct. For low-power or QRP work, or for receive only applications, low-cost RG-58 52 ohm coax is fine. The formula for dipole antennas: 234/(frequency in MHz) = ¼-wavelength in feet will give the length of wire needed for each leg of the dipole. For example, if you wanted to build a 40-meter dipole for the CW portion of the band you would divide 234 by 7.040 MHz and the result would be 33.238 feet per leg on either side of the center insulator: see Figure 9-4.

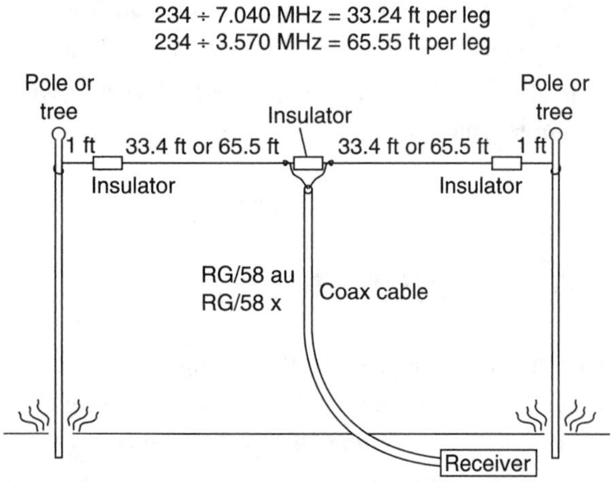

$$234 \div 7.040 \text{ MHz} = 33.24 \text{ ft per leg}$$
$$234 \div 3.570 \text{ MHz} = 65.55 \text{ ft per leg}$$

Figure 9-4 *80/40 meter dipole antenna*

Tuning notes

While trying to tune CW stations, you should tune across the band until a CW signal is heard. Set the gain control at R1 for a comfortable listening level. Then, carefully adjust the tuning control until the "beat note," or tone, sounds best for you. Most experienced CW operators prefer a lower beat note of about 300 or 400 Hz.

When trying to tune-in single-side band (SSB) stations, you may need a little practice to tune-in the signal properly. Direct conversion receivers offer no upper or lower sideband rejection, so you will have to slowly tune across the SSB signal until the voices become natural sounding. Use the gain control to prevent overload and to set a comfortable listening level.

In order to tune-in AM stations, you will need to realize that, unlike SSB signals, AM signals have a carrier frequency. That means you will hear a beat note as you tune across an AM signal. Tune the signal so the beat note becomes lower in frequency and stop when "zero beat" (beat note disappears) is reached.

Possible overload problems

An unfortunate fact of life is that low-power amateur signals are surrounded by high-power SW broadcast signals, so there may be a possibility of overload problems. In order to compensate for this possible problem, you may need to lower the gain control. Simply lowering the gain setting will resolve many overload problems. Another approach to an overload problem is to use a **preselector**. A preselector provides protection from strong out-of-band signals. They are most effective when the interfering signal is far from the operating frequency.

A resonant antenna favors signals in the band it was cut for, so when faced with a severe overload problem, you elect to use a resonant antenna. The final approach to solving an overload problem is to change the direction the dipole faces; this may lower unwanted pickup of undesired SW signals. Lowering the antenna will change its angle of pickup; a low antenna may favor local stations over the low-angle of distant broadcast signals. A shorter antenna can also be tried; it will be less efficient and signals will be weaker.

Table 9-5

WIAW on-air code practice sessions

Pacific	Mtn	Central	East	Mon	Tues	Weds	Thur	Fri
6 am	7 am	8 am	9 am		fast code	slow code	fast code	slow code
7-9 am	8-10 am	9-11 am	10-noon	visiting	operator	time	————	————
10 am-12:45 pm	11 am-1:45 pm	noon-2:45 pm	1-3:45 pm					
1 pm	2 pm	3 pm	4 pm	fast	slow	fast	slow	fast
2 pm	3 pm	4 pm	5 pm	————	code	bulletin	————	————
3 pm	4 pm	5 pm	6 pm	————	teleprint	bulletin	————	————
4 pm	5 pm	6 pm	7 pm	slow	fast	slow	fast	slow
5 pm	6 pm	7 pm	8 pm	————	code	bulletin	————	————
6 pm	7 pm	8 pm	9 pm	————	teleprint	bulletin	————	————
6:45 pm	7:45	8:45	9:45	————	voice	bulletin	————	————
7 pm	8 pm	9 pm	10 pm	fast	slow	fast	slow	fast
8 pm	9 pm	10 pm	11 pm	————	code	bulletin	————	————

Frequencies are 1.8175, 3.5815, 7.0475, 14.0475, 18.0975, 21.0675, 28.0675 and 147.555 MHz.

Slow Code = practice sent at 5, 7½, 10, 13 and 15 words per minute (wpm).

Fast Code = practice sent at 35, 30, 25, 20, 15, 13 and 10 wpm.

The chart shown in Table 9-5 illustrates the W1AW code practice sessions for Ham Radio operators or those interested in obtaining a Ham Radio license. The 80-meter frequency is 3.5815 MHz and the 40-meter frequency is 7.0475. The Amateur Radio Relay League (ARRL) offers these code practice sessions every day to help amateur radio enthusiasts increase their code speed. The ARRL is a great resource for those interested in Amateur Radio and offers many books, magazines and free literature in including videos to help you understand what Amateur Radio is all about. Take a few minutes and explore their web-site.

Chapter 10

WWV 10 MHz "Time-Code" Receiver

Parts list

WWV Time Receiver

R1   22k ohm, $\frac{1}{4}$w,
   5% resistor

R2,R6,R7,R10   100 ohm,
   $\frac{1}{4}$w, 5% resistor

R16,R18,R30   100 ohms,
   $\frac{1}{4}$w, 5% resistor

R3,R21   1 megohm, $\frac{1}{4}$w,
   5% resistor

R4   3k ohm, $\frac{1}{4}$w,
   5% resistor

R5,R11,R12   2k ohm, $\frac{1}{4}$w,
   5% resistor

R8,R13   47k ohm, $\frac{1}{4}$w,
   5% resistor

R9   220k ohm, $\frac{1}{4}$w,
   5% resistor

R14,R24,R29   100k ohm,
   $\frac{1}{4}$w, 5% resistor

R19,R22   5.6k ohm, $\frac{1}{4}$w,
   5% resistor

R20   27k ohm, $\frac{1}{4}$w,
   5% resistor

R17,R23,R28   4.7k ohm,
   $\frac{1}{4}$w, 5% resistor

R31,R33,R34   4.7k ohm,
   $\frac{1}{4}$w, 5% resistor

R25,R26,R27   10k ohm,
   $\frac{1}{4}$w, 5% resistor

R32   12k ohm, $\frac{1}{4}$w,
   5% resistor

R35,R36   22 ohm, $\frac{1}{4}$w,
   5% resistor

R37   2.7k ohm, $\frac{1}{4}$w,
   5% resistor

C1,C2,C6,C7,C15   .001 µF,
   35v ceramic disk
   capacitor

C3,C17,C19,C22   .01 µF,
   35v ceramic disk
   capacitor

C4,C5,C8,C9,C14   .1 µF,
   35v ceramic disk
   capacitor

C16,C23,C28,C31   .1 µF,
   35v ceramic disk
   capacitor

C10,C11,C12,C13   20 pF,
   35v plastic capacitor

C18   1000 pF, 35v
   plastic capacitor

C20,C21   220 pF, 35v
   plastic capacitor

C24,C27   1.0 µF, 35v
   electrolytic capacitor

C25,C32   22 µF, 35v
   electrolytic capacitor

C26   2.2 µF, 35v
   electrolytic capacitor

C29,C30   390 pF, 35v
   plastic capacitor

C33 100 µF, 35v electrolytic capacitor

C34 2200 µF, 35v electrolytic capacitor

C35 10 µF, 35v electrolytic capacitor

C36,C37 1000 µF, 35v electrolytic capacitor

C38 330 µF, 35v electrolytic capacitor

C39 470 µF, 35 volt electrolytic capacitor

CV1 10-40 pF trimmer capacitor

CV2 10-70 pF trimmer capacitor

VR1 10k ohm RF gain potentiometer

VR2 1M ohm Sensitivity potentiometer

VR3 10k ohm volume control potentiometer

L1 4.1 µH coil - 27 turns on T-28-2 ferrite core using 22 AWG wire tap 2-turns from ground side

L2,L4 12 turns on FT37-43 ferrite toroid core

L3 3.0 µH - 25 turns #26 AWG wire on T50-2 ferrite core

L5 12 turns on FT37-43 ferrite toroid core

L6 1000 µH coil

D1 1N5711 silicon Schottky diode NTE 112

D2 optional LED - power "on" indicator

Y1,Y2,Y3 10 MHz crystal

(F) 3-ferrite beads - (see front-end)-run wire through beads

Q1,Q2,Q3 Dual Gate MOSFET 40673 or ECG222

Q4,Q5 nJFET MPF102

Q6,Q7 2N3904 NPN transistor

Q8 2N3906 PNP transistor

U1 Signetics NE5532 Op-Amp

J1 RCA jack - antenna

J2 Coaxial power jack

SPK 8-ohm 3" speaker

S1 SPST power switch

Misc PC board, wire, hardware, sockets, etc.

Does anybody really know what time it is? The WWV receiver, shown in Figures 10-1 and 10-2, is designed to tune the National Institute of Standards and Technology (NIST) atomic clock for moment by moment time accuracy! This is a great receiver project that actually serves a dedicated purpose. Put one in the boat or on a desk and you'll always know what time it is!

The WWV receiver is quite sensitive, and with a good antenna you should be able to receive time signals most of the time! The receiver features easy tuning and small size, it can be packed up and taken anywhere and runs on a 12 volt DC power supply, or a 12 volt battery for portable operation. An earphone jack is provided for headset operation so you can use it night or day.

Figure 10-1 *WWV time code received*

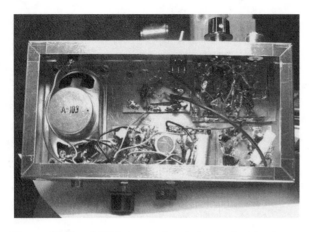

Figure 10-2 *WWV time code received – inside view*

People have always been fascinated with the measurement and properties of time. What happened to that last second that you just spent reading this sentence? Perhaps this is the reason that so many of us have tuned in over the years to the National Institute of Standards and Technology (NIST) "clock" operating at a transmitting frequency of 10 MHz. The cesium standard NIST-7 has an accuracy of $5 \times 10\text{-}15$ seconds. Now that's accurate! In fact, the definition of a "second" is the duration of 9,192,631,770 periods of the radiation corresponding to the transition between the two hyperfine levels of the ground state of the cesium-133 atom.

We set our clocks, watches, and chronometers to the all-important "beep" of the radio signal that denotes the correct time to the second. In addition; avid listeners of the NIST clock are also treated to the following (from the NIST web site).

The OMEGA Navigation System status reports are voice announcements broadcast on WWV at 16 minutes after the hour. The OMEGA Navigation System consists of eight radio stations transmitting in the 10- to 14-kHz frequency band. These stations serve as international aids to navigation. The status reports are updated as necessary by the U.S. Coast Guard. Current geophysical alerts (Geoalerts) are broadcast in voice from WWV at 18 minutes after the hour. The messages are less than 45 s in length and are updated every 3 hours (typically at 0000, 0300, 0600, 0900, 1200, 1500, 1800, and 2100 UTC). Hourly updates are made when necessary.

Marine storm warnings are broadcast for the marine areas that the United States has warning responsibility for under international agreement. The storm warning information is provided by the National Weather Service. Storm warnings for the Atlantic and eastern North Pacific are broadcast by voice on WWV at 8, 9, and 10 minutes after the hour. An additional segment (at 11 minutes after the hour on WWV) is used occasionally if there are unusually widespread storm conditions. The brief voice messages warn mariners of storm threats present in their areas. The storm warnings are based on the most recent forecasts. Updated forecasts are issued by the National Weather Service at 0500, 1100, 1700, and 2300 UTC for WWV. Since March 1990 the U.S. Coast Guard has sponsored two voice announcements per hour on WWV giving current status information about the GPS satellites and related operations. The 45-s announcements begin at 14 and 15 minutes after each hour.

WWV radiates 10,000 watts on 5, 10, and 15 MHz. The WWV antennas are half-wave dipoles that radiate omnidirectional patterns. The station uses double sideband amplitude modulation. The modulation level is 50 percent for the steady tones, 25 percent for the BCD time code, 100 percent for the seconds pulses and the minute and hour markers, and 75 percent for the voice announcements. That's a lot of information being broadcast, and you can easily pickup these signals with the high frequency 10 MHz Time Receiver shown in this project. The WWV Time receiver has excellent sensitivity, selectivity, and dynamic range. (from the NIST web site http://www.nist.gov/)

Circuit description

The 10 MHz WWV Receiver is shown in the block diagram of Figure 10-3 and in the main schematic diagram in Figure 10-4. The WWV Time Code receiver is an AM TRF or Tuned Radio Frequency or TRF type receiver. This receiver project is designed specifically for 10 MHz time signal operation. The block diagram of the WWV receiver illustrates the four major sections of the receiver. The receiver's front-end, as shown in the schematic diagram, features two sensitive dual-gate MOSFETs in the front-end circuit with a single-pole filter. An RF gain control is shown at VR1 which adjusts the input sensitivity, via a 10k ohm potentiometer. Note inductor L1 is 4.1 μH coil, which consists of 27 turns of 22 ga. AWG wire wound on a

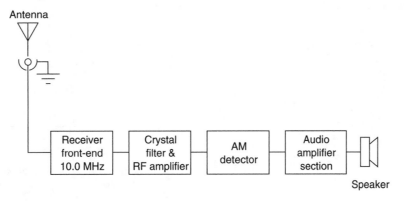

Figure 10-3   *10 MHz WWV receiver block diagram*

T-68 ferrite core. Inductors L2 and L4 consist of 12 turns of 22 ga. AWG wire on an FT37-43 ferrite core, and finally L3 consists of 25 turns of #26 ga. AWG wire on a T50-2 ferrite core. For even greater selectivity (but greater insertion loss), you can consider moving the L1 tap to 1 turn from ground. Fine tuning the WWV receiver is accomplished by adjusting CV1 and CV2 trimmer capacitors. The front-end MOSFETs are extremely sensitive to static electricity, so be very careful in handling them. Note the two ferrite beads, one is placed in series with the Drain lead of Q1, while the second ferrite bead is in series with the Drain of Q2. These ferrite beads are important, so be sure to use them. The receiver front-end, at J1, should be connected to a dipole antenna "cut" for 10 MHz. TRF receivers are especially sensitive and are subject to overloading a BCB or Broadcast Band Interference, so internal lead lengths should be kept short between components and the front-end should be shielded and the receiver placed in a metal enclosure for best results.

The output impedance of Q2 is 2000 ohms, which nicely matches the input impedance of the Cohn crystal filter, in the crystal filter section of the receiver. This filter utilizes three 10 MHz crystals. Choose 10 MHz crystals that are marked for a 20 pF or 32 pF load capacitance if possible. Using a 10 MHz crystal oscillator, find three that are closest to one another in frequency. You may substitute 2.2k ohm resistors instead of the specified 2k with a slight penalty in pass band shape. The crystal filter section utilizes a single dual-gate MOSFET transistor at Q3. Coil L5 is 12 turns on an FT37-43 ferrite core. The output of the Cohn crystal filter is next fed to the input of the AM detector stage.

The AM detector has three distinct advantages; it has high bandwidth, low distortion and incredible (and variable) sensitivity. The variable bias control allows the listener to adjust the bias to maintain detected audio fidelity even when the RF signal is weak. The detector uses a Schottky UHF mixer diode at D1. Increasing the diode bias from zero volts to maximum causes three things to happen: increased sensitivity, increased audio high frequency response, and a slight increase in receiver noise.

The AM detector section matches the 2000 ohm impedance of the crystal filter stage quite nicely. A sensitivity control at VR2 is a 1 megohm potentiometer, which feeds directly to L6, a 1000 μH coil. Coil L6 is connected directly to a Schottky diode at D1. The resultant signal is next sent through a 5.6k ohm resistor and on to a 1 μF capacitor which feeds the nJFET at Q4. The Source lead of Q1 is connected to the Drain lead of Q5, and the Source lead of Q5 is connected to ground via a 5.6k ohm resistor. Note the two 22 μF electrolytic capacitors on each of the Drain leads of the nJFETs.

When the WWV RF signal is weak, turning the bias off may result in the detected WWV signal disappearing. Increasing the bias will bring the WWV signal back in. You run the bias control pot about half way and, of course, higher as WWV fades out. The fidelity that the bias adds even when the WWV signal is strong is quite pleasing to the ear. For the 1 μF and 2.2 μF capacitors, I used polyester film types which sounded better than electrolytic capacitors.

The audio amplifier section features low noise audio amplifier. Distortion is very low as long as the input is

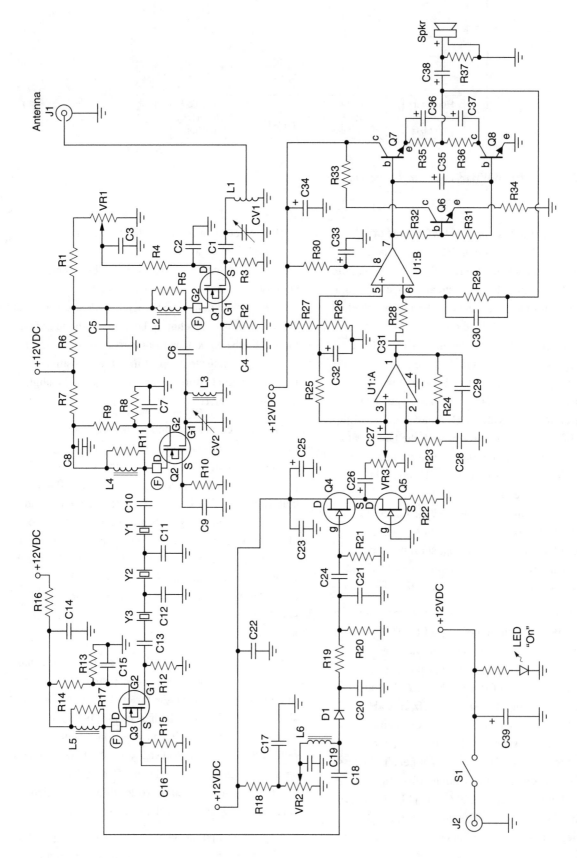

Figure 10-4 *WWV time code receiver schematic. Courtesy of VC7BFO – Todd Gale*

not over driven. The audio amplifier section consists of a dual NE5532 op-amp and three transistors. Transistors Q6 and Q7 are NPN types, while transistor Q7 is a PNP type. The input signal from the detector is fed directly to the volume control at VR3, a 10k ohm potentiometer. The input signal is first sent to the non-inverting stage of the first op-amp at pin 3. The output of the first op-amp section at pin 1 is next sent to the second stage of the op-amp through a .1 μF capacitor in series with a 4.7k ohm resistor to pin of U1:b. The output of U1, at pin 7 is then fed to the output transistor stage. The output of the transistor pair is two 1000 μF capacitors which coupled to a 330 μF capacitor which feeds the 8 ohm speaker.

The entire WWV receiver is powered from a 12 volt "wall-wart" power or can be powered from a 12 volt battery for portable operation.

Receiver assembly

You are probably quite anxious to begin the assembly of the receiver project; it is best to follow the assembly steps in order. Try to avoid the urge to "jump ahead" installing components. Locate your layout and schematic diagram. Place all the components in front of you and locate a 27-30 watt pencil tip soldering iron.

Make sure you have a large work table in order to spread out all the parts and diagrams. A printed circuit board is highly recommended for this project, so you should have your circuit board in front of you oriented correctly, in order to begin populating the PC board. Remember that the components will be mounted on the "component" side of the circuit board and soldered on the "solder" side of the circuit board, the side with the printed circuit traces. Locate a spool of 60/40 tin/lead rosin core solder for this project.

First we will begin by locating all of the resistors for the project. Most all of the resistors are ¼ watt carbon composition type resistors except for the potentiometer. Before we begin building the Time receiver, take a moment to refer to the resistor color chart shown in Table 10-1. Note that each resistor will have three or four color bands beginning at one end of the resistor body. The first color code represents the resistor's first digit value, while the second color band denotes the resistor's second digit value. The third color band represents the multiplier value of the resistor and the fourth band denotes the tolerance value of the resistor. If the fourth color band is silver then the resistor has a tolerance of 10%, while a gold band denotes a 5% tolerance. If there is no fourth color band then the resistor has a tolerance of 20%. The first color band of a 10,000 ohm resistor, for example,

Table 10-1

Resistor color code chart

Color Band	1st Digit	2nd Digit	Multiplier	Tolerance
Black	0	0	1	
Brown	1	1	10	1%
Red	2	2	100	2%
Orange	3	3	1,000 (K)	3%
Yellow	4	4	10,000	4%
Green	5	5	100,000	
Blue	6	6	1,000,000 (M)	
Violet	7	7	10,000,000	
Gray	8	8	100,000,000	
White	9	9	1,000,000,000	
Gold			0.1	5%
Silver			0.01	10%
No color				20%

would be brown with a value of (1). The second color band would be black with a value of (0). The third color band would be orange with a value of (000). Read the color codes on the resistors and compare them to the parts list and layout diagram so that you place the correct resistor in the correct hole on the printed circuit board. You can install the resistors in groups of four or five. Solder the first group of resistors onto the circuit board, then cut the excess component leads flush to the edge of the PC board using a pair of small end-cutters. Then move on to the next group of resistors until all of the resistors have been installed.

Next we will move on to installing the capacitors. Most all of the capacitors will have some sort of marking on them, so compare the value or marking on the component with the parts list before installing the components on the PC board. Note that capacitors C24, C25, C26, C27 are all electrolytic capacitors; these capacitors will have additional polarity markings on them and they must be installed with respect to these polarity markings in order for the circuit to work properly. Each of these capacitors will have a white or black band with either a plus or minus marking on one side or edge of the component. You must orient the capacitor correctly, so you should refer to the layout diagram and schematic diagram when installing these parts. Also note that capacitors C32 through C38 are also all electrolytic types. All the other capacitors have no polarity marking so they can be installed in either direction. Now, you will want to refer to the capacitor code chart shown in Table 10-2, which illustrates the codes used for small size capacitors. Often very small capacitors will not have their actual value printed on them, because there may not be enough room, so a three-digit code was developed. Look at the chart and compare the numbers on the chart with the actual capacitors to help identify them. You can go ahead and install all of the capacitors at this time. It is best to install the capacitors in groups of four or five at a time, then solder them on to the circuit board before moving on to the next group of capacitors; this helps keep some order to the building process. It is also advisable to check off each of the components against the parts list, so you can keep tabs on your progress and make sure that all of the parts are accounted for. Solder a group of four or five components, cut the excess leads and then move on to the next group of components. This will keep things in an orderly fashion.

Next, go ahead and locate and install the Schottky diode at D1. Note that the diode has a polarity marking on the glass enclosure. At one end of the diode body you should see a black or white band; this denotes the cathode side of the diode. Look at your schematic diagram to observe the direction of the cathode band in the circuit before you install the diodes.

The WWV receiver incorporates six inductors; five of the inductors are wound on ferrite cores, while L6 is a ready-made 1000 µH choke coil. Refer to the chart in Table 10-3 for details on winding each of the coils. When winding the coil you will have to scrape the insulation off each of the wire ends in order to solder the coil to the circuit board. You will need to secure the coil to the circuit board using a couple of "dabs" of rubber cement. Once the coils have been secured to the PC board, you can solder the coil leads to the circuit board, and then remove the excess lead lengths.

Locate the three crystals, Y1, Y2 and Y3 and place them on the PC board. Now solder the crystals to the PC board and then trim the excess leads flush to the edge of the circuit board.

Before we attempt to install the semiconductors, take a quick look at the semiconductor pin-out diagram shown in Figure 10-5. Now locate the MOSFETs Q1, Q2 and Q3. These devices are very static sensitive and must be handled very carefully using an anti-static wrist-band to avoid damage before they are soldered into the circuit. Transistor sockets are highly recommended for all of the transistors, see the pin-out diagram for the MOSFETs and make sure you can identify each of the leads before placing the MOSFETs on to the circuit board. The MOSFETs have two Gate leads, G1 and G2, as well as a Source lead (S) and a Drain lead (D). Once you are sure of each of the leads, you can go ahead and install the MOSFETS onto the circuit board. The AM detector portion of the receiver utilizes two nJFETS at Q4 and Q5, these semiconductors are also static sensitive and must be handled with extreme care to avoid damage. The nJFETS have three leads, like most conventional transistors, but they have a Gate, Source and Drain lead,

Table 10-2

Capacitance code information

This table provides the value of alphanumeric coded ceramic, mylar and mica capacitors in general. They come in many sizes, shapes, values and ratings; many different manufacturers worldwide produce them and not all play by the same rules. Some capacitors actually have the numeric values stamped on them; however, many are color coded and some have alphanumeric codes. The capacitor's first and second significant number IDs are the first and second values, followed by the multiplier number code, followed by the percentage tolerance letter code. Usually the first two digits of the code represent the significant part of the value, while the third digit, called the multiplier, corresponds to the number of zeros to be added to the first two digits.

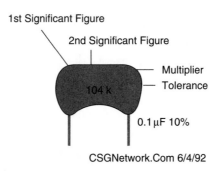

CSGNetwork.Com 6/4/92

Value	Type	Code	Value	Type	Code
1.5 pF	Ceramic		1,000 pF /.001 µF	Ceramic / Mylar	102
3.3 pF	Ceramic		1,500 pF /.0015 µF	Ceramic / Mylar	152
10 pF	Ceramic		2,000 pF /.002 µF	Ceramic / Mylar	202
15 pF	Ceramic		2,200 pF /.0022 µF	Ceramic / Mylar	222
20 pF	Ceramic		4,700 pF /.0047 µF	Ceramic / Mylar	472
30 pF	Ceramic		5,000 pF /.005 µF	Ceramic / Mylar	502
33 pF	Ceramic		5,600 pF /.0056 µF	Ceramic / Mylar	562
47 pF	Ceramic		6,800 pF /.0068 µF	Ceramic / Mylar	682
56 pF	Ceramic		.01	Ceramic / Mylar	103
68 pF	Ceramic		.015	Mylar	
75 pF	Ceramic		.02	Mylar	203
82 pF	Ceramic		.022	Mylar	223
91 pF	Ceramic		.033	Mylar	333
100 pF	Ceramic	101	.047	Mylar	473
120 pF	Ceramic	121	.05	Mylar	503
130 pF	Ceramic	131	.056	Mylar	563
150 pF	Ceramic	151	.068	Mylar	683
180 pF	Ceramic	181	.1	Mylar	104
220 pF	Ceramic	221	.2	Mylar	204
330 pF	Ceramic	331	.22	Mylar	224
470 pF	Ceramic	471	.33	Mylar	334
560 pF	Ceramic	561	.47	Mylar	474
680 pF	Ceramic	681	.56	Mylar	564
750 pF	Ceramic	751	1	Mylar	105
820 pF	Ceramic	821	2	Mylar	205

Table 10-3

Inductor coil winding information

Coil	Value	Ferrite Core	Description
L1	4.1 µH	T-28 toroid (*)	27 turns #22 AWG wire on ferrite core
L2		FT-37-43 (**)	12 turns #22 AWG wire on ferrite core
L3	3.0 µH	T50-2 toroid *	25 turns #26 AWG wire on ferrite core
L4		FT37-43 (**)	12 turns #22 AWG wire on ferrite core
L5		FT37-43 (**)	12 turns #22 AWG wire on ferrite core
L6	1000 µH		pre-wound choke coil; Digi-Key

* Toroids can be obtained from Circuit Specialists; see Appendix.

** FT-37-43 toroid available from RP electronics; see Appendix.

unlike conventional transistors, with a Base, Emitter and Collector. Once the nJFETS have been identified, they can be installed on the circuit board. Finally, the audio amplifier section utilizes three conventional transistors. Transistors Q6 and Q7 are NPN types, while Q8 is a PNP transistor. Transistors have three "legs" and must be oriented correctly. Notice that the part contains a

"flat" side with the writing imprinted on it. Be sure to follow the parts diagram for correct placement. To install, slide the center legs through the circuit board and push the component as close to the board as possible without "straining" the leads. Identify the transistor pin-outs and then go ahead and install the transistors onto the circuit board. Remember to trim the

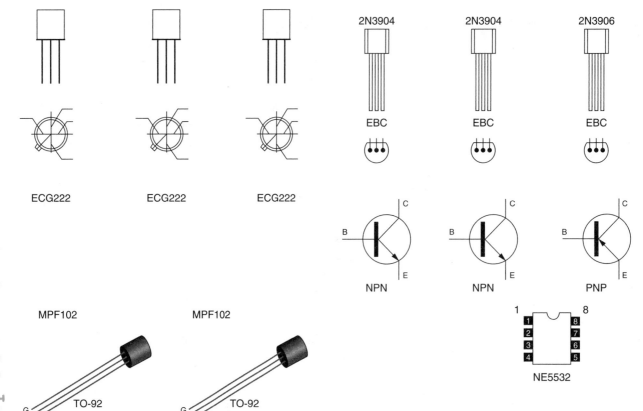

Figure 10-5 *Semiconductor pin-outs*

excess transistor leads after all of the transistors have been installed on the PC board.

It is a good idea to use an IC socket for each of the integrated circuits in the event of a component failure at a possible later date.

Leads length from the board

Now that your soldering skills are at their peak, we'll install the integrated circuit. An integrated circuit socket is highly recommended in the event of a possible circuit failure. It is much easier to simply unplug a "burnt" IC, rather than trying to un-solder a defective IC from the circuit board. Most hobbyists have considerable trouble trying to un-solder multi-pin integrated circuits from a PC board successfully, so it is easier to install a socket from the beginning. Most IC sockets will have a notch or cut-out and one end of the plastic package. Pin 1 will be just to the left of the notch or cut-out. Make sure you orient the socket, so that pin 1 will be close to C29, C31 and R22. When installing the IC into the socket make sure you align pin of the IC with pin 1 of the IC socket.

That's a lot of parts! Take a short break and rest your weary eyes. Before we start up again, we will check your solder joints for "short" circuits and "cold" solder joints. First, pick up the circuit board with the foil side of the board facing upwards toward you. Take a close look at the solder joints in front of you. The solder joints should all look clean, bright and shiny. If any of the solder joints look dark, dull or "blobby," then you need to remove the solder from that particular joint, re-apply solder and re-solder the joint all over again. Next, we will inspect the circuit board for possible "short" circuits. Look closely for small solder "blobs" that often bridge circuit traces. Also look carefully for any "stray" components leads which may have stuck to the foil side of the PC board. Sometimes sticky rosin from solder can leave a residue that will "hold" pieces of wire on the board, often causing "short" circuits between circuit traces.

Identify and install the three potentiometer; they can now be mounted on the metal enclosure that you chose to use for the receiver. Make up three or four inches

leads for each of the potentiometer leads, so you can connect the potentiometer to the circuit board. All three potentiometers can be mounted on the front panel of the receiver enclosure. Using a length of speaker wire, connect the speaker to the circuit board. The speaker can be mounted on the top or front of the enclosure.

If you desire you can customize your new receiver with an LED power indicator. Install a red LED on the front panel. Being a diode, this component is polarized and must be oriented correctly. Examine the LED and notice how one lead is longer than the other. The longer of the two leads is the anode, or (+) connection. Most diodes also have a flat side molded in the component body. This corresponds to the cathode or (−) side of the part. This flat should face in the direction of the band marking of the diode. Leave the diode's leads as long as possible as this component will also mount to the front panel as a power indicator. You can connect up the LED across the power input of the circuit's +12 volt lead and ground.

You can also customize the receiver by installing a closed circuit earphone jack if desired, so that the jack will bypass the speaker when the headphones are plugged into the jack.

Moving to the rear of the enclosure, install antenna connector J1, you can use an RCA jack or a BNC connector as desired. Next, go ahead and mount a coaxial power plug on the rear panel of the chassis and then solder the connectors to the circuit board using a length of insulated wire. Now you will need to decide how you will power your new WWV receiver. You elect to power the receiver from a 12 volt battery or re-chargeable power pack or from a 12 volt DC "wall-wart" power supply.

Congratulations, you have successfully completed your WWV receiver; now let's move on to final testing of the receiver!

Initial receiver testing

This project was designed to be pretty much "plug and play," with minimal testing upon completion. Turn all three potentiometers to their mid-point position. Connect up you power supply or battery and depress the power switch. The power on LED should illuminate. Connect up your outdoor antenna via connector J1.

Now increase the RF gain control, more clockwise. You can also adjust the volume control for more gain as well. You should begin to hear the "beep" signal from the WWV Colorado transmitting station. You may have to adjust VR2 until the signal is sound best in the speaker. If the signal is too strong, you may have to adjust the RF gain and volume control as needed.

In the event that you do not hear any sounds from your new receiver, you will have to turn off the receiver and remove the battery and do a close inspection of the circuit. The most common reasons for circuit problems is due to placing the wrong component in a particular circuit board location or possible reverse placement of a component. Take a look at the Time-Code Receiver circuit board very carefully. Sometimes it is better to get a pair of "fresh eyes" or have a friend take a look at your circuit to find a possible error. Look at each of the resistors and make sure you have placed the correct resistor in each resistor location; read the color code carefully. Check the placement of the diodes and electrolytic capacitors to make sure you have installed them with respect to their polarity marking. Next, take a close look to make sure that you have installed the transistors correctly, looking closely at the pin-outs with respect to the manufacturer's pin-out diagram. Finally, examine the circuit to make sure that the integrated circuit has been oriented properly in its IC socket, since this is a common cause for error. Once you have fully inspected the circuit board and located your error, you can re-connect the battery and switch on the receiver to test it once again.

Antenna considerations

In order to obtain the best reception on your new WWV Time-Code receiver you will want to construct an optimum antenna. Using a random length of wire will give poor results. A very efficient method to use with the WWV Time-Code receiver is a dipole antenna "cut" for the proper frequency of 10 MHz. Half-wave dipoles are very effective antennas and are simple to construct. For low-power work or receiver use, you can use low-cost RG-58 52 ohm coax from the dipole antenna to your new receiver. The formula for dipole antennas: 234/(frequency in MHz) = ¼-wavelength in feet will give the length of wire needed for each leg of the dipole. For example, if you wanted to build a

10 MHz dipole for your Time-Code receiver, you would divide 234 by 10 MHz and the result would be 23.4 feet per leg on either side of the center insulator. The diagram in Figure 10-6 illustrates a dipole antenna which can be used with the Time-Code receiver. Depending upon your location, you will want to orient the antenna for best reception. The antenna should be placed broad-side or parallel towards Colorado for best reception.

Using your new WWV receiver

You will have fun learning about other interesting things on WWV besides the time function feature, such as accurate tone frequencies, geophysical alerts (solar activity reports), marine storm warnings, the global positioning system (GPS) and Omega navigation system status reports.

Table 10-4 shows the broadcast schedule for WWV in Fort Collins, CO. A similar schedule is available for WWVH in Hawaii. This diagram is extracted from a National Institute of Standards and Technology (NIST) publication. If you want to order a copy, it is available for a modest charge from the Government Printing Office, Washington DC 20402. Write and ask for a current list of available publications. The two features of most interest are precise frequency of the carrier signal, to calibrate radio and test equipment, and precise time information given each minute as a voice transmission. But there are many other interesting parts of their transmissions.

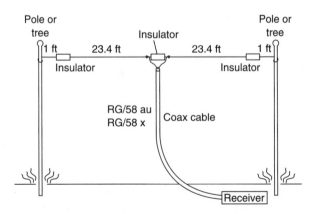

Figure 10-6 *MHz dipole antenna*

Table 10-4

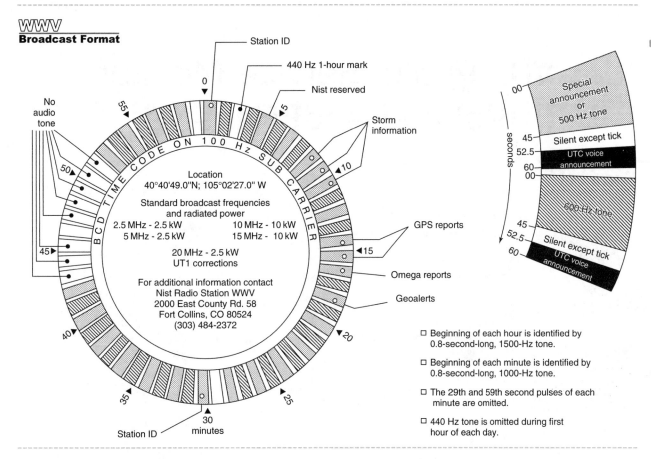

There are various tones which can be used to check audio equipment or musical instruments and they can be captured to detect the start of each hour or each minute. At certain intervals, voice announcements are made for the benefit of other government agencies.
The predominant ones are as follows:

Marine storm warnings are prepared by the National Weather Service and broadcast for areas of the Atlantic and Pacific Oceans which are of interest to the US Government.

Global Positioning System status announcements are prepared by the Coast Guard to give current status information about the GPS satellites.

Omega Navigation System reports are prepared by the Coast Guard to give the status of the 8 Omega transmitting stations in the 10–l4 kHz frequency range. These serve as navigation aids.

Geophysical alerts are prepared by the Space Environment Service Center of NOAA. They are broadcast on the 18th minute of the hour and give

information of interest to amateur radio operators, as well to as various scientific organizations regarding solar activity, geomagnetic fields, solar flares, and other geophysical statistics. This propagation information can help you decide when the DX will be good.

- Beginning of each hour is identified by 0.8 second-long, 1500 Hz tone.

- Beginning of each minute is identified by 0.8 second-long, 1000 Hz tone.

- The 29th and 59th second pulses of each minute are omitted.

- 440 Hz tone is omitted during the first hour of each day.

You will have fun learning about other interesting things on WWV at certain times besides just the time function feature. Features such as accurate tone frequencies, geophysical alerts (solar activity reports), marine storm warnings, and global positioning system (GPS) and Omega navigation system status reports.

Chapter 11

VHF Public Service Monitor (Action-Band) Receiver

Parts list

VHF Public Service Monitor Receiver:

R1,R2,R3   10k ohm potentiometer

R4,R8,R13,R21   47k ohm, $\frac{1}{4}$w, 5% resistor [yellow-violet-orange]

R5   470 ohm, $\frac{1}{4}$w, 5% resistor [yellow-violet-brown]

R6,R9,R20   270 ohm, $\frac{1}{4}$w, 5% resistor [red-violet-brown]

R7,R17,R19   10k ohm, $\frac{1}{4}$w, 5% resistor [brown-black-orange]

R10   1 megohm, $\frac{1}{4}$w, 5% resistor [brown-black-green]

R11,R12   1k ohm, $\frac{1}{4}$w, 5% resistor [brown-black-red]

R14   33k ohm, $\frac{1}{4}$w, 5% resistor [orange-orange-orange]

R15   470k ohm, $\frac{1}{4}$w, 5% resistor [yellow-violet-yellow]

R16   100k ohm, $\frac{1}{4}$w, 5% resistor [brown-black-yellow]

R18   18k ohm, $\frac{1}{4}$w, 5% resistor [brown-gray-orange]

R22   2 ohm, $\frac{1}{4}$w, 5% resistor [red-black-gold]

C1,C6,C22   100 pF, 35 vdc disk capacitor [marked 100, 101, or 101K]

C2,C5   10 pF, 35 vdc disk capacitor

C3,C4   56 pF, 35 vdc disk capacitor

C7,C8,C11,C20   .001 µF, 35 vdc disk capacitor [.001 or 102 or 1 nF]

C9,C10   15 pF, 35 vdc disk capacitor

C13,C16,C17,C21   .01 µF, 35 vdc disk capacitor [marked .01 or 103 or 10 nF]

C12   8.2 pF, 35 vdc disk capacitor

C14,C32,C33,C35   100 to 220 µF, 35 vdc electrolytic capacitor

C15,C30,C36   4.7 µF, 35 vdc electrolytic capacitor

C18   220 pF, 35 vdc disk capacitor [marked 220 or 221]

C19   22 pF, 35 vdc disk capacitor

C23,C24,C25   .001 µF, 35 vdc disk capacitor [.001 or 102 or 1 nF]

C26,C28,C29,C31   .01 µF, 35 vdc disk capacitor [marked .01 or 103 or 10 nF]

C34   .1 µF, 35 vdc disk capacitor [marked .1 or 104]

C27   1.0 µF, 35 vdc electrolytic capacitor

L1,L2   2-turn coil 5/32" OD dia - #24 ga. Insulated magnet wire

L3   slug-tuned 3 ½ turn coil, see Table 11-3

L4   450 kHz shielded quadrature coil, see Table 11-3

D1   Varactor diode, BB505 [orange body marked BB505]

D2   1N4148 signal diode

Q1   2SC2498 or 2SC2570A transistor

Q2,Q3,Q4   2N3904 transistor

U1   SA602 8-pin IC

U2   MC3359 18-pin FM receiver IC

U3   LM386 8-pin audio amplifier IC

Y1   10.24 MHz Crystal

FL1   10.7 MHz ceramic filter [brown, molded, 3 leads]

FL2   450 kHz ceramic filter [black, square]

S1   DPDT push switch

J1   PC mount RCA jack

J2   PC mount

subminiature speaker jack

B1   9 volt transistor radio battery

ANT   telescoping antenna

Misc   PC board, battery holder, battery clip, speaker, etc.

Keep an ear on the local action from your easy chair with the VHF Public Service Receiver project, shown in Figure 11-1. The VHF Public Service Radio will allow you to tune the VHF high-band Police and Fire band, marine band, weather band, as well as railroad frequencies. The receiver tunes any 5 MHz portion of the 136 to 175 MHz band. Excellent performance, less than 1 uV sensitivity. The receiver features: tuned input, low noise pre-amp stage with true dual-conversion superhet design with 2 pole Ceramic High Intermediate Filter or I-F and 6 pole ceramic low I-F filters! The receiver also features front panel volume, squelch, and tuning controls. This is a great inexpensive receiver to put on your desk at home or to "watch" local radio activity while you are away from home on vacation or business trip.

Circuit description

The heart of the VHF Public Service Receiver shown in the block diagram of Figure 11-2 is the MC3359

VHF Public Service Receiver

Figure 11-1   *VHF public service receiver*

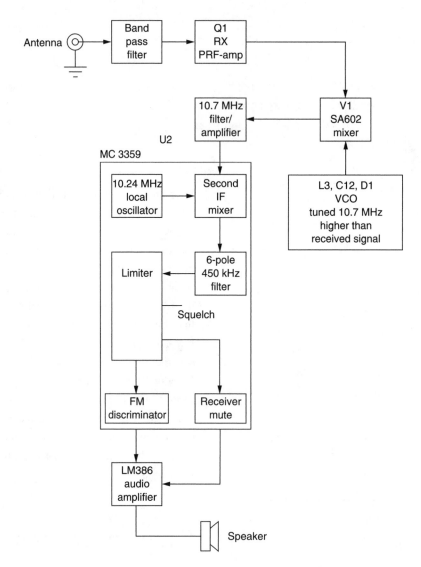

Figure 11-2 *VHF monitor receiver block diagram*

integrated circuit. The integrated circuit is a complete high gain low power FM IF sub-system which includes an oscillator, a mixer, a limiting amplifier, AFC, quadrature discriminator, op-amp, squelch, scan control and mute switch all combined in a single IC. The MC3359 was designed to detect narrow-band FM signals using a 455 KHz ceramic filter.

Let's get into more details of the VHF Public Service Receiver by examining the schematic diagram shown in Figure 11-3. VHF signals from the antenna are amplified through the tuned input circuit (L1, C3 and L2, C4) by Q1, a microwave bipolar transistor. Q1's output is fed to the input of the SA602 IC, an efficient single-package (8-pin DIP) mixer-product detector-oscillator. The tuneable oscillator section of the SA602

is aligned to operate at 10.7 MHz higher than the signal fed and amplified by Q1. For example, to receive 144-148 MHz signals, the oscillator must tune 154.7 to 158.7 MHz in order for the SA602's mixing capability to produce a steady 10.7 MHz output signal to the rest of the circuit. The oscillator frequency is determined by L3 and its associated capacitors, and varied by the varactor tuning network using D1 and varied by R1.

The output from pin 4 of the SA602 passes through a ceramic 10.7 MHz filter, amplified by transistor Q2 and applied to input pin 18 of U2. Q3 provides AFC (Automatic Frequency Control) by keeping the local oscillator of U1 from drifting away from an incoming signal. This is accomplished by tuning the varactor circuit in the direction opposite the drift.

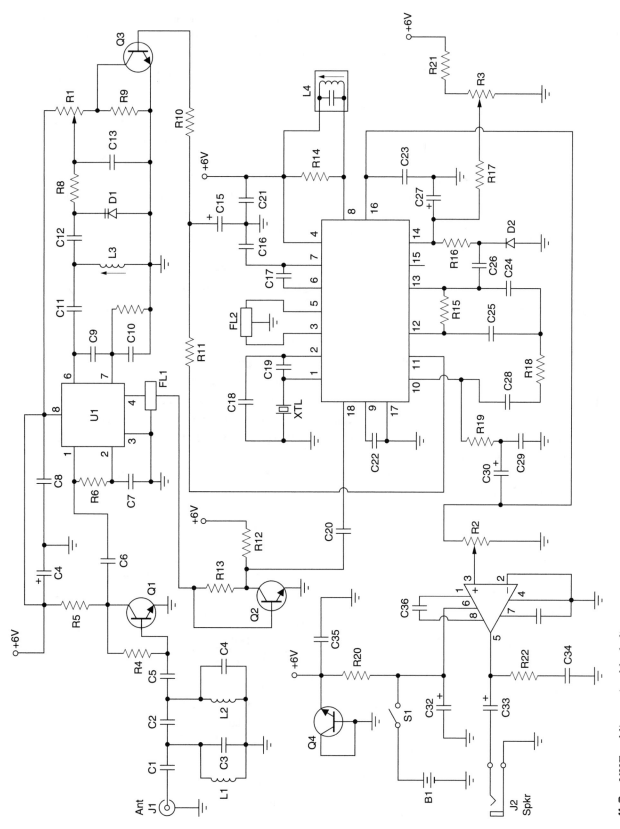

Figure 11-3 *VHF public service block diagram*

The MC3359 IC has an internal oscillator controlled by the 10.24 MHz crystal. The 10.24 MHz signal is mixed with the 10.7 MHz input from U1 to convert down to the low IF of 450 kHz. The 450 kHz IF is filtered by FL2 and then amplified by a limiting amplifier in U2. Audio demodulation takes place in the quadrature detector, with L4 adjusting the detector. The audio modulation is detected in U2, and the background noise is used to control the squelch. U3 amplifies the audio output from pin 10 of U2 to a practical level for speaker or headphone operation. Pin 16 of U2 will ground the input of U3 when the squelch is closed. L4, a 450 kHz IF coil, permits adjustment of the 90° voltage-current phasing ("quadrature") of FL2's output to the FM detector demodulator (pin 8 of the MC3359).

Circuit assembly

Look for a clean well lit area in which to assembly your shortwave receiver. You will need a 27 to 30 watt pencil tip soldering iron, some 60/40 tin/lead rosin core solder, a few hand tools, including a pair of end-cutters, a needle nose pliers, a few screwdrivers and a pair of tweezers to pick up small parts. Locate your schematic and layout diagrams and we will begin to construct your new shortwave receiver. Let your soldering iron tip heat both the component lead wire and PC board trace enough so that the wire itself AND the foil trace BOTH become hot enough TOGETHER to melt a bit of solder so that it flows smoothly from the pin to the PC board trace. Locate the printed circuit board and all the components and we will start.

Locate all of the resistors for the project and place them in front of you. Each resistor has a color code which describes its value. The first color band at one edge of the resistor body represents the resistor's first digit value, while the second color band denotes the second digit of the resistor value. The third color band represents the multiplier value of the resistor. A fourth color band is the tolerance value. A silver band denotes a 10% tolerance, while a gold band denotes a 5% tolerance. The absence of a fourth color band denotes a 20% tolerance value. Check the color code against the parts list and schematic before installing them onto the circuit board. Refer to the resistor color chart in Table 11-1. Place about four or five resistors on the PC board and place these resistors in their respective locations carefully making sure each one is in the correct location. Next, solder each of the resistors onto the circuit board. Trim off the excess component leads with a small pair of end-cutters. Pick up and identify the next grouping of resistors and install them on the circuit board, follow up by soldering this next batch of resistors

Table 11-1

Resistor color code chart

Color Band	1st Digit	2nd Digit	Multiplier	Tolerance
Black	0	0	1	
Brown	1	1	10	1%
Red	2	2	100	2%
Orange	3	3	1,000 (K)	3%
Yellow	4	4	10,000	4%
Green	5	5	100,000	
Blue	6	6	1,000,000 (M)	
Violet	7	7	10,000,000	
Gray	8	8	100,000,000	
White	9	9	1,000,000,000	
Gold			0.1	5%
Silver			0.01	10%
No color				20%

to the circuit board; remember to trim the extra lead lengths form the PC board.

Once all the resistors have been installed onto the circuit board, we can move on to identifying all of the capacitors for the project. Note that there will be different looking capacitors for this project, some will be small disks, while others will be larger body devices with minus and plus marking on them. These larger capacitors with polarity marking are electrolytic capacitors and they have polarity which must be observed for the circuit to work correctly. Look through the parts pile and locate the capacitors and place them in front of you. You will notice a number of small disk capacitors, we will install these first. Look closely at each of the disk capacitors at their markings. Sometimes the capacitors are very small and they may not have their actual value printed on them but will instead have some sort of code marked on them. Refer to the parts list for these codes or the capacitor code chart in Table 11-2. Locate four or five of these small disk capacitors, identify them and install them on the circuit board, while referring to the schematic and layout diagrams. After inserting the disk capacitors onto the circuit board, you can go ahead and solder them to the circuit board. Remember to cut the excess component leads. Next, move on and install another grouping of small disk capacitors, then solder them in place on the printed circuit board. Cut the extra component leads after soldering in the capacitors. When you are finished installing the small capacitors we will move on to the larger electrolytic types.

Capacitors C14, C32, C33 and C35 are all 100 uF to 200 μF electrolytic types. These capacitors will have either a plus or minus marking and/or black band denoting polarity. You must orient these capacitors with respect to their polarity markings. Check both the layout diagram and the schematic diagram for polarity references. Note also that capacitors C15, C30 and C36 are also electrolytic capacitors with a 4.7 μF value. These too must be oriented correctly for the circuit to operate correctly when power is applied. It is possible to actually damage the circuit upon power-up if these capacitors have been installed incorrectly, so pay careful attention when mounting these capacitors. Install the electrolytic capacitors in groups of two or three. Once they have been placed on the circuit board, you can move on to soldering them in place. Remember to trim the excess component leads from the PC board after soldering.

Next we will move to installing the two diodes. Diode D1 is the varactor "tuning" diode while D2 is a 1N4148 silicon diode. When mounting the diodes be careful to observe polarity markings on the diodes. Usually each diode will have a black or white band at one end of the diode body. This color band denotes the cathode end of the diode. Refer to the schematic diagram and layout diagrams in order to see how the diodes should be mounted with respect to polarity considerations.

Before we go ahead and install the semiconductors, take a look at the semiconductor pin-out diagram in Figure 11-4 which will help you orient the components. Now locate the transistors from the components. Transistor Q1 is a 2SC2498 or 2SC2507A, while transistors Q2, Q3 and Q4 are 2N3904 NPN types. Transistors have three leads, a Base lead, a Collector lead and an Emitter lead. The symbol for the Base is a vertical line, while the Collector and Emitter are slanted lines pointing to the Base lead. The Emitter lead is shown with a small arrow pointing towards or away from the Base lead. Refer to the schematic and layout diagrams to determine the orientation of the transistors. Clearly identify Q1, the 2SC2498 or 2SC2570A transistor. Do not confuse it with the other transistors supplied. Position Q1 as shown on the PC board layout, with the flat side facing to the right, toward the middle of the board. Press the transistor snugly into the PC board so that only a minimum amount of wire lead is exposed above the board. In soldering, don't be afraid to use enough heat to make good clean connections.

Once the transistors are in place on the PC board, you can go ahead and solder them to the foil side of the PC board. Don't forget to trim the extra component leads.

Now we are going to install FL1, the ceramic filter. This component looks like a capacitor with three leads and may be installed either way. Next go ahead and locate FL2, the 450 kHz filter. Its three leads are delicate and fit in only one way. At this time you can locate Y1, the 10.24 MHz crystal. No special procedure is required. Simply press the crystal firmly into its holes as far as it will go. Now go ahead and solder in the crystal and ceramic filters, remove the excess leads if any.

Table 11-2
Capacitance code information

This table provides the value of alphanumeric coded ceramic, mylar and mica capacitors in general. They come in many sizes, shapes, values and ratings; many different manufacturers worldwide produce them and not all play by the same rules. Some capacitors actually have the numeric values stamped on them; however, many are color coded and some have alphanumeric codes. The capacitor's first and second significant number IDs are the first and second values, followed by the multiplier number code, followed by the percentage tolerance letter code. Usually the first two digits of the code represent the significant part of the value, while the third digit, called the multiplier, corresponds to the number of zeros to be added to the first two digits.

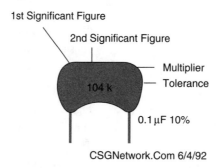

CSGNetwork.Com 6/4/92

Value	Type	Code	Value	Type	Code
1.5 pF	Ceramic		1,000 pF /.001 µF	Ceramic / Mylar	102
3.3 pF	Ceramic		1,500 pF /.0015 µF	Ceramic / Mylar	152
10 pF	Ceramic		2,000 pF /.002 µF	Ceramic / Mylar	202
15 pF	Ceramic		2,200 pF /.0022 µF	Ceramic / Mylar	222
20 pF	Ceramic		4,700 pF /.0047 µF	Ceramic / Mylar	472
30 pF	Ceramic		5,000 pF /.005 µF	Ceramic / Mylar	502
33 pF	Ceramic		5,600 pF /.0056 µF	Ceramic / Mylar	562
47 pF	Ceramic		6,800 pF /.0068 µF	Ceramic / Mylar	682
56 pF	Ceramic		.01	Ceramic / Mylar	103
68 pF	Ceramic		.015	Mylar	
75 pF	Ceramic		.02	Mylar	203
82 pF	Ceramic		.022	Mylar	223
91 pF	Ceramic		.033	Mylar	333
100 pF	Ceramic	101	.047	Mylar	473
120 pF	Ceramic	121	.05	Mylar	503
130 pF	Ceramic	131	.056	Mylar	563
150 pF	Ceramic	151	.068	Mylar	683
180 pF	Ceramic	181	.1	Mylar	104
220 pF	Ceramic	221	.2	Mylar	204
330 pF	Ceramic	331	.22	Mylar	224
470 pF	Ceramic	471	.33	Mylar	334
560 pF	Ceramic	561	.47	Mylar	474
680 pF	Ceramic	681	.56	Mylar	564
750 pF	Ceramic	751	1	Mylar	105
820 pF	Ceramic	821	2	Mylar	205

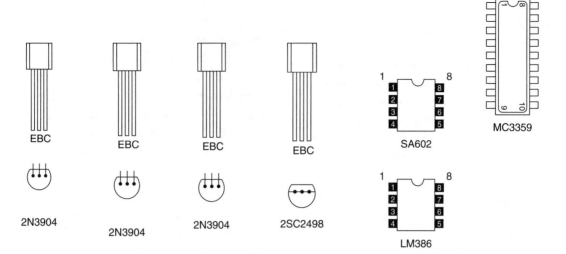

Figure 11-4 *Semiconductor pin-outs*

Look through the remaining components and locate the inductors L1 through L4. Coils L1 and L2 are small 1 and ½ turn coils. Go ahead and install coils L1 and L2, they are both the same. Now, locate and install L3, the slug tuned coil. It is important that the body of this coil be snug against the PC board for mechanical stability, which directly affects the frequency stability of the receiver. Make good, solid solder connections, when installing the above coils. Finally locate coil L4, the 455 kHz quadrature coil, soldering the two pins and the two mounting tabs. Remove any excess lead lengths if any; see Table 11-3 for coil winding information.

Let's move on to installing the SA602 IC, you may wish to use an 8-pin DIP socket rather than soldering the IC directly to the board. Reasons for doing this might include the peace of mind of beginners afraid of damaging the IC, or experienced builders testing the merits of the SA602's cousin, the SE602. However,

please be aware that we have seen more service problems with improper socket insertion than from soldering heat damage to ICs. Even if this is your first IC, don't be afraid to use enough heat to make eight clean connections, but DO be sure to correctly orient the end marked by a band, dot, or notch. Before soldering, make sure that the IC or socket is perfectly flat against the top of the PC board. Install U1, the SA602, in accord with the preceding directions.

Now, install MC3359 IC, U2. As in the case of the SA602 IC, you could choose to install an 18-pin IC socket rather than soldering the IC directly. Re-read the discussion of IC sockets offered for the installation of U1. Larger ICs such as the MC3359 require considerably more care in socket insertion. Notice that the end of the IC marked by a band, dot, or notch must be oriented correctly as shown on the parts layout diagram. Solder each of the 18 connections carefully.

Table 11-3
Coil winding information

L1,L2	2-turn coil - #24 ga. Insulated magnet wire - 5/32″ OD dia.
	Air-core coil.
L3	3 ½ turn coil ferrite slug-tuned coil, unwind original coil from form and rewind 3½ turns of #24 ga. insulated magnet wire on form - use series 49A coil
	Circuit Specialists (49A127MPC).
L4	450 kHz shielded quadrature coil
	TOKO America Coils (7MC-81282).

Make sure you have good lighting as well as good technique to make sure that no solder "bridges" flow between the connections.

Finally locate and install U3, the LM386 IC audio amplifier integrated circuit. Once again use of an IC socket is highly recommended. Solder the IC socket to the PC board and insert the LM386 paying particular attention to the small notch or indented circle on the top of the IC body. Pin 1 is to the left of the notch, cut-out or indented circle.

Take a short well-deserved break and when we return, we will install the remaining components on the PC board and finish up building the receiver. Now locate power switch S1. Press S1 firmly into its six holes and solder all six pins. The switch fits the board only one way. Next, install J1, the RCA-style antenna jack and solder all four points. Now, install J2, the subminiature headphone jack. Solder all three points. Be gentle and patient in inserting, so as not to damage the solder tabs.

Locate all three potentiometers and we will install them now. Potentiometer controls R1, R2 and R3 are all mounted in a row next to switch S1. Insert the three PC mount potentiometers into their positions. Check that the controls are pressed in firmly and straight against the top of the board. Solder the three center pins and then the two larger mechanical mounting tabs. Use enough solder for a solid connection.

The circuit board may require a few jumpers; "cut" component leads make good jumpers. Insert the jumpers and solder into place, remember to trim the excess leads. We are now down to the home-stretch. Install the battery snap terminal in the two holes below S1, making sure that the positive (red) lead is inserted into the (+) hole on the PC board.

Finally, go ahead and mount the battery bracket, it may be attached in a variety of ways. A wire jumper can be passed through the two holes on the PC board, then both ends soldered on the underside of the board. You may prefer to use very small screws or even a double-sided mounting adhesive strip or glue. In using such methods, make sure that the battery will not block the PC board's mounting hole in its vicinity.

Wiring of the PC board is now complete. If you have the patience, we suggest a short break. Then take a magnifying lens and a bright light and examine all your solder joints, touching up any connection which appears less than perfect. Make sure all excess leads have been trimmed, and that one is not bent back flat against the board, possibly causing a short. Brush the solder side of the board with a stiff brush to make sure that no loose wire trimmings or solder drippings are lodged between connections. Connect up a suitable antenna and connect up the 9 volt battery to the receiver and you are ready to try out the receiver for the first time.

After power is first applied and you have connected up your antenna, you will need to adjust the gain control to midrange. Next, you should adjust the squelch control just below the threshold point, so that you can hear the normal background hiss. Now begin turning the tuning control and you should soon come across a public service transmission from a local police or fire broadcast.

In the event that you do not hear any hiss coming from the speaker or headphones, or you do not hear any stations after about 20 to 30 minutes, then you may have to troubleshoot the receiver for any errors that may have occurred during the building process. First, disconnect the battery and antenna and un-mount the receiver circuit board from the enclosure. Have a friend assist you in inspecting the circuit board; a different set of eyeballs can often help to find errors quickly.

The most common causes for circuit failure are resistors placed in the wrong locations, the second possible cause for errors are improper installation of diodes and electrolytic capacitors. Make sure that you re-check the orientation of these components one more time and refer to the schematic, and parts layout diagrams. Another possible cause for mistake is the improper installation of transistors and integrated circuits. Be sure to recheck the orientation of transistors. Each transistor has three leads, an Emitter, a Collector and a Base lead. Closely inspect and compare the schematic diagram along with the parts layout diagram and manufacturer's specifications sheets to ensure that you have oriented the transistors correctly. Lastly, integrated circuits are often placed backwards in their sockets. After you have checked all the above tips, make sure that the battery is connected correctly with the black battery or minus lead (−) connected to the circuit ground and the plus (+) or red battery lead connected to the power input regulator of the circuit. Reconnect the battery and the antenna and then switch the receiver to the "On" position and retest the receiver once again to make sure that it is now working.

Testing, alignment and adjustment

In order to prepare the VHF FM Receiver for reliable monitoring operation, you will need a few basic tools, such as useful VHF signal source, a hexagonal, non-metallic coil slug alignment tool for L3, and a small screwdriver to adjust L4. To align your new receiver, you will need to locate some sort of a signal source. This signal can be from your test bench equipment or from a "live" source such as a local repeater. The "live" signals are best for fine-tuning, but they also have that VHF communications character of being fast and to the point. Good test bench signal sources include your own 2-meter transceiver, if you are licensed to use it, a signal generator or grid-dip oscillator. If you do not have a steady test signal source for the band within 135–175 MHz of primary interest to you, and still wish to verify immediately the proper operation of your receiver, a good and steady VHF signal source will be your local NOAA Weather Service station, broadcasting around 160 MHz. Once you have verified reliable reception of your regional NOAA station, adjustment to your desired listening range will become easier.

If you do not already have a set of plastic or nylon coil alignment tools and do expect to try further ham radio or electronic hobby projects, such tools are worth having and can be found inexpensively at any electronics store including Radio Shack. While a metal hex key wrench can fit the coil slug, the metal itself would damage the coil inductance drastically and therefore should not be used. With patience and sandpaper, a useable tool might be formed from a wood or plastic rod.

The alignment procedure should begin as follows: first make sure the power switch is off. Next, connect the antenna, earphone or speaker and battery, then turn the slug of L3 until it is even with the top of the coil form. Then turn the slug of L3 clockwise seven turns back into the form.

Now, turn the slug of L4 until it is flush with the top of the coil and then turn it two turns back into the coil. Next turn all three controls to the left, fully counterclockwise.

Now, you can turn the power switch to "On." Now turn R2 (volume control) until you hear some noise. Finally, adjust L4 for maximum noise from the speaker.

Further alignment now consists of adjusting the oscillator coil L3 to permit the tuning control (R1) to cover the 5 MHz segment between 135 and 175 MHz of primary interest to you. Assuming you wish to adjust for the 2-meter 144–148 MHz Amateur Band, adjust L3 until you hear your intended test signal. If you are a beginner with no license or other equipment, any ham operator with a 2-meter transceiver should be willing to give you the test signal and extra help that you need. Your new VHF receiver is very sensitive, so operate the transceiver on low power on a simplex frequency from a distance of at least across the room. An 8″ piece of wire will be a sufficient receiving antenna for such tests. If you don't know any hams, visit a friendly two-way radio service center to get close to the test signal you need!

Receiver sensitivity

Your FM receiver features sensitivity under 1 uv. Radio hams constantly marvel at how an FR146 displayed at hamfests tunes in dozens of hand-held QSOs on the premises without an antenna connected! You can expect to monitor local repeater and simplex transmissions easily, using a simple ground-plane style antenna.

Customization

For many people, a pilot lamp to indicate "power on" is more than a nice touch. They expect it and depend on it. Adding a simple LED power-on indicator to your VHF receiver is easy. All you need is the LED itself and a small 1k to 2.2k resistor. Study the PC traces between the positive battery supply wire and the on-off switch. The unused connectors on top of your switch are an ideal point to get the + DC voltage needed for the anode (longer lead) of the LED. Plan where and how you wish to install the LED in your enclosure.

Locating the LED immediately above the on-off switch is logical and ideal. The simplest way to make a good installation is to drill a neat hole just slightly smaller than the diameter of the LED. Then, enlarge the hole a little bit at a time just enough to let the LED be pressed in and held firmly. The resistor may be connected to either the anode or cathode of the LED, but the anode MUST go to + DC, with the cathode connected to the nearest common ground point.

Antenna considerations

For local VHF reception, the VHF receiver will operate fine just using a whip antenna. If you wish to receive stations which are further away than just the local stations, you may want to install some sort of an outdoor antenna such a ground plane or Discone antenna for VHF operation. Since most of the transmissions in this band are of vertical polarization, vertical antennas, ground plane and Discone antennas will work best in this band. Various VHF scanner type antennas from your local Radio Shack store will work fine. The chart illustrated in Table 11-4 shows the utilization of the VHF Public Service band. Have fun and enjoy you new VHF "action" band receiver.

Table 11-4

VHF high-band frequency band utilization

150.995 MHz to 151.130 MHz	Highway maintenance
151.145 MHz to 151.475 MHz	Forestry conservation
152.030 MHz to 152.240 MHz	Mobile telephone
152.270 MHz to 152.450 MHz	Taxi services
152.510 MHz to 152.810 MHz	Mobile telephone
153.410 MHz to 154.115 MHz	Utilities/light/power/water and local gov't
153.830/153.890/154.010/154.070 MHz	Fire
154.130 MHz to 154.445 MHz	Fire
154.650 MHz to 155.700 MHz	State & Local police
155.715 MHz to 156.255 MHz	Local Gov't / police/ highway maintenance
156.275 MHz to 157.425 MHz	Marine Frequencies
157.530 MHz to 158.265 MHz	Taxi & Mobile telephones
158.280 MHz to 158.460 MHz	Utilities water/power/industrial
158.730 MHz to 159.210 MHz	Highway Maintenance & police
159.225 MHz to 159.465 MHz	Forestry
159.495 MHz to 160.200 MHz	Motor carrier trucks
160.215 MHz to 161.565 MHz	Railroad
161.640 MHz to 161.760 MHz	Broadcast pickups & Studio to Xmiter Links
161.800 MHz to 162.00 MHz	Maritime Shore Stations
162.400 MHz to 162.550 MHz	NOAA weather broadcasts
162.250 MHz to 173.500 MHz	Gov't sonobuoys
169.00 MHz to 172.000 MHz	Low power wireless mikes
173.2037 MHz to 173.396 MHz	Fixed/mobile industrial & remote control
173.400 MHz to 173.5000 MHz	Remote control & telemetry

Chapter 12

6 & 2-Meter Dual-Band Amateur Radio Receiver

Parts list

Parts Bin

6N2 Meter Amateur Radio Receiver

R1   68 k, $\frac{1}{4}$w, 5% resistor

R2   47 k, $\frac{1}{4}$w, 5% resistor

R3, R7   3.3k, $\frac{1}{4}$w, 5% resistor

R5   20 k 5% $\frac{1}{4}$W

R8   8.2k, 5% $\frac{1}{4}$W

R9   Jumper wire (use a discarded resistor lead)

R10, R12   1k, $\frac{1}{4}$w, 5% resistor

R11   10 k, $\frac{1}{4}$w, 5% resistor

R15   10 ohm, $\frac{1}{4}$w, 5% resistor

VR1, VR3   100 k potentiometer (chassis mount)

VR2   5k, 20% linear potentiometer (chassis mount)

C1   120 pF, 25v capacitor (121)

C2   .005 µF, 35v 20% capacitor (502)

C3,C4,C9   .1 µF +80, −20% 50v capacitor (104)

C5   3.9 pF, 50v disk capacitor

C6   33 pF, 50v, 10% disk capacitor

C7   470 pF, 50v, 10% 50v disk

C8   7 pF, 25v, 20% disk capacitor

C10   1000 pF, 50v, 10% disk capacitor (102)

C11,C13,C16   .01 µF, 50v, +80, −20% capacitor (103)

C17   51 pF, 25v, 10% capacitor (51K)

C12   68 pF, 25v, 10% capacitor (68K)

C14, C21   4.7 µF, 50v Radial capacitor

C15   220 µF, 16v Radial capacitor

C18,C19,C20   .1 µF, 50v, +80, −20% capacitor (104)

C22   .047 µF, 50v, +80, −20% capacitor (473)

D1, D2 Diode   1N914 silicon diode

U1   IC MC3362P FM Receiver IC

U2   IC LM386N-1 Audio Amp IC

U3 IC 78L05 Regulator +5v 78L05

L1 Coil (4.5 turns coated wire- #24 ga.) -air-core

L2, L3 Coil (on form with core)

L4 Coil (1.5 turns coated wire- #24 ga.) -air-core

L5 .64 mH Coil (yellow) I-F Transformer (42IF301)

F1 455 kHz Ceramic Filter

F2 10.7 MHz Ceramic Filter

Y1 10.245 MHz Crystal

SW1, SW2 Switch Slide DPDT

B1 9 volt transistor radio battery

SPKR 8 ohm speaker

J1 RCA jack - antenna

Misc PC Board, battery holder, battery clip,

The 6N2 receiver is a dual conversion, dual band amateur radio receiver; it combines the excitement of two very active amateur radio bands in one small receiver, see Figure 12-1. The 6N2 receiver will allow radio enthusiasts, both the young and old, to get a

Figure 12-1 *6N2 meter amateur radio receiver*

glimpse of the active world of VHF amateur radio. This receiver combines both the 2-meter and the 6-meter bands which are popular with radio enthusiasts around the country. The 2-meter band is probably the most active ham radio band especially for newcomers with a technician class license. Two-meter band coverage is usually quite good in most parts of the country. The 2-meter band features many ham radio repeaters, and in some have wide-area long-range repeaters in most US states and in many Canadian provinces. These ham radio repeaters allow amateur radio operators to communicate over long distances. The 6-meter band has often been called the "magic band." This amateur band is usually only used for fairly local coverage but when the band is "open" you can often hear hams on the opposite side of the US or in Cuba or the Caribbean, which is really exciting!

You can listen-in to these exciting two ham radio bands and hear all the action, day or night with the 6N2 portable receiver. You may just decide the action sounds like something you would like to be part of, and you can easily get your own ham radio "ticket" or license. Contact the Amateur Radio Relay League (ARRL) at http://www.arrl.org; they have many books, publications as well as introductory videos. They are a great resource in locating local ham radio clubs, where you can make friends and learn a lot about ham radio and communication.

The 6N2 radio is known as a narrow band FM VHF receiver. FM of course is a type of modulation and is denoted as Frequency Modulation. Frequency modulation means the data or voice changes the frequency of the radio wave. The other popular forms of modulation are AM (Amplitude Modulation) and PM (Phase Modulation). We also mentioned narrow band, which notes that the selectivity of the entire system is limited to only enough frequencies to pass voice or low frequency data. A commercial broadcast FM receiver would have a bandwidth large enough to pass music and high frequency data transmissions.

The 6N2 receiver is a dual conversion, two-band radio receiver. In a dual conversion receiver, the original radio frequency is converted first to a 10.7 MHz intermediate frequency (I-F) and amplified in (Block 4). The 10.7 MHz signal is then converted to 455 kHz and amplified in (Block 3). Noise is removed and the

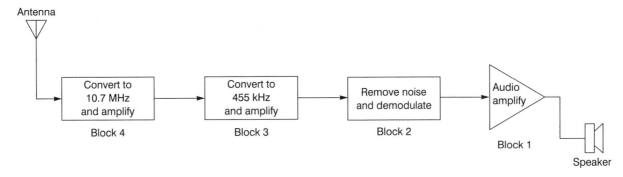

Antenna

| Convert to 10.7 MHz and amplify | Convert to 455 kHz and amplify | Remove noise and demodulate | Audio amplify |
| Block 4 | Block 3 | Block 2 | Block 1 |

Speaker

Figure 12-2 *½ meter amateur radio receiver block diagram*

modulated signal is recovered from the 455 kHz I-F signal in (Block 2). The audio signal is amplified to drive a speaker in (Block 1), see Figure 12-2. The 6N2 receiver specifications are illustrated in Table 12-1.

So you ask, why dual conversion? To answer this question, take a look at the diagram shown in Figure 12-3, and it may become more clear. In order to obtain the desired intermediate frequency (I-F) signal by mixing the local oscillator with the desired radio frequency, an unwanted output may result due to a transmission spaced one intermediate frequency or (I-F) on the opposite side of the oscillator (image). By using a large I-F frequency, this image is moved further out of the band of desired frequencies. The second conversion

provides the selectivity to filter out the desired narrow band transmission.

The 6N2 receiver operates on two of the most popular VHF amateur radio bands. The 6N2 VHF amateur radio receiver receives both the 6-meter ham band and the 2-meter ham bands, and you can select the band with a band-switch.

To help you visualize the relationship between frequency and wavelength, take a look at the diagram shown in Figure 12-4. If the speed of a wave (meters per second) is divided by the number of waves that pass a given point (cycles per second), the seconds cancel and you obtain the wavelength λ = meters per cycle. The speed of radio waves is approximately 300,000,000 meters per second. If the frequency is 50 to 54 MHz, the wavelength becomes 300,000,000/50,000,000 or 6 meters. So for the 2-meter band, which is approximately 150 MHz, the wavelength becomes 300,000,000/150,000,000 or 2 meters. The actual bands are: 2-meter band which covers 144 MHz to 148 MHz, and the 6-meter band which covers 50 MHz to 54 MHz.

The 6N2 dual band VHF receiver is composed of four major blocks, i.e. blocks 1 through 4. It is a bit easier to follow how the receiver works if we break it down into smaller defined blocks, see Figure 12-5. The heart of the receiver is the integrated circuit at U1, an almost complete FM receiver on-a-chip. The FM receiver chip only requires timing components, a crystal and RF tuning components, a filter. The output of the FM radio chip is sent to the audio amplifier chip at U2. Power for the receiver is regulated via the regulator chip at U3.

Table 12-1
The 6N2 amateur radio specifications

2-meter band covers 144 MHz to 148 MHz

6-meter band covers 50 MHz to 54 MHz

- Single supply voltage (4-12 v)
- Idle current - 4 milliamps
- Inputs referenced to ground
- Input resistance - 50 k
- Self-centering output voltage
- Total harmonic distortion less than 0.2%
- Output power with 9 volt supply voltage
- Voltage gain with 10 µF from pin 1 to 8 - 200 or 46 dB
- Voltage gain with pins 1 and 8 open - 20 or 26 dB
- Bandwidth with pins 1 and 8 open - 300 kHz

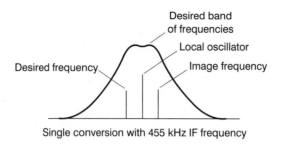

Single conversion with 455 kHz IF frequency

vs.

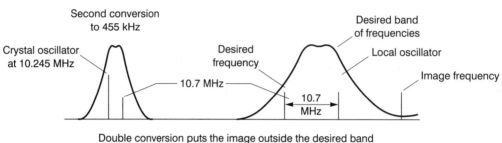

Double conversion puts the image outside the desired band
and still allows narrow band output

Figure 12-3 *Single vs. double conversion receiver*

The receiver is relatively straightforward and can therefore be constructed in stages, one through four. We will begin with Stage 1, which can be built and tested separately. You can progress step by step and build and test each stage before moving on to the next stage. Let's begin with Block 1, the audio amplifier stage.

The audio amplifier – block 1

The audio section in this receiver is amplified by using an integrated circuit audio power amplifier. The output

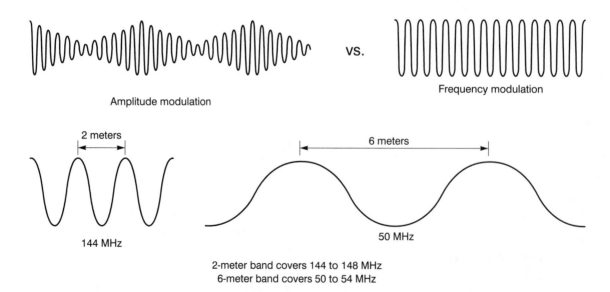

Figure 12-4 *Frequency coverage diagram*

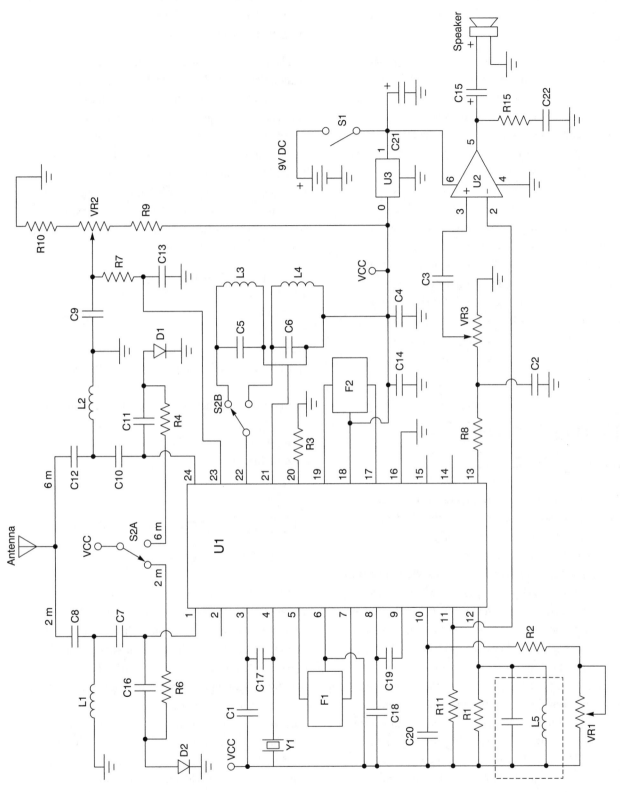

Figure 12-5   *VHF public service receiver schematic. Courtesy of Elenco Electronics*

impedance of the amplifier is low enough to drive an 8 ohm speaker directly. The coupling capacitor value is picked to pass audio signals down to 100 cycles by matching the reactance of the capacitor with the speaker impedance.

Due to the high input resistance of the amplifier (50 k), the audio coupling capacitor C3 can be as small as 0.1 µF. The equivalent resistance at the junction of R8 and VR3 is approximately 6.6 k (the parallel combination of R8, VR3 and the 50 k input impedance of the LM-386). The capacitor C2 and this equivalent resistance sets the 3 dB corner used to attenuate any IF voltage at pin 13. A simple RC filter attenuates at a rate of 6 dB per octave (an octave is the same as doubling the frequency). By using 6.6 k as the equivalent resistance and 0.005 µF as the capacitance, we get a 3 dB corner at approximately 4.8 kHz. To get to 455 kHz, you must double 4.8 kHz approximately 6.6 times. This equates to a reduction of the IF voltage at the R8 - VR3 junction of 39.6 dB (6 dB per octave times 6.6 octaves), or 95 times.

Recovering the data – block 2

In working with FM receivers, there are some terms that need to be defined. First, let's determine the term **deviation** as the frequency swing of the incoming FM signal. When no modulation is present, the incoming signal is a fixed frequency **carrier wave (Fc)**. **Positive deviation (Fp)** is the increase in Fc due to modulation, and **negative deviation (Fn)** is the decrease in Fc due to the modulation. The detector must be linear over the maximum total deviation (Fp max. - Fn max.) produced by the transmitter or distortion will occur. Before the second I-F signal reaches the detector, it is applied to a **limiting amplifier** inside the integrated circuit. A limiting amplifier is designed to remove any amplitude variations in the signal by amplifying the signal well beyond the limit of the amplifier. The frequency variations (FM) are not affected by the limiter. The limiter removes the affects of fading (driving through a tunnel) and impulse noise (lightning and ignition). These affects produce significant

unwanted amplitude variations in the received signal, but minor frequency variations. Noise immunity is one of the great advantages of frequency modulation over amplitude modulation. The remaining signal contains only the frequency modulation.

The 6N2 receiver uses a quadrature detector to demodulate the FM signal. After the noise is removed by the limiter, the signal is internally connected to the quadrature detector. A parallel tuned circuit must be connected from pin 12 to the power supply. This circuit produces the 90° phase needed by the quadrature detector. A resistor shunting this coil sets the peak separation of the detector. If the value of the resistor is lowered, it will increase the linearity, but decrease the amplitude of the recovered audio.

The quadrature detector combines two phases of the I-F signal that are 90° apart, or in quadrature, to recover the modulation. The shifted signal is used to gate the non-shifted signal in a synchronous detector. When the frequency increases above the carrier frequency (positive deviation), the phase shift increases causing a decrease in the width of the gated impulse output. In a similar manner, when negative deviation occurs, the gated impulse output will widen. The gated output is then filtered to remove the pulses and recover the modulating signal.

455 kHz conversion – blocks 3 & 4

The first local oscillator should be set at 133.3 MHz to 137.3 MHz for the 2-meter band and 39.3 MHz to 43.3 MHz for the 6-meter band. This oscillator is free-running varactor-tuned. The first mixer converts the RF input to an I-F frequency of 10.7 MHz. This I-F signal is then filtered through a ceramic filter and fed into the second mixer. If the oscillator of the second mixer is precisely set at 10.245 MHz, it will produce an output containing the sum (20.945 MHz) and the difference (0.455 MHz or 455 kHz) frequencies. This 455 kHz signal is then sent to the limiter, detector, and audio circuits. The mixers are doubly balanced to reduce spurious (unwanted) responses. The first and second mixers have conversion gains of 18 to 22 dB respectively.

Conversion gain is the increase in a signal after the signal has been converted to a new frequency and amplified. For both converters, the mixers are designed to allow the use of pre-tuned ceramic filters. After the first mixer, a 10.7 MHz ceramic bypass filter is used. This eliminates the need for special test equipment for aligning I-F circuits. The ceramic filter also has a better aging and temperature characteristic than conventional LC tuned circuits.

Assembling the 6N2 receiver

Before constructing your new 6N2 receiver, you will need to locate a clean well lit, well ventilated work space. Locate a small pencil tipped 27 to 33 watt soldering iron. You should also locate a spool of 22 ga. rosin core solder, a small container of "Tip Tinner" from your local Radio Shack store. "Tip Tinner" helps to clean and dress the soldering iron tip. Place the schematics and parts layout diagram in front of you along with all of the project components. A few small tools would be helpful to constructing your project, so find a small pair of end-cutters, a pair of tweezers, a magnifying glass, a small Phillips and flat-blade screwdrivers, and we can begin. When building a

project with integrated circuits it is wise to consider installing integrated circuit sockets on the PC board, as a form of insurance in the event of a possible circuit failure somewhere down the road. It is much more easy to simply unplug a damaged IC and simply replace it by plugging in a new one.

Before we get some momentum on the project, you may also want to refer to the chart in Table 12-2. This table lists the color codes for the resistors, and will greatly aide you in constructing the receiver. Go ahead and locate resistor R1, a 68 k resistor, its color code is (blue-gray-orange). Now locate resistor R2, a 47 k ohm resistor with (yellow-violet-orange) color code. Place these resistors on the circuit board in their respective locations and solder them in place. Now you can identify the remaining resistors and place them on the circuit board and solder them in their proper locations. Use your end-cutters to trim the excess component leads. Cut the extra component leads flush to the edge of the circuit board.

Now locate and install capacitors; first locate and identify capacitors C1 and C2. Capacitor C1 is a small disk capacitor marked (121) or120 pF. Small capacitors often do not have their actual values printed on them but use a three-digit code to represent their value. Refer to the chart in Table 12-3 to help you identify the small capacitors. Now look for capacitor C2, its value

Table 12-2

Resistor color code chart

Color Band	1st Digit	2nd Digit	Multiplier	Tolerance
Black	0	0	1	
Brown	1	1	10	1%
Red	2	2	100	2%
Orange	3	3	1,000 (K)	3%
Yellow	4	4	10,000	4%
Green	5	5	100,000	
Blue	6	6	1,000,000 (M)	
Violet	7	7	10,000,000	
Gray	8	8	100,000,000	
White	9	9	1,000,000,000	
Gold			0.1	5%
Silver			0.01	10%
No color				20%

Table 12-3
Capacitance code information

This table is designed to provide the value of alphanumeric coded ceramic, mylar and mica capacitors in general. They come in many sizes, shapes, values and ratings; many different manufacturers worldwide produce them and not all play by the same rules. Some capacitors actually have the numeric values stamped on them; however, some are color coded and some have alphanumeric codes. The capacitor's first and second significant number IDs are the first and second values, followed by the multiplier number code, followed by the percentage tolerance letter code. Usually the first two digits of the code represent the significant part of the value, while the third digit, called the multiplier, corresponds to the number of zeros to be added to the first two digits.

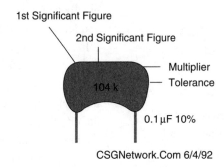

Value	Type	Code	Value	Type	Code
1.5 pF	Ceramic		1,000 pF /.001 µF	Ceramic / Mylar	102
3.3 pF	Ceramic		1,500 pF /.0015 µF	Ceramic / Mylar	152
10 pF	Ceramic		2,000 pF /.002 µF	Ceramic / Mylar	202
15 pF	Ceramic		2,200 pF /.0022 µF	Ceramic / Mylar	222
20 pF	Ceramic		4,700 pF /.0047 µF	Ceramic / Mylar	472
30 pF	Ceramic		5,000 pF /.005 µF	Ceramic / Mylar	502
33 pF	Ceramic		5,600 pF /.0056 µF	Ceramic / Mylar	562
47 pF	Ceramic		6,800 pF /.0068 µF	Ceramic / Mylar	682
56 pF	Ceramic		.01	Ceramic / Mylar	103
68 pF	Ceramic		.015	Mylar	
75 pF	Ceramic		.02	Mylar	203
82 pF	Ceramic		.022	Mylar	223
91 pF	Ceramic		.033	Mylar	333
100 pF	Ceramic	101	.047	Mylar	473
120 pF	Ceramic	121	.05	Mylar	503
130 pF	Ceramic	131	.056	Mylar	563
150 pF	Ceramic	151	.068	Mylar	683
180 pF	Ceramic	181	.1	Mylar	104
220 pF	Ceramic	221	.2	Mylar	204
330 pF	Ceramic	331	.22	Mylar	224
470 pF	Ceramic	471	.33	Mylar	334
560 pF	Ceramic	561	.47	Mylar	474
680 pF	Ceramic	681	.56	Mylar	564
750 pF	Ceramic	751	1	Mylar	105
820 pF	Ceramic	821	2	Mylar	205

is .005 μF, but may be marked (502). Go ahead and install these capacitors in their respective locations on the circuit board. Remember to trim the excess component lead lengths with your end-cutters. Next, locate the remaining small capacitors and install them on the circuit board.

Next, we are going to locate and install the electrolytic capacitors. Electrolytic capacitors are generally larger in size and higher in value than the small value capacitors and usually they are vertical or horizontally mounted. Remember, electrolytic capacitors have polarity and must be installed with respect to this polarity if the circuit is going to work properly. The capacitor's value will be printed on the body of the capacitor along with a white or black polarity marking, either a plus or minus marking. Pay attention to this plus or minus marking and align it with the proper holes on the PC board, you will have to refer to the schematic and or parts layout diagram when placing the electrolytic capacitors. Locate and install the remaining electrolytic capacitors onto the circuit board. Remember to trim the excess electrolytic capacitor leads after they are soldered in place.

In this project there are two silicon diodes. Diodes are generally rectangular in shape like a resistor but they are often clear glass or painted black. You will see a black or white band at one edge of the diode body, this marking denotes the diode's polarity. The colored band is the diode's cathode lead. When installing the diode make sure that you observe the polarity marking by referring to the parts layout and the schematic diagram. Polarity is important and must be observed in order for the circuit to work properly.

The 6N2 receiver has two crystal filters, a 455 kHz one and a 10.7 MHz one. These devices are small three-lead devices and look much like a small capacitor with three leads. Make sure that you can identify them properly before installing them on the circuit board. The 455 kHz unit connects to U1 at pins 5, 6 and 7, while the 10.7 MHz crystal filter is connected to pins 17, 18 and 19. The receiver also employs a single 10.24 MHz crystal which is connected between pins 4 and 6. This is a two-lead device and is usually a small metal can.

Now locate the air-wound coils L1 and L4. Coil L1 is 4½ turns of wire, it will be placed on the right edge of

the PC board, and coil L4, a 1½ turn air-wound coil, will be mounted at the bottom center of the board. Coils L2 and L3 are 9 turn coils wound on a coil form with a ferrite tunable slug in the center of the form. Coil L5 is a 0.64 mH coil in a small metal can with five leads. Coil L5 is a tunable slug tuned coil with a yellow ferrite core, see Table 12-4 for coil winding details.

The 6N2 receiver utilizes three integrated circuits. Before we go ahead and install the integrated circuits, let's take a brief look at the semiconductor pin-out diagram shown in Figure 12-6. The main receiver chip U1 is the Motorola MC3362, a 24 pin dual conversion FM receiver on-a-chip which contains oscillators, mixers, quadrature discriminator, and meter drive/carrier detect circuitry all in one chip. The audio amplifier at U2 is an LM386 chip, while U3 is 5 volt regulator, an LM78L05. When installing the integrated circuits, you will need to identify the pin-outs of each of the integrated circuits. It is advisable to install IC sockets prior to installing the ICs on the circuit board, to avoid the rare event of circuit failure at a late date. It is much easier to simply unplug an IC and insert another without un-soldering a 24 pin chip from the circuit board. Integrated circuits will generally have a small cut-out at one end of the IC package or a small indented circle at one end of the chip. Pin one (1) of the integrated circuit will be to the left of these markings. Be sure to orient the IC before placing into its proper socket. Note that pin 1 of U1 will connect to the junction of capacitors C7 and C16.

Table 12-4
Coil Winding Information

L1	4.5 turns insulated magnet wire - #22 ga. Air-core coil wound on ¼″ form then remove coil.
L2, L3	8½ turns #24 ga. insulated magnet wire coil-wound on 49A Series form - ferrite slug adjustable (49A127MPC) Circuit Specialists. Remove original coil from form and rewind 8½ turns.
L4	1.5 turns insulated magnet wire - #22 ga. Air-core wound on ¼″ form then remove coil.
L5	.64 mH Coil (yellow) 455 kHz I-F Transformer Circuit Specialists - (42IF301).

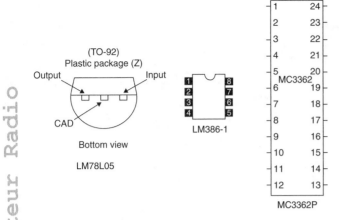

Figure 12-6   *Semiconductor pin-outs*

Finishing the PC board assembly

Now cut three 3″ pieces of wire from the roll of 22 gauge solid wire and strip ¼″ of insulation off each end. Using these wires, solder the 5k potentiometer (VR2) to the PC board. Be sure to mechanically attach each wire. Note that VR3 is the tuning control potentiometer, the center lead is connected to the PC board hole marked (D), while the left-most pot terminal gets connected to the hole marked (N). The final potentiometer wire on the right side of the potentiometer goes to the hole marked (P).

Finally, connect up a short piece of coax to the terminals on the PC board marked ANT and GND. The center wire of the coax goes to the terminal marked antenna (ANT) and shield wire from the coax goes to the terminal marked ground or (GND). The other free end of the coax goes to RCA antenna jack which gets mounted on the front panel of the chassis box.

After completing the circuit board, take a short break and when we return, we will examine the circuit board for possible "cold" solder joints and "short" circuits. Pick up the circuit board with the foil side of the circuit board facing upwards toward you. First, we will look at the circuit board for possible "cold" solder joints. Take a close look at the solder joints, they should all look clean, shiny and bright. If any of the solder joints look

dark, dull or "blobby," then you should un-solder that particular joint and clean it and then re-solder the joint so that it looks clean, shiny and bright. Next we will examine the circuit board for possible "short" circuits.

The front panel assembly

The speaker was mounted on the receiver's front panel. Using the three black 4-40 screws, 4-40 nuts and #4 internal tooth washers, mount the three small right-angle brackets to the speaker mounting holes. The short side of the brackets should be against the panel. Use a blunt tool to bend the brackets over the speaker. Using the hardware shown, mount the phono jack. Using the hardware shown, mount the On-Off switch, 2M/6M switch, squelch pot, volume pot, and tuning pot to the front panel.

Orient the speaker terminals, so they don't interfere with the nearby battery mounting. Solder the loose end of the shielded cable to the phono jack, center conductor to the phono jack terminal and shield to the GND terminal. Peel the backing off one side of the ½″ wide, double-sided tape and stick it to the battery clamp. Peel the backing off the other side of the double-sided tape and stick it to the inside bottom of the plastic case. With the power switch OFF, connect a 9V alkaline battery to the battery snap and put the battery into the battery clamp. When the final test and alignment (next section of manual) is completed, place the front plate on the plastic case and insert the four nylon plugs into the holes in each corner of the plate.

Connecting the front panel parts to the PC board

To wire the speaker, pot and switch, cut the indicated length of wire from the roll of 22 gauge solid wire and strip ¼″ of insulation off each end. Next, go ahead and mount the phono jack at J1. Finally mount the On/Off switch, the 2 meter/6 meter band switch, the squelch potentiometer, the volume potentiometer, and tuning potentiometer to the front panel as shown in Figure 12–7. For the jumper wire, cut 2″ of 22 gauge solid wire, strip ¼″ of insulation off each end and solder to the points indicated. Attach two 4″ long pieces of insulated wire to the speaker terminals and solder them. Now solder two 4″ pieces of insulated wire to the On-Off switch, use the center terminal and one of the outside terminals on the switch. Next attach three 3″ long insulated wires to the volume control on VR3.

Next locate the Squelch control (VR1) and solder two 3″ long insulated wires to the potentiometer. Take the two free wires and connect them to the PC board. The center terminal from the potentiometer goes to the PC board hole marked (Q), while the other potentiometer lead goes to the hole marked (R).

Take the free ends of the two speaker wires and solder them to the circuit board at their respective terminal strip on the bottom left side of the PC board. Now take the two free ends of the On-Off switch leads and solder them to the terminals on the left side of the PC board. Next take the three potentiometer leads and solder them to the PC board. The center terminal of VR3 goes to the terminal marked (D) on the circuit board. The left-most potentiometer terminal goes to the hole marked (C) on the PC board, while the right-most potentiometer terminal

goes to the hole marked (E) on the circuit board. Finally locate the 9 volt snap battery terminal stip and connect it to the battery terminals on the far left marked BATT.

Now that your 6N2 receiver has been completed, it is time to test out your receiver to see if it is going to work. Connect up the 9 volt battery clip to the 9 volt battery. Connect an antenna to the antenna terminal. For testing purposes you can simply connect a 19″ piece of wire to the antenna terminals, or if you have an outdoor antenna you can connect it up to the receiver. Position the band switch to the 2-meter band position. Now adjust the volume control to the mid-way position, and adjust the squelch control counterclockwise so that you just break the squelch and you should hear the receiver noise coming from the speaker. If all is well, you should hear a hissing sound or receiver noise. Adjust the tuning control around the band, back and forth, and you should eventually hear a ham radio operator calling another friend.

In the event that you hear no receiver noise and the receiver appears "dead," you will have to disconnect the antenna and battery clip and examine the receiver's circuit board for any errors which might have occurred during construction. Some of the most common causes for construction errors include installing diodes and electrolytic capacitors backwards in the circuit; remember that these components have polarity and it must be observed in order for the circuit to work properly. Another common problem is having resistors installed in the wrong place on the circuit board. Finally, when installing semiconductors such as transistors and FETs, you must refer to the manufacturer's pin-out diagram and the parts layout diagram along with the schematic when constructing the circuit. Often times transistors are installed incorrectly; also check the orientation of all integrated circuits to make sure that you have oriented them correctly with respect to their marking and pin numbers. Hopefully you have found your error by now and are ready to reassemble the circuit in order to test once again.

Reconnect the antenna and battery leads and turn the On-Off switch to the "On" position. Turn up the volume control to the midway position, and set the BAND switch to the 2-meter position. The 2-meter band will

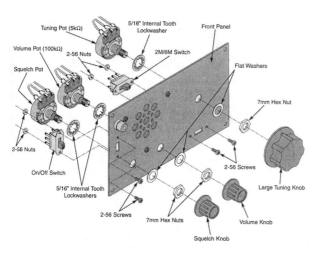

Figure 12-7 *6N2 meters*

generally be much more active than the 6-meter band unless there is a 6-meter band "opening." Now, you will need to adjust the squelch control. The correct adjustment of the squelch control is to advance the control until the receiver noise is silenced. When the receiver is tuned across the band and a signal is found, the squelch will release and the transmission will be heard. As you continue tuning away from this signal, the squelch should again come into effect and "squelch" the receiver noise. Remember that the function of the squelch control is to eliminate static when there is no signal present. This control sets the level of passable signals. The squelch control works by raising the voltage to the (–) input to the audio amplifier and thus disabling the amplifier output. Turn the squelch control fully counterclockwise. The audio should produce a hissing sound. Turning the squelch control clockwise should quiet the audio. Check the leads going to the control.

Now that the receiver seems to be functioning, you will need to calibrate the receiver using an RF frequency generator. You will need to borrow an RF signal generator from a friend or local radio enthusiast in order to calibrate your receiver for best operation.

Frequency measurements

Attach a two foot or longer length of wire to the male RCA plug and connect it to the radio antenna jack. Install a fresh battery and turn the power ON. Make sure that the squelch control is fully counterclockwise and the volume control is set at mid-position. A "rushing" noise should be heard from the speaker when no signal is present, which is normal. Slowly tune the radio on each band and listen for activity. Reception will only be possible if someone is transmitting. If you know someone with a 2M or 6M transmitter, you should test your receiver with known frequencies of transmission and adjust the high frequency oscillator for proper tuning of desired bands.

A calibrated RF generator may also be used to adjust the oscillators. Start with the 2-meter band. Loosely couple the output of the RF generator to the receiver input. Turn the tuning control on the receiver to the full counterclockwise position (lowest frequency). With the squelch control tuned fully counterclockwise (OFF), a "rushing" noise will be heard. Tune the RF generator until the receiver noise disappears and note the frequency on the generator dial. Next, rotate the receiver tuning control to the full clockwise position (higher frequencies). Repeat the above procedure to determine the upper frequency of your tuning range. If the range is too low, decrease L4 by spreading the turns. If the range is too high, substitute the 39 pF capacitor with the 33 pF capacitor in C6. Then, repeat both procedures to determine the upper and lower frequency limits. With a little perseverance, you should arrive at a range covering the 2-meter band (144 to 148 MHz). Set the generator at different points (144, 145, 146, etc.) and tune the radio for quieting at each setting. Mark the dial with the appropriate values. Using the plastic alignment tool to adjust L3, follow the same procedure for the 6-meter band (50 to 54 MHz). **Note:** Before the receiver is properly adjusted, you may receive a broadcast from an FM station or TV station. These signals will be distorted because they are wide band FM transmissions and the 6N2 is a narrow band amateur receiver.

If a frequency counter is available, it can be used to measure the frequency of the local oscillator at pin 20 of U1. On the 2-meter band, the counter should read 133.3 MHz to 137 MHz, and on the 6-meter band, it should read 39.3 MHz to 43.3 MHz.

Antenna considerations

For local VHF reception, the VHF receiver will operate fine just using a whip antenna. If you wish to receive stations which are further away than just the local stations, you may want to install some sort of outdoor antenna such as a ground plane or discone antenna for VHF operation. Since most of the transmissions in this band are of vertical polarization, vertical antennas, ground plane and discone antennas will work very well in this band. Various VHF scanner type antennas from your local Radio Shack store will work fine as well.

Connect up a whip antenna or an outdoor Discone antenna and you can now begin to listen-in to the exciting world of VHF ham radio communication. The charts in Table 12-5 illustrate the band plan for the two popular ham radio VHF bands. Note the SSB satellite and long-range portion of the 2-meter band from 144.1 MHz to 144.3 MHz and the 6-meter DX windows around 51.1 and 52.05 MHz.

Table 12-5

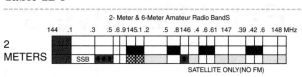

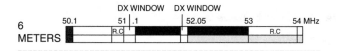

Chapter 13

Active and Passive Aircraft Band Receivers

Parts list

Parts Bin

Active Aircraft Receiver

R1,R2,R3   10k
   potentiometers

R5,7,11,18,25,27   1k ohm
   [brown-black-red]

R6,R28   270 ohm
   [red-violet-brown]

R8,12,17,23   10k ohm
   [brown-black-orange]

R26   22k ohm
   [red-red-orange]

R13,R22   33k ohm
   [orange-orange-orange]

R4,9,15,16,20,21,24
   47k ohm
   [yellow-violet-orange]

R19   100k ohm
   [brown-black-yellow]

R10,R14   1 megohm
   [brown-black-green]

C3,C5   3.9 pF ceramic
   capacitor

C11   10 pF ceramic
   capacitor

C12,C14   27 pF ceramic
   capacitor

C2,C4,C6   82 pF ceramic
   capacitor

C1,7,8,13,16   .001 µF
   disk capacitor
   (marked.001 or 102)

C9,17,19,20,28,30   .01 µF
   disk capacitor
   (marked .01 or 103 or
   10 nf)

C23,C24   .1 µF disk
   capacitor (marked .1
   or 104)

C22   .47 µF electrolytic
   capacitor

C10,15,21,25,26,31   4.7
   to 10 µF electrolytic
   capacitor

C18,27,29   100 to 220 µF
   electrolytic capacitor

Q1   2SC2498 or 2SC2570A
   NPN UHF transistor

Q2   2N3904 NPN
   transistor

U1   SA602 8 pin IC

U2   MC1350 8 pin IC

U3   LM324 14 pin IC

U4   LM386 8 pin IC

D1   BB505 varactor diode
   (marked BB505)

D2   1N270 germanium diode

D3   1N4148 silicon diode

L1,L3,L5   See Table 13-3

L2,L4   See Table 13-3

L6   See Table 13-3

L7   See Table 13-3

FL1   10.7 MHz ceramic
   filter - (Murata/
   Digi-Key)

S1 PC mount DPDT switch

J1 PC mount RCA jack

J2 subminiature phone jack

B1 9 volt battery

Misc PC board, battery clip, battery holder

*** Ramsey Electronics Inc.

Passive Aircraft Monitor

R1,R2 10 megohm, $\frac{1}{4}$ watt, 5% resistor

R3 100 k ohm potentiometer - volume (chassis mount)

R4,R5 3.3k ohm, $\frac{1}{4}$ watt, 5% resistor

R6 100 k ohm, $\frac{1}{4}$ watt, 5% resistor

R7 100 ohm $\frac{1}{4}$ watt, 5% resistor

C1 1.5 pF, 35 volt mica/poly capacitor

C2 5-18 pF trimmer capacitor (tuning)

C3 .01 μF, 35 volt ceramic disk capacitor

C4 330 pF, 35 volt polyester capacitor

C5,C8 100 μF, 35 volt electrolytic capacitor

C6,C7 1 μF, 35 volt electrolytic capacitor

D1 1N34 germanium diode

L1 .15 μH choke coil - (Digi-Key)

U1 LM358 dual op-amp

S1 SPST toggle switch

B1 9 volt transistor radio battery

PH high-impedance (2 k ohm) headphone

Misc battery clip, battery holder, wire, enclosure, etc.

Eavesdroping in on the pilots as they trace their way across the sky can be very fascinating and exciting listening for aircraft enthusiasts of all ages. An aircraft receiver will allow you to listen-in to high flying commercial aircraft from jumbos to pipers, as well as control towers passing instructions to the airplanes overhead. It will provide many hours of interesting listening.

In this chapter, we will present two different types of aircraft receivers: a passive aircraft receiver and an active aircraft receiver. A passive aircraft receiver is a more simple receiver with no local oscillator to interfere with the on-board aircraft electronics. A passive aircraft receiver is defined as a tuned detector with an amplifier which can be carried aboard an airplane in order to listen-in to the pilot while you are on a flight. You will be able to listen-in to the pilot in flight as well as the control tower when in fairly close range. This type of receiver does not interfere with the airplane's electronics and in most instances you will be permitted to use this type of receiver on-board your aircraft. The active receiver is a more sensitive active receiver which can be used at home to listen-in to high-flying aircraft as well as control towers over longer distances. Listening to aircraft communications can be very interesting but very cryptic, with lots of bursts of information passing in a short period of time.

Active aircraft band receiver

Ideal for arm-chair pilots, student pilots and the seasoned pilots, this sensitive receiver will pick up planes up to 100 miles away. The active aircraft band receiver tunes the aircraft band between 118 and 136 MHz and will allow you to listen to control towers, centers, and planes en-route to their destination. The active aircraft band receiver, shown in Figure 13-1 has good sensitivity, image rejection, signal-to-noise ratio and stability. It is designed for casual "listening in" both

Figure 13-1 *Active aircraft band receiver*

ground and air communication, for both commercial airlines and general aviation, and it will provide you with many years of easy sky-monitoring enjoyment.

This project is especially good for people with an interest in learning more about aviation and electronics and radio. The active aircraft receiver can be built by folks of all ages and skill levels.

What you can hear

A basic fact about the VHF Aviation Band which even licensed pilots can overlook or forget is that communications are in the AM mode, not FM, as in the case of the FM broadcast band immediately below it, and the VHF public service and ham bands immediately above it. No matter where you live, you will be able to receive at least the airborne side of many air traffic communications. If you know where to tune, you'll hear any aircraft you can see, PLUS planes a hundred miles away and more, since VHF signals travel "line of sight". An airliner at 35,000 feet altitude in the next state is still line of sight to your antenna. Similarly, whatever ground stations you may hear are also determined by this "line of sight" character of VHF communication. If there are no major obstacles between your antenna and an airport (tall buildings, hills, etc.), you'll be able to hear both sides of many kinds of aviation communication. Be prepared for them to be fast and to the point, and for the same airplane to move to several different frequencies in the span of a few minutes! The most common types of ground services with which

pilots communicate is the control tower, ground control, clearance delivery and ATIS or "Automatic Terminal Information System."

At most metropolitan airports, a pilot communicates with the FAA on a frequency called "Clearance Delivery" to obtain approval or clearance of the intended flight plan. This communication is done before contacting ground control for taxi instructions. From the control tower, ground movements on ramps and taxiways are handled on the "Ground Control" frequency, while runway and in-flight maneuvers near the airport (takeoffs, local traffic patterns, final approaches and landings) are on the "Control Tower" frequency. ATIS, or "Automatic Terminal Information System" is a repeated broadcast about basic weather information, runways in use, and any special information such as closed taxiways or runways. Such a broadcast offers an excellent steady signal source for initial adjustment of your receiver. If you are close enough to the airport to receive ATIS, you will hear the Approach Control and the Departure Control transmissions. These air traffic radar controllers coordinate all flight operations in the vicinity of busy metropolitan airport areas, and are called the ATC Center.

When you hear a pilot talking with "Jacksonville Center" or "Indianapolis Center," you know the aircraft is really enroute on a flight rather than just leaving or just approaching a destination. A pilot will be in touch with several different "Regional Centers" during a cross-country flight. These smaller centers are called "Unicom" centers.

Airports without control towers rely on the local "Unicom" frequency dedicated only to advisory communications between pilots and ground personnel such as fuel service operators. The people on the ground can advise the pilot on the status of incoming or outgoing aircraft, but the pilot remains responsible for landing and takeoff decisions. Typical Unicom frequencies are 122.8 and 123.0 MHz.

The FAA's network of Flight Service Stations keeps track of flight plans, provides weather briefings and other services to pilots. Some advisory radio communication takes place between pilots and a regional "FSS." If there is an FSS in your local area, but no airport control towers, the FSS radio frequency will stay interesting.

Fast talking pilot and controllers

Aviation communication is brief, but it is clear and full of meaning. Usually, pilots repeat back exactly what they hear from a controller so that both know that the message or instructions were correctly interpreted. If you are listening in, it is hard to track everything said from a cockpit, particularly in big city areas. Just to taxi, take off and fly a few miles, a pilot may talk with six or eight different air traffic control operations, all on different frequencies, all within a few minutes! Here are the meanings of a few typical communications:

"Miami center, delta 545 heavy out of three-zero for two-five."

Delta Flight 545 acknowledges Miami Center's clearance to descend from 30,000 feet to 25,000 feet altitude. The word "heavy" means that the plane is a jumbo jet such as 747, DC-10, etc.

"Seneca 432 lima cleared to outer marker. contact Tower 118.7."

The local Approach Control is saying that the Piper Seneca with the N-number (tail number) ending in "432L" is cleared to continue flying an instrument approach to the outer marker (a precision radio beacon located near the airport) and should immediately call the airport radio control tower at 118.7 MHz. This message also implies that the controller does not expect to talk again with that aircraft.

"Cessna 723, squawk 6750, climb and maintain five thousand."

A controller is telling the Cessna pilot to set the airplane's radar transponder to code 6750, climb to and fly level at an altitude of 5000 feet.

"United 330, traffic at 9 o'clock, 4 miles, altitude unknown."

The controller alerts United Airlines flight #330 of radar contact with some other aircraft off to the pilot's left at a 9 o'clock position. Since the unknown plane's altitude is also unknown, both controller and pilot realize that it is a smaller private plane not equipped with altitude-reporting equipment.

Active receiver circuit description

Now, take a look at the schematic diagram shown in Figure 13-2. The aircraft band antenna is first coupled through C1 to a three-section tuned LC filter input network. The aircraft band signals ranging from 118 to 135 MHz signals are amplified by VHF transistor Q1 and fed to the input of U1, the SA602 mixer-oscillator. Inductor L6 and its associated capacitor network establish the LO (local oscillator) frequency at 10.7 MHz higher than the incoming 118–135 MHz signals. The local oscillator frequency may be tuned across about 15 MHz by the varactor tuning network formed by diode D1 and resistor R1. The 10.7 MHz difference between the local oscillator and the received signal is fed through the 10.7 MHz ceramic filter FL1, and then amplified by Q2 and the signal is then applied to U2, the MC1350 IF amplifier IC with AGC input. The 10.7 MHz IF is peaked by inductor L7, and the AM audio signal is then demodulated by diode D2 and fed through the four op-amps of U3, the LM324, where volume control, AGC output, audio filtering and squelch functions are managed. The LM386 (U4) is the audio amplifier, and is capable of driving simple communications speakers to excellent volume levels.

Circuit construction

The aircraft band receiver is best constructed on a printed circuit board for best results. Place your schematic diagram in front of you, heat up your 27 to 30 watt pencil tipped soldering iron and prepare to insert

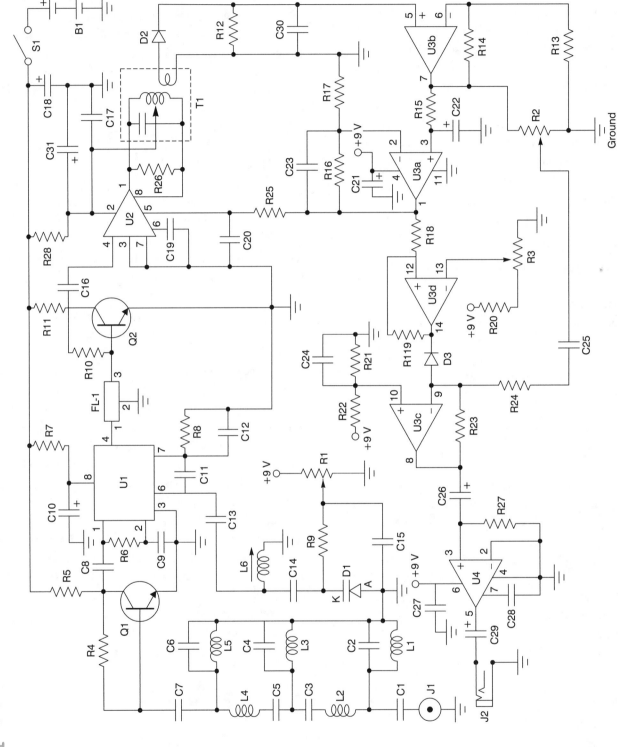

Figure 13-2 *Active aircraft receiver schematic. Courtesy of Ramsey Electronics*

Table 13-1

Resistor color code chart

Color Band	1st Digit	2nd Digit	Multiplier	Tolerance
Black	0	0	1	
Brown	1	1	10	1%
Red	2	2	100	2%
Orange	3	3	1,000 (K)	3%
Yellow	4	4	10,000	4%
Green	5	5	100,000	
Blue	6	6	1,000,000 (M)	
Violet	7	7	10,000,000	
Gray	8	8	100,000,000	
White	9	9	1,000,000,000	
Gold			0.1	5%
Silver			0.01	10%
No color				20%

the components into the circuit board. You can fabricate your own circuit board from scratch if you are an experienced builder or you could elect to build the receiver from the kit supplier, Ramsey Electronics, listed in the Appendix. We will build the receiver in sections and will begin first with the antenna and input pre-amplifier section.

Before we go ahead and build the aircraft receiver, let's take a few moments to look at Table 13-1, which illustrates the resistor color chart. Resistors have three or four color bands, which begin at one edge of the resistor body. The first color band represents the resistor's first digit value, while the second color band denotes the second digit of the resistor value. The third color band represents the multiplier value of the resistor. The fourth color band denotes the tolerance value of the resistor. A silver band represents a 10% tolerance, while a gold band notes a 5% tolerance value. The absence of a fourth color band states that the resistor has a 20% tolerance value. Therefore, a 10,000 ohm or 10 k ohm resistor would have a brown band (1) a black band (0) and a multiplier of (000), and an orange band.

Go ahead and install resistors R4, R5 and R6, then solder them in place on the PC board. Then cut the excess leads from the PC board with a pair of

end-cutters. Cut the excess leads flush to the PC board.

Next we are going to install the capacitors for the project. Capacitors are described as electrolytic and non-electrolytic types. Electrolytic types are large sized and larger in value than non-electrolytic types. Electrolytic type capacitors have polarity, i.e. both a positive and negative terminal, often with color band and/or (+) or minus (−) marking on them. These electrolytic capacitors must be installed with respect to these polarity markings if you expect the circuit to work correctly. Take your time installing them and refer regularly to both the schematic and parts layout diagrams. Non-electrolytic capacitors can sometimes be very small in size, and often their actual value will NOT be printed on them. A three-digit code is used to help identify these capacitors; refer to Table 13-2. For example, a .001 µF capacitor would be marked with (102), while a .01 µF capacitor would be marked with (103).

Go ahead and install capacitors C1 through C8 into their respective PC holes on the PC board, being sure to observe polarity on the capacitors before installing them on the PC board. Most electrolytics will have a black band or a plus or minus marking next to a pin. Solder the capacitors to the PC board, then remove the excess leads. Next install inductors L1, L2, L3 and L4 into

Table 13-2
Capacitance code information

This table provides the value of alphanumeric coded ceramic, mylar and mica capacitors in general. They come in many sizes, shapes, values and ratings; many different manufacturers worldwide produce them and not all play by the same rules. Some capacitors actually have the numeric values stamped on them; however, many are color coded and some have alphanumeric codes. The capacitor's first and second significant number IDs are the first and second values, followed by the multiplier number code, followed by the percentage tolerance letter code. Usually the first two digits of the code represent the significant part of the value, while the third digit, called the multiplier, corresponds to the number of zeros to be added to the first two digits.

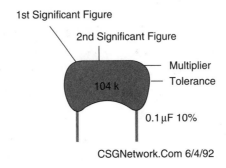

1st Significant Figure
2nd Significant Figure
Multiplier
Tolerance
104 k
0.1 µF 10%
CSGNetwork.Com 6/4/92

Value	Type	Code	Value	Type	Code
1.5 pF	Ceramic		1,000 pF /.001 µF	Ceramic / Mylar	102
3.3 pF	Ceramic		1,500 pF /.0015 µF	Ceramic / Mylar	152
10 pF	Ceramic		2,000 pF /.002 µF	Ceramic / Mylar	202
15 pF	Ceramic		2,200 pF /.0022 µF	Ceramic / Mylar	222
20 pF	Ceramic		4,700 pF /.0047 µF	Ceramic / Mylar	472
30 pF	Ceramic		5,000 pF /.005 µF	Ceramic / Mylar	502
33 pF	Ceramic		5,600 pF /.0056 µF	Ceramic / Mylar	562
47 pF	Ceramic		6,800 pF /.0068 µF	Ceramic / Mylar	682
56 pF	Ceramic		.01	Ceramic / Mylar	103
68 pF	Ceramic		.015	Mylar	
75 pF	Ceramic		.02	Mylar	203
82 pF	Ceramic		.022	Mylar	223
91 pF	Ceramic		.033	Mylar	333
100 pF	Ceramic	101	.047	Mylar	473
120 pF	Ceramic	121	.05	Mylar	503
130 pF	Ceramic	131	.056	Mylar	563
150 pF	Ceramic	151	.068	Mylar	683
180 pF	Ceramic	181	.1	Mylar	104
220 pF	Ceramic	221	.2	Mylar	204
330 pF	Ceramic	331	.22	Mylar	224
470 pF	Ceramic	471	.33	Mylar	334
560 pF	Ceramic	561	.47	Mylar	474
680 pF	Ceramic	681	.56	Mylar	564
750 pF	Ceramic	751	1	Mylar	105
820 pF	Ceramic	821	2	Mylar	205

Table 13-3

Coil winding information

L1,L3,L5	1.5 turns #28 ga. magnet wire, 5 mm inner diameter air core - about 1 mm spacing (2-turns works fine)
L2,L4	.33 μH, molded inductor (Digi-Key M9R33-ND)
L6	.1 μH, 3.5 turns, slug tuned coil (Digi-Key TK2816-ND)
L7	10.7 MHz shielded I-F transformer (Mouser 42FIF122) or Circuit Specialists

their respective holes, followed by L5; see Table 13-3 for winding instructions. When you have inserted the inductors, you can solder them in place and then cut the excess component leads flush to the edge of the circuit board.

Before you go ahead and install the transistors and integrated circuits, refer to the semiconductor pin-out diagram shown in Figure 13-3, which will help you orient the components. Now locate and install transistor Q1, being sure you have the correct component before installing it on the PC board. Be sure you can identify all three leads as to their correct pin-outs before inserting it into the circuit board to avoid damage to the circuit once power is later applied. Solder Q1 to the board and then remove the excess leads.

Next we are going to move on to installing the Local Oscillator components. First install resistors R1, R7, R8, and R9 followed by resistors R10, R11, R12 and then resistors R25, R26 and R28. Solder the resistors into their respective locations and be sure to remove the excess lead lengths. Next we will install capacitors C9, C10, C11 and C12 followed by capacitors C13, C16, C17, C19 and C20 followed by C30 and C31. Be sure to observe the polarity of the capacitors when installing them on the board to prevent circuit damage when power is first applied. Install the capacitors into their respective location on the PC board. Once inserted, solder the capacitors to the circuit board and then remove the excess leads.

Now locate diode D1 and D2, and note that the cathode or black band side of the diode points to the battery. Diode D1 is the varactor "tuning" diode. Go ahead and install these diodes, then solder them to the board. Remove the excess leads. Locate transistor Q2 and install it on the PC, once you have identified all the leads. Solder Q2 to the board, and then remove the excess leads with your end-cutter. Now locate and install inductor L6, the 3.5-turn slug tuned coil and L7, the 10.7 I-F transformer. Remember to trim the extra lead lengths as needed. Locate the FL1, the 10.7 MHz ceramic filter, it will have three leads and is usually a small square component. Solder it in place and remove the excess leads. Finally install the 8-pin SA602 local oscillator-mixer at U1. It is advisable to install an IC socket into place in the proper location and then insert the IC into the socket. This is good insurance in the event of a possible component failure at some later point. When inserting the IC into the socket be sure to first identify which pin is #1. Usually each IC will have a small circle or cut-out on the top or top left side of the IC. Usually the pin is just to the right of this indented circle.

Next we will install the LM324 IC circuit components. First locate resistors R13, R14, R15, R16, R17, R18 and R19 followed by resistors R20, R21, R22, R23 and R24. Install these resistors in their respective locations and then solder them in place on the

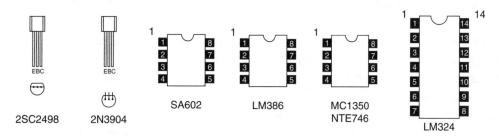

Figure 13-3 *Semiconductor pin-outs*

PC board. Remember to cut the extra leads. Now locate diode D3, remember to observe the polarity when installing D3. Solder the diode in place, then remove the excess lead lengths. Now locate and install capacitors C18, C21, C22 followed by capacitors C23, C24 and C25. Remember electrolytic capacitors have polarity markings and they must be observed for the circuit to operate correctly. You will see either a plus or minus marking at one end of the capacitor. Once installed, solder the capacitors in place and then remove the extra leads by cutting them flush to the circuit board. Finally install an integrated circuit socket for U3 and then place U3 into the socket, but be sure to install it correctly, observing the indented circle as near pin 1 on the IC.

Finally we are going to install the audio amplifier components. First locate resistor R27 and install it on the PC board. Next solder it in place and cut the extra leads. Now, locate and install capacitors C26, C27, C28, and C29. Remember C26 and C27 are electrolytic types, so polarity must be observed before installing them. Finally install an IC socket for U4 the LM386 amplifier IC. When installing U4 be sure to observe the correct orientation before inserting it into the socket to prevent damage when power is applied.

Now you finish up the circuit by installing the 9 volt battery holder, the 9 volt battery clip, followed by switch S1. Once these components have been installed you can locate and install jacks J1 and J2 onto the circuit board.

If you haven't already installed the potentiometers, R1, R2 and R3, go ahead and install them and solder the controls firmly in place. Be sure the mounting tabs are completely soldered for a good mechanical connection.

Let's take a short break and when we return we will inspect the PC board for poor solder joints and "short" circuits before applying power to the circuit board. Pick up the circuit board and bring the foil side of the board up towards you to inspect the board. First, we will look for "cold" or poor solder joints on the PC board. Take a look at each of the solder joints. You want to make sure each of the solder joints appears to look clean, smooth and shiny. If you see any solder joints that look dull or "blobby" then you will need to remove the solder from the joint with a solder-sucker or wick and then re-solder the joint, so that it appears to be clean, shiny and smooth.

Next we will inspect the PC board for "short" circuits. Often "stray" or "cut" component leads will "bridge" across or between circuit traces on the PC board. These possible "shorts" can cause damage to the circuit when power is first applied, so it is best to remove any "bridging" leads or solder blobs or "shorts" before power is sent to the circuit board. Look over the board carefully for any signs of "bridging" wires or blobs of solder. Once you are comfortable with your inspection, you can lay the circuit board back down on the table.

Enclosure recommendations

If your first goal is economy and rugged portability, you will find that the circuit board can be nicely mounted in a standard VHS videotape storage box, which also gives room for a speaker or earphone storage, and even a roll of antenna wire. The controls are easily mounted at one end of such a box. It may be necessary to cut away the molded posts which secure the tape cassette itself. These storage boxes come in several styles, so pick one which truly looks practical as a project enclosure.

The most economical metal enclosure nicely suited for Ramsey PC board kits is the Radio Shack 270-253A. This metal utility cabinet can accommodate one or two different receiver boards plus a speaker, with room for various refinements you might want to add, such as a bigger battery pack, fine tuning control, and so forth. When you have selected your enclosure, you can place the circuit board into the case and mount it. Be sure to allow the extension of the squelch control, the volume control and the tuning control on the front panel of the chassis. The speaker can be mounted on the top of the case, and the earphone and antenna jack can be mounted on the rear of the receiver.

Congratulations, you have successfully completed your active aircraft band receiver. Now you are ready to test and adjust the receiver before setting down and enjoying it. Now, you can attach a speaker or pair of headphones to jack J2. Next, attach a whip antenna to jack J1, and then connect a 9 volt battery to the battery clip leads and we can go ahead and test the active aircraft receiver. Now, turn the On-Off switch to the "ON" position. Be sure to turn down the squelch control to below the threshold point, so you can hear a hissing sound coming from the speaker or headphone. Now, turn up the volume control to about midway. Take a few minutes to tune up and down the band with the main tuning control. If after about 15 minutes, you do not hear any transmissions on your radio or if you do not hear any hissing sounds from the speaker, then you know there could be a problem with your receiver.

In the event of a problem with your new aircraft band receiver, you will have to disconnect the speaker, antenna and battery and visually inspect the circuit board for any errors. Often a second or new pair of eyes is very helpful in locating problems on circuit boards. The most likely causes of circuit failure are misplaced components or components such as diodes and electrolytic capacitors installed backwards. The second most likely cause for problems are semiconductors, such as transistors installed backwards. Be sure to re-check the manufacturer data sheets for proper pin-out information when installing the transistors. Finally, check the orientation of the integrated circuits in their respective sockets; sometimes the ICs can be put in backwards in their sockets causing the circuit to malfunction. Once you have fully inspected and hopefully located your problem, you can reconnect the battery, headphones and antenna and then retest the receiver once again.

Antenna considerations

An antenna for your new aircraft receiver can be as simple as a 21″ piece of wire, an extendable whip antenna or a roof-mounted ground-plane aviation antenna. Most folks near an airport will get plenty of in-the-air action from a wire or whip antenna, but if you're more than a few miles away, a decent roof-mount antenna is the way to go. Radio Shack sells an ideal antenna designed for scanners which covers the aircraft band nicely and it costs around $30. A low cost TV antenna works well, even better if rotated 90 degrees (remember aircraft antennas are vertically polarized).

Adjustment and alignment

Alignment of the air-band receiver consists of simply adjusting L6 for the desired tuning range and peaking the IF transformer (L7). If you are using a signal generator, frequency counter or other VHF receiver for calibration, remember that you want to set the local oscillator frequency 10.7 MHz higher than the desired signal or range to be received. Adjustment of the L6 oscillator coil MUST be made with a non-metallic alignment tool. The use of a metal tool of any kind will detune the coil drastically, making alignment almost impossible. Also, if you're receiving FM broadcast stations, you have the slug tuned too far down in the coil form. Turn it until it is higher in the form and try again. One other thing that you can do to improve the operation of your kit is to spread out the three 2-turn coils at the antenna input, L1, L3, and L5. Stretching these out will give you greater sensitivity.

Once you know you are receiving aircraft or airport transmissions, adjust the IF transformer (L7) for best reception. Typically, L7 is adjusted 2–3 turns from the top of the shield can. If you don't have any signal reference equipment at home, and are not yet hearing airplanes, your best bet is to pack up your AR1C and needed tools and head for the nearest airport! If there is no control tower, don't hesitate to visit a general aviation service center on the airport grounds. If you've never done this before, you will probably find it to be a fun and interesting experience. Ask which are the most active frequencies and adjust L6 and your front panel tuning control until you hear the action. A ground service operator or private pilot may be willing to give you a brief test transmission on the 122.8 Unicom frequency. Remember, also, that if your airport has ATIS transmissions, you can get a steady test signal as soon as you are line-of-sight with its antenna.

The aircraft receiver does not produce a loud hiss when no signal is being received (unlike an FM receiver or expensive AM receiver); this is due to the somewhat limited amount of IF or Intermediate Frequency gain. Increasing the IF gain would produce a hiss and marginally better sensitivity (about a microvolt) but also require much more alignment, AGC circuitry, and builder ability—far beyond the intention of this kit.

Tuning the aircraft receiver

With the varactor tuning control capable of going across 10–15 MHz, and with pilots and controllers talking so briefly, you will need to get used to tuning your receiver! You'll find that ANY knob gives smoother tuning than the bare control shaft, and that a "vernier" dial will make the procedure even easier—but at the expense of being able to check up and down the band quickly, which you might like to do if you're tracking the same airplane. The air-band receiver is designed to let you explore the entire communications section of the aviation band. If you become really interested only in being able to check a certain frequency such as a nearby FAA control tower, or Unicom, it will be helpful to mark that spot on a dial template such as is reproduced for your convenience below.

"VOR" or "OMNI" transmissions

While driving around the countryside you may have come across a tall white "cone" structure near your airport or in the middle of a big farming field; it is useful for you to know that these are VHF navigational aids operating in the 118–135 MHz frequency range, just below the air-ground communications range, sending a steady signal which may be helpful in initial alignment. If you tune in such a signal on your receiver, remember that you will have to increase the local oscillator frequency later in order to listen to air traffic communications.

Passive aircraft receiver

The passive air-band receiver is a type of "crystal" radio which contains no local oscillator which might interferes with the on-board aircraft's sensitive electronics. The passive aircraft receiver broadly tunes from 118 MHz to 136 MHz, and was designed to listen-in to in-flight communications between your pilot and the control tower. The passive aircraft receiver is shown in Figure 13-4. The passive aircraft receiver can be built small enough to place inside your vest pocket and it operates from an ordinary 9 volt battery.

The passive aircraft receiver is basically an amplified type of "crystal radio" designed to receive AM aircraft transmissions. The "passive" design uses no oscillators or other RF circuitry capable of interfering with aircraft communications. This receiver utilizes a coil/capacitor tuned "front-end," which feeds an RF signal to a detector diode. The tuning capacitor may be any small variable with a range from about 5 pF to about 15 or 20 pF. The 0.15 μH inductor may be a molded choke or a few turns wound with a small diameter. Experiment with the coil to get the desired tuning range. The aircraft frequencies are directly above the FM band so a proper inductor will tune FM stations with the capacitor set near maximum capacity. (The FM stations will sound distorted since they are being slope detected.) A 1N34 germanium detector diode or a Schottky diode like the 1N5711 or HP2835 should be used as the detector diode in this receiver circuit. The 10 megohm resistors provide a small diode bias current for better detector efficiency.

The LM358 dual op-amp amplifier draws under 1 ma so the battery life should be quite long. Potentiometer R3 is used to adjust the gain to the second stage of the dual op-amp. The second op-amp stage drives a 100 ohm resistor in series with a 100 μF capacitor. You must use a high impedance "crystal" headphone, since the op-amp output will not drive a speaker directly. A speaker amplifier may be added to drive a speaker or low-z earphone, if desired, but the power consumption will increase sharply, as will the size of the receiver.

The passive aircraft receiver is powered from an ordinary 9 volt transistor radio battery. The entire aircraft band receiver can be built inside a small plastic box. Simply mount an SPST slide switch on the side of the case to apply power to the circuit. If you elect, you can install a ⅛″ phone jack for the "crystal" headphone.

Construction

The passive air-band receiver can be constructed on a regular circuit board or it could be built on a

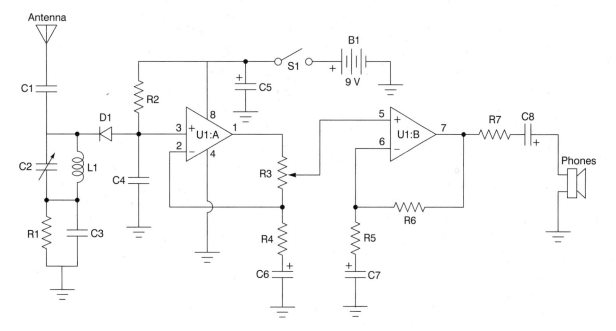

Figure 13-4 *Passive aircraft receiver schematic*

"quick-board" which contains ready-made circuit traces, these are available at Radio Shack stores.

You may want to lay out the circuit so that the tuning control is at one edge of the board, which will make things a little easier when installing the circuit in an enclosure. Building the circuit is pretty straightforward and should only take a few hours to build. Refer to Table 13-1 for identifying the resistors and Table 13-2 for identifying the small capacitors in the circuit. Remember it is very important to install the diode correctly with respect to its polarity in order for the circuit to work properly. The passive air-band receiver also contains a few electrolytic capacitors, and you will remember that these devices have polarity and it must be observed for the circuit to work. The aircraft receiver uses a dual op-amp to amplify the signals and you will have to ensure that it is installed correctly. An integrated circuit is highly recommended for the op-amp. In the event of a circuit failure at some later date, it is much easier to simply unplug an IC rather than trying to un-solder one from a PC board to replace it with a new one. When installing the IC into its socket, you will have to make sure that you orient it correctly. An IC will typically have a notch or small cut-out at one end of the IC case. Pin one (1) will be to the left of the cut-out of notch when viewed from the top of the IC. In this project pin 1 of the IC connects to potentiometer R3, while pin 8 of the IC is connected to the

power source. Note that pin 4 of the IC is connected to the circuit ground.

You can elect to use a small speaker in this circuit, but in order to make a miniature receiver you may want to simply use a small ⅛″ mini jack and just plug into a pair of small earphones. The circuit is powered from a standard 9 volt battery; remember the black or minus (−) battery clip is connected to ground, while the plus (+) or red battery clip lead is connected to capacitor C5 at the plus (+) end.

You will want to have the tuning capacitor C2 near the edge of the circuit board, so that you can allow the tuning knob to protrude through the case so you can easily tune the receiver. Once the receiver has been completed, you should take a few minutes rest.

After the break, you should inspect the foil side of the circuit board for any "cold" solder joints. Cold solder joints often look dark, dull or "blobby." If you see any solder joints that look this way, you should re-solder the connections. You should also take a moment to examine the circuit board for any "short-circuits" which may be the result of a "stray" component lead shorting between circuit traces.

You can make yourself an antenna from a piece of stiff 20 ga. wire as a whip antenna to capacitor C1, or install a miniature jack in order to plug in a whip antenna. The antenna can be a 6″ piece of #20 ga. stiff solid copper wire or a small telescoping aluminum antenna connected to capacitor C1. Radio Shack has a good selection of whip antennas. You could elect to solder the wire antenna directly to C1 and just bring the antenna out the top of the enclosure, or you could install a small ½″ mini jack for the antenna on the rear of the enclosure. The selectivity is reduce as the antenna length is increased so best performance is achieved with the shortest acceptable antenna. Try increasing the 1.8 pF capacitor value when using very short antennas and decreasing it for long antennas. In order to "tune" the passive aircraft receiver you can either leave a small hole on the side of the plastic enclosure to adjust the frequency using a plastic tuning tool, or you need to find a way to attach a plastic knob onto the tuning screw; you will have to drill a larger hole in the case to bring the knob outside the case for easy tuning.

To test the passive aircraft receiver, park near your local airport. Turn on the passive air-band receiver, attach your wire antenna or pull up your telescoping antenna and tune the receiver. You should try to listen when you see an aircraft take-off or land or when you see an aircraft taxiing along the runway. If you can get close to the tower in a small airport, that should ensure that you will be able to test your receiver adequately. As you see a plane arrive or leave, "tune" around the band and you should hear either the pilot or the control tower. If you plan on taking the passive aircraft receiver on-board an aircraft to listen-in to the pilot during your flight, you should ask the pilot or stewardess if it is permitted aboard your aircraft. Generally you are given permission without a problem. Enjoy building and using your new aircraft receivers!

Chapter 14

VLF or Very Low Frequency Radio Receiver

Parts list

Parts Bin

VLF Whistler Receiver

R1,R4 10k ohm ¼w, 5% resistor

R2,R3 22k ohm ¼w, 5% resistor

R5 6.2 megohm, ¼w 5% resistor

R6 1 meg ohm potentiometer (trimpot) - single turn

R7 10 megohm, ¼w 5% resistor

R8 1k ohm, ¼w 5% resistor

R9 1k ohm, ¼w 5% resistor

R10 3.3k ohm, ¼w 5% resistor

R11,R12 820 ohm, ¼w 5% resistor

R13 10k ohm potentiometer (panel)

R14,R17 10 ohm, ¼w, 5% resistor

R15,R16 2.2k ohm, ¼w, 5% resistor

C1,C3 47 pF, 35 volt mica capacitor

C2 100 pF, 35 volt mica capacitor

C4,C5,C6 3.3 nF, 35 volt mylar capacitor

C7,C15 .01 µF, 35 volt ceramic disk capacitor

C8 27 pF, 35 volt mica capacitor

C9 1 µF, 35 volt electrolytic capacitor

C10 .18 µF, 35 volt tantalum capacitor

C11 .12 µF, 35 volt tantalum capacitor

C12 1.8 µF, 35 volt tantalum capacitor

C13 .68 µF, 35 volt tantalum capacitor

C14 .22 µF, 35 volt tantalum capacitor

C16 .068 µF, 35 volt tantalum capacitor

C17,C18 .22 µF, 35 volt tantalum capacitor

C19 .05 µF, 35 volt ceramic disk capacitor

C20,C22,C24 100 µF, 35 volt electrolytic capacitor

C21,C23 .1 µF, 35 volt
   ceramic disk capacitor

L1 120 mH coil (Mouser
   electronics)

L2 150 mH coil (Mouser
   electronics)

L3 18 mH coil (Mouser
   electronics)

L4 56 mH coil (Mouser
   electronics)

Q1a,Q1b U401 Siliconix
   Dual matched N-Channel
   FETs

Q2 MPSA56 transistor

U1 LM386 op-amp

S1 DPDT toggle switch
   (trap)

S2 DPDT toggle switch
   (high-pass)

S3 DPDT toggle switch
   (low-pass)

B1 9 volt transistor
   radio battery

J1,J2 ⅛ mini phone
   jack w/switch

J3 SO-238 UHF chassis
   jack

P1 PL259 UHF plug

P2 binding post
   (ground)

Misc PC board, IC
   socket, solder, wire,
   antenna, battery
   holder

Battery clip, hardware,
   chassis

"Natural Radio" is a name sometimes used for describing radio noise with natural origins, mainly due to but not confined to lightning. At first suggestion, you may think that listening to lightning crackles would be dull or uninteresting, but electromagnetic radiation from lightning can travel great distances and undergo strange modifications along the way. The frequencies of the original pulse can be spread out in time (a process called "dispersion") since the higher frequencies travel a little faster than the lower. The result is that the short impulse from a lightning strike in South America can sound like a chirp in Texas. Slower sweeping tones are called "whistlers" and they are a bit of a mystery. The energy from a lightning bolt streams out into space into a region called the "magnetosphere," magnetized plasma created by the interaction of solar wind with the Earth's magnetic field. The lightning pulse is reflected or "ducted" back down to Earth after a very long trip during which time the frequencies are spread out by a dispersion-like process. Short whistlers might be due to dispersion, but some whistlers last five seconds so ordinary dispersion is probably inadequate an explanation. A radio wave can travel a million miles in five seconds, so to accumulate that much difference in arrival times, the signal would have to travel hundreds of millions of miles, assuming a pretty steep dispersion curve. More likely, the whistler is an emission from the magnetosphere triggered by the lightning pulse. When conditions are just right, numerous lightning strikes combine with numerous reflections to give an eerie chorus that sounds a bit like a flock of geese.

Most people have never heard of Natural Radio produced by several of natures's processes, including lightning storms and aurora, aided by events occurring on the Sun. The majority of Earth's natural radio emissions occur in the extremely-low-frequency and very-low-frequency (ELF/VLF) radio spectrum. Whistlers, one of the more frequent natural radio emissions to be heard, are just one of many natural radio "sounds" the Earth produces at all times in one form or another, and these signals have caught the interest and fascination of a small but growing number of hobby listeners and professional researchers for the past four decades.

Whistlers are amazing sounding bursts of ELF/VLF radio energy initiated by lightning strikes which "fall" in pitch. A whistler, as heard in the audio output from a VLF or "whistler" receiver, generally falls lower in pitch, from as high as the middle-to-upper frequency range of our hearing downward to a low pitch of a couple hundred cycles-per-second (Hz). Measured in frequency terms, a whistler can begin at over 10,000 Hz and fall to less than 200 Hz, though the majority are heard from 6000 down to 500 Hz. Whistlers can tell scientists a great deal of the space environment between the Sun and the Earth and also about Earth's magnetosphere.

The causes of whistlers are generally well known today though not yet completely understood. What is clear is that whistlers owe their existence to lightning storms. Lightning stroke energy happens at all electromagnetic frequencies simultaneously. The Earth is literally bathed in lightning-stroke radio energy from an estimated 1500 to 2000 lightning storms in progress at any given time, triggering over a million lightning strikes daily. The total energy output of lightning storms far exceeds the combined power output of all man-made radio signals and electric power generated from power plants.

Whistlers also owe their existence to Earth's magnetic field (magnetosphere), which surrounds the planet like an enormous glove, and also to the Sun. Streaming from the Sun is the Solar Wind, which consists of energy and charged particles, called ions. And so, the combination of the Sun's Solar Wind, the Earth's magnetosphere surrounding the entire planet, and lightning storms all interact to create the intriguing sounds and great varieties of whistlers.

How do whistlers occur from this combination of natural solar-terrestrial forces? Some of the radio energy bursts from lightning strokes travel into space beyond Earth's ionosphere layers and into the magnetosphere, where they follow approximately the lines-of-force of the Earth's magnetic field to the opposite polar hemisphere along "ducts" formed by ions streaming toward Earth from the Sun's Solar Wind. Solar-Wind ions get trapped in and aligned with Earth's magnetic field. As the lightning energy travels along a field-aligned duct, its radio frequencies become spread out (dispersed) in a similar fashion to light shining into a glass prism. The higher radio frequencies arrive before the lower frequencies, resulting in a downward falling tone of varying purity.

A whistler will often be heard many thousands of miles from its initiating lightning stroke—and in the opposite polar hemisphere! Lightning storms in British Columbia and Alaska may produce whistlers that are heard in New Zealand. Likewise, lightning storms in eastern North America may produce whistlers that are heard in southern Argentina or even Antarctica. Even more remarkably, whistler energy can also be "bounced back" through the magnetosphere near

(or not-so-near) the lightning storm from which it was born!

Whistlers are descending tones. Their duration can range from a fraction of a second to several seconds, with the rate of frequency shift steadily decreasing as the frequency decreases. A whistler's note may be pure, sounding almost as if it was produced using a laboratory audio signal generator; other whistlers are more diffuse, sounding like a breathy "swoosh" or composed of multiple tones. On occasion, some whistlers produce echoes or long progressions of echoes, known as "echo trains," that can continue for many minutes.

Whistlers can occur in any season. Some types are more likely to be heard in summer than winter and vice-versa. Statistically, the odds are especially high for good whistler activity between mid-March and mid-April.

Whistlers and the sounds of the dawn chorus are not heard equally well everywhere in the world. Reception of these are poor in equatorial regions, with best reception at geomagnetic latitudes above 50°. Fortunately the continental United States and Canada are well positioned for reception of whistlers and other signals of natural radio.

Considered by many listeners to be the "Music of Earth", whistlers are among the accidental discoveries of science. In the late 19th century, European long-distance telegraph and telephone operators were the first people to hear whistlers. The long telegraph wires often picked up the snapping and crackling of lightning storms, which was mixed with the Morse code "buzzes" or voice audio from the sending station. Sometimes, the telephone operators also heard strange whistling tones in the background. They were attributed to problems in the wires and connections of the telegraph system and disregarded.

The first written report of this phenomenon dates back to 1886 in Austria, when whistlers were heard on a 22-km (14 mile) telephone wire without amplification. The German scientist, H. Barkhausen, was eavesdropping on Allied telephone conversations during World War I. In order to pick up the telephone conversations, he inserted two metal probes in the ground some distance apart and connected these to the input of a sensitive audio amplifier. He was surprised to hear whistling tones which lasted for one or two seconds and which glided from a high frequency in the audio range to a lower frequency where they disappeared into the amplifier or background noise level. On occasion, the whistlers were so numerous and loud that he could not detect any type of telephone.

Later, three researchers from the Marconi Company in England reported in 1928 on the work they had done in whistler research. Eckersley, Smith, and Tremellen established a positive correlation between whistler occurrence and solar activity, and they found that whistlers frequently occur in groups preceded by a loud click. The time between the click and the whistlers was about three seconds. They concluded that there must have been two paths of propagation involved; one for the click which preceded the whistler, and a second, much longer path, over which the whistler and its echoes propagated.

Eckersley was able to show in 1931 that Earth's magnetic field permits a suitably polarized wave to pass completely through the ionosphere. This was in accordance with the magneto-ionic theory. Tremellen had on one occasion observed that during a summer thunderstorm at night every visible flash was followed by a whistler. This served as the first definite evidence of the relationship between lightning discharges and whistlers—the click being the lightning discharged which caused the whistler. As the Marconi workers continued their research, they also discover a new type of atmospheric which sounded somewhat like the warbling of birds. Since the sound tended to occur most frequently at dawn, they gave it the name *dawn chorus*.

L.R.O. Storey of Cambridge University began an intensive study of whistlers in 1951. He confirmed Eckersley's law which showed that most whistlers originate in ordinary lightning discharges, and found the path of propagation of the whistler to be along the lines of the Earth's magnetic field.

Other interesting sounds include "tweeks," which have been described by one listener as a cross between chirping birds and a hundred little men hitting iron bars with hammers. The ephemeral "dawn chorus," is a cacophony of sounds that resembles nothing else on Earth.

Tweeks are abrupt descending notes that resemble "pings." They are usually heard at night during the winter and early spring.

The dawn chorus has been variously described as sounding like birds at sunrise, a swamp full of frogs, or seals barking. In fact, it varies constantly. "Hooks," "risers," and "hiss" are all part of the dawn chorus, but sometimes these sounds are also heard as solitary events. A hook starts out like a whistler but then abruptly turns into a rising tone. Risers increase in frequency from beginning to end. And hiss sounds just like its name. Appropriately enough, dawn chorus is best heard around dawn, but it may occur at any time of the day or night. The dawn chorus is most likely to be heard during and shortly after geomagnetic storms.

A whistler receiver

You cannot easily buy a radio capable of tuning 1 to 10 kHz, but it is possible to build a sensitive receiver to permit you to hear the sounds of natural radio. We will explore what causes these sounds and see how you can study them yourself.

A whistler receiver, even though whistlers and related emissions occur at acoustical frequencies, is a receiver of radio signals. To hear whistlers, you must intercept their electromagnetic energy with an antenna and transform it to the mechanical vibrations to which our ears respond.

The "classic" whistler receiving system consists of an antenna for signal collection, an amplifier to boost the signal level, and headphones or a speaker to transform the signal to sound waves. (A magnetic tape recorder can be substituted for the headphones or speaker.) A whistler receiver can simply be an audio amplifier connected to an antenna system. However, powerful man-made interference immediately above and below the frequencies of interest tend to seriously overload receivers of this sort and make reception of whistlers and related phenomena difficult.

To overcome these problems, a good whistler receiver circuit includes a circuit called a low-pass filter that attenuates all signals above 7 kHz. This greatly reduces interference from such sources. To escape interference from AC power lines and other forms of non-natural radio noise, the receiver is designed for portable operation "in the field" away from such interference sources.

A sensitive dual FET whistler receiver is shown in the photo in Figure 14-1, and in the schematic diagram in Figure 14-2. The antenna is first fed to an RF filter

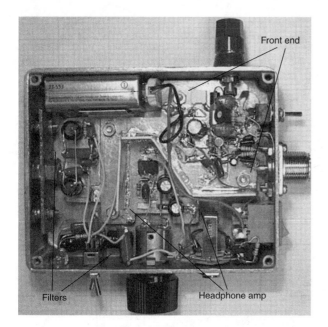

Figure 14-1 *VLF whistler receiver*

which is composed of resistors R1 through R4 and capacitors C1 through C3. This filter only allows the desired frequencies to be sent to the next section of the receiver. The RF filter is next fed to the 60 Hz filter or "trap" which filters out 60 Hz power line frequencies before the RF signals are amplified. The 60 Hz filter can be switched in and out of the circuit with switch S1:a. The dual matched FET transistor Q1 along with transistor Q2 are used to amplify the RF signals of the desired band of frequencies. The amplified signals are then fed to the input of a set of selectable filters which can be switched in and out of the circuit. Toggle switch S2 is used to switch in the high-pass filter circuit, while switch S3 is used to switch-in the low-pass filter. You can select either or both filters if desired.

The output of the low-pass filter circuit is coupled to the final audio amplifier stage via capacitor C17. Potentiometer R12 controls the audio level entering the LM386 audio amplifier at U1; also note that R12 also contains a switch S5, which is used to supply power to the LM386 audio amplifier. The output of the LM386 is coupled to an audio output jack via capacitor C20. Note that capacitor C18 is coupled to a recorder output jack to allow a chart recorder or analog to digital converter to be used to monitor and record the output from the whistler receiver. The whistler receiver is powered from a 9 volt battery for both stationary and field receiving applications.

Building a whistler receiver

Before we start building the Whistler receiver, you will need to locate a clean, well lit and well ventilated area. A workbench or large table top to spread out all you diagrams, components and tools for the project. First, you will need a small pencil tipped 27 to 33 watt soldering iron, a spool of 22 ga. 60/40 tin/lead solder, and a small jar of "Tip Tinner." "Tip Tinner" is a soldering iron tip cleaner/dresser and is available at your local Radio Shack store. You will also want to locate a few small tools such as a pair of end-cutters, a pair of needle-nose pliers, a pair of tweezers and a magnifying glass. Look for a small flat-blade screwdriver as well as a small Phillips screwdriver to round out your tools for the project. Next locate all the components for the Whistler project as well as all of the diagrams including the schematic, parts layout diagram and the necessary charts and tables and we will begin.

In order to construct a stable well performing radio receiver, it is recommended practice to build the Whistler receiver on a printed circuit board, with large ground plane areas. You will need to use good RF building practices, such as keeping leads as short as possible, clean solder joints, etc. With your circuit board in hand we will begin building the circuit.

Take a look at the schematic diagram and you will notice that there are dotted or dashed lines. These dashed lines represent shielded compartments on the circuit board. These shielded areas are used to separate different portions of the circuit from one another. You can create a shielded area by soldering a vertical piece of scrap circuit board material onto the circuit board ground plane areas on the circuit board. You could also modularize the circuit building three small circuit boards, one for the RF filter and "trap" circuit, one for the high-pass and low-pass filter, and a third board for the audio amplifier and power circuit. Small metal shield covers could be built using scrap circuit board material.

First, you will need to locate the resistors for the project, and you will also want to refer to the resistor color code chart in Table 14-1, which will help you to identify each of the resistors. Resistors generally have three or four color bands which are used to denote the

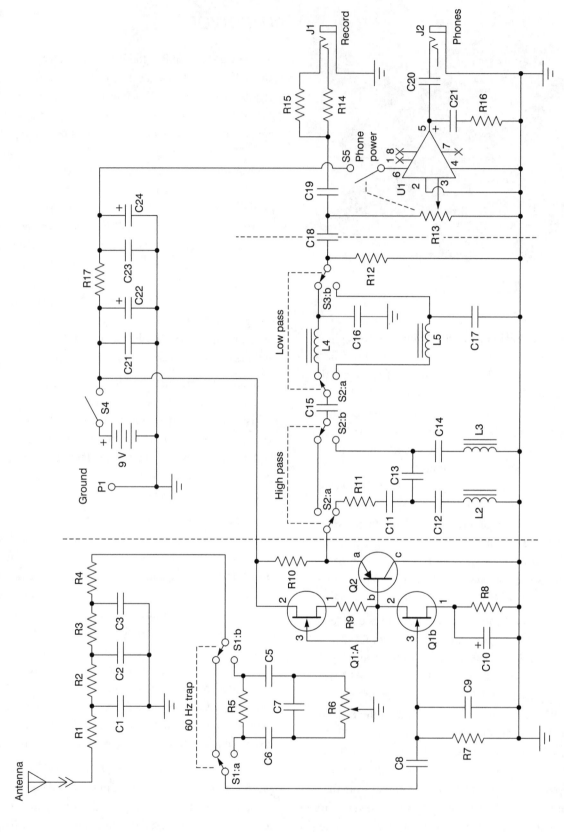

Figure 14-2 *FET VLF receiver schematic. Courtesy of Eric Vobel*

Table 14-1

Resistor color code chart

Color Band	1st Digit	2nd Digit	Multiplier	Tolerance
Black	0	0	1	
Brown	1	1	10	1%
Red	2	2	100	2%
Orange	3	3	1,000 (K)	3%
Yellow	4	4	10,000	4%
Green	5	5	100,000	
Blue	6	6	1,000,000 (M)	
Violet	7	7	10,000,000	
Gray	8	8	100,000,000	
White	9	9	1,000,000,000	
Gold			0.1	5%
Silver			0.01	10%
No color				20%

resistor's value. The first color band will be closest to one edge of the resistor. The first color band represents the first digit of the value. Resistor R1, for example, is a 10k ohm resistor. Its first color band would be brown (1), the second color band would be black (0) and the third color band which denotes the resistor's multiplier value would be orange or (000), so 10,000 or ten thousand ohms. A fourth color band represents the resistor's tolerance value. The absence of a fourth color band denotes a 20% tolerance, while a silver band represents a 10% tolerance and a gold denotes a 5% tolerance resistor. Use the color code chart to identify resistors R1, R2 and R3. When you are confident that you know the color codes and can identify these resistors, you can go ahead and solder these resistors onto the circuit board. Use your end-cutters to cut the excess component leads flush to the edge of the circuit board. Now locate another grouping of resistors and install them to the circuit board, then solder them in place on the PC board. Remember to trim the extra component lead lengths from the PC board. Install the remaining resistors onto the circuit board, until all have been mounted to the board, solder them in place and trim the leads as necessary.

Next, we are going to install the capacitors for the Whistler receiver. Capacitors come in two major types, i.e. electrolytic and non-electrolytic. Generally the non-electrolytic capacitors are the small types such as ceramic disks, mylar or polyester types. These small capacitors may have their actual value printed on them, or if the capacitors are very small may instead have a three-digit code printed on them. Refer to the chart in Table 14-2, which will help you identify these capacitors. A .001 µF capacitor, for example, would have (102) marked on the body of the device to denote its value of .001 µF. Electrolytic capacitors are generally the larger capacitors with polarity marking on them. You will find either a black or white color band on them or a plus (+) or minus (−) marking on them. Electrolytic capacitors have polarity, so this means that you have to orient them correctly while installing them in order for the circuit to work properly. Reversing electrolytic capacitors can often cause the circuit to malfunction, and it can destroy the component and/or the circuit itself, so be careful when installing them. Refer to the schematic and the parts layout diagram when installing electrolytic capacitors.

Let's begin installing the capacitors now. Try and separate the electrolytic capacitors from the other types of capacitors. Locate a few of the non-electrolytic capacitors, identify them and install a few of the capacitors on the circuit board. Go ahead and solder these capacitors onto the PC board, once you are sure you have correctly identified them. Trim the excess

Table 14-2
Capacitance code information

This table provides the value of alphanumeric coded ceramic, mylar and mica capacitors in general. They come in many sizes, shapes, values and ratings; many different manufacturers worldwide produce them and not all play by the same rules. Some capacitors actually have the numeric values stamped on them; however, many are color coded and some have alphanumeric codes. The capacitor's first and second significant number IDs are the first and second values, followed by the multiplier number code, followed by the percentage tolerance letter code. Usually the first two digits of the code represent the significant part of the value, while the third digit, called the multiplier, corresponds to the number of zeros to be added to the first two digits.

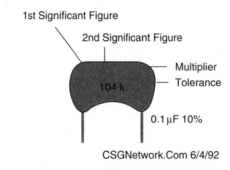

Value	Type	Code	Value	Type	Code
1.5 pF	Ceramic		1,000 pF /.001 µF	Ceramic / Mylar	102
3.3 pF	Ceramic		1,500 pF /.0015 µF	Ceramic / Mylar	152
10 pF	Ceramic		2,000 pF /.002 µF	Ceramic / Mylar	202
15 pF	Ceramic		2,200 pF /.0022 µF	Ceramic / Mylar	222
20 pF	Ceramic		4,700 pF /.0047 µF	Ceramic / Mylar	472
30 pF	Ceramic		5,000 pF /.005 µF	Ceramic / Mylar	502
33 pF	Ceramic		5,600 pF /.0056 µF	Ceramic / Mylar	562
47 pF	Ceramic		6,800 pF /.0068 µF	Ceramic / Mylar	682
56 pF	Ceramic		.01	Ceramic / Mylar	103
68 pF	Ceramic		.015	Mylar	
75 pF	Ceramic		.02	Mylar	203
82 pF	Ceramic		.022	Mylar	223
91 pF	Ceramic		.033	Mylar	333
100 pF	Ceramic	101	.047	Mylar	473
120 pF	Ceramic	121	.05	Mylar	503
130 pF	Ceramic	131	.056	Mylar	563
150 Pf	Ceramic	151	.068	Mylar	683
180 pF	Ceramic	181	.1	Mylar	104
220 pF	Ceramic	221	.2	Mylar	204
330 pF	Ceramic	331	.22	Mylar	224
470 pF	Ceramic	471	.33	Mylar	334
560 pF	Ceramic	561	.47	Mylar	474
680 pF	Ceramic	681	.56	Mylar	564
750 pF	Ceramic	751	1	Mylar	105
820 pF	Ceramic	821	2	Mylar	205

component leads from the PC board. Install a few more non-electrolytic capacitors and solder them to the board, remember to cut the extra leads from the board. Install the remaining non-electrolytic capacitors to the circuit board and solder them in place.

Next, locate the electrolytic capacitors and we will install them on the PC board. Refer to the schematic and parts layout diagrams to make sure you can orient the capacitors correctly before soldering them. Place a few electrolytic capacitors on the board and solder them in, trim the excess leads as necessary. Add the remaining electrolytic capacitors on to the PC board and solder them in place.

The Whistler receiver also has a number of coils in the circuit, most of them are low cost and readily available from Mouser Electronics, see Appendix. Identify the coils for the project and place them in their respective locations on the PC board. Now, solder them in place and remember to trim the excess component leads flush to the edge of the circuit board.

Looking at the circuit you will notice three transistors, more correctly one dual FETs and one conventional transistor. Before we go ahead and install the semiconductors, take a quick look at the semiconductor pin-out diagram shown in Figure 14-3. The FETs are very sensitive to static electricity damage, so be extremely careful when handling them, use a grounded anti-static wrist-band. The FETs have three leads much like a regular transistor but their pin-outs are a bit different. An FET will usually have a Drain, Source and Gate lead; see pin-out diagram for each particular FET, when installing them on the PC board. The conventional transistor at Q2 is a PNP type which has a Base, Collector and Emitter lead, once again refer to the transistor pin-out diagram, the schematic and parts layout

diagrams when installing the transistor and FET. Installing the FET and transistor incorrectly will prevent the circuit from working correctly and may damage the component as well, so be careful when installing the FET and transistor.

The receiver has a single integrated circuit at U1, an LM386 audio amplifier IC. As an insurance policy against a possible circuit failure in the present or possible future date, it is recommended that you use an integrated circuit socket. Integrated circuit sockets will usually have a notch or cut-out at one end of the socket; this is supposed to represent pin 1 of the IC socket. You will have to align the socket correctly so that pin 2 of the IC socket connects to the ground bus. Note that pin 1 is not used in this application. When installing the IC into the socket, you will note that the IC itself will have either a small indented circle, a cut-out or a notch at one end of the plastic IC body. Pin 1 of the IC is just to the left of the notch. Insert the IC into the socket making sure pin of the IC is aligned with pin of the socket.

Take a short break and when we return we will check over the printed circuit board for "cold" solder joints and possible "short" circuits. Pick up the PC board with the foil side facing upwards toward you. First we will inspect the PC board for possible "cold" solder joints. Take a look at all of the solder joints. All the solder joints should appear to look clean, shiny and bright, with well formed solder connections. If any of the solder joints appear dull, dark or "blobby," then you should un-solder and remove the solder from the joint and re-solder the joint all over again, so that it looks good. Next, we will inspect the PC board for possible "short" circuits. "Short" circuits can be caused by small solder balls or blobs which stick to the underside of the board. Many times rosin core solder leaves a sticky residue on the PC board. Sometimes "stray" component leads will adhere to the underside of the board and often they can form a bridge across the PC traces on the PC board causing a "short" circuit, which can damage the circuit upon power-up, so be careful to remove any component leads or solder blobs from the underside of the board.

The VLF Whistler receiver was housed in a metal chassis box to prevent "stray" RF from affecting the circuit operation. The main circuit board can be mounted to the bottom of the enclosure with $\frac{1}{8}''$

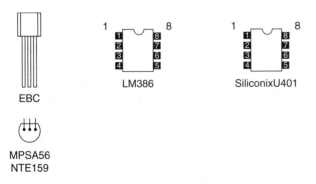

Figure 14-3 *Semiconductor pin-outs*

plastic standoffs. The 60 Hz "trap" filter switch, the high-pass filter and low-pass filter switches, along with the power on-off switch, were mounted on the front panel of the chassis box. The headphone power amplifier power switch, the headphone jack and volume control at R13 were also mounted on the front panel of the chassis box. The SO-238 antenna jack, the ⅛″ record jack and the ground binding post were all mounted on the rear panel of the chassis box. A 9 volt battery holder was used to secure the battery to the left rear corner of the chassis box. Once the circuit has been mounted in the selected chassis, you can move on to applying power and testing the VLF receiver. Connect a fresh 9 volt battery to the VLF receiver, connect a pair of headphones and a whip antenna to the antenna terminals. Turn the power switch to the "ON" position and you should hear a hissing sound from the headphones. If the receiver is totally quiet and you hear nothing, you have a problem! First, you will need to disconnect the battery, the headphones and the antenna, and you will need to carefully examine the circuit board. The most common cause for circuit failure is improper installation of electrolytic capacitors and diodes. Remember that these components have polarity and it must be observed for the circuit to work correctly. Look over the circuit very carefully. It is often advisable to have a second pair of eyes. Have a knowledgeable electronics friend help you examine the circuit board for mistakes. Another cause for possible errors are the incorrect placement of resistors. Often colors are mistaken and resistors are installed in the wrong location. Another possible problem can be the installation of the semiconductors, namely in this circuit the transistors or FETs. Check the manufacturer's pin-out diagrams for the particular device and carefully re-examine the installation. Hopefully at this point you have found your error and corrected it and you can then re-connect the battery, antenna and headphones and re-test the VLF receiver to make sure it is now working.

Using your VLF receiver

The whistler receiver works best out doors in open areas, with a vertical antenna of roughly 3 to 12 feet in height. It can also be used in wooded or obstructed areas with wire antennas between roughly 50 to 200 feet

in length connected instead of the vertical antenna. Wire antennas should be insulated and supported off the ground as high as you can place them. The performance of a wire antenna improves the more you can get it in the clear and the more vertically it can be positioned. If you do use a wire antenna, you may have to reduce the value of the input resistor R1 for best results.

The whistler receiver also needs a ground connection, also called a counterpoise. A simple but often effective ground can be provided by your body in contact with the ground terminal (BP2). I have used an improvised finger ring attached to a wire connected to the ground terminal. Other places to make a ground connection include car bodies, wire fences, or other large metal objects not connected in any way to an AC power circuit. Another approach is to use short sections of copper tubing as ground stakes which can be driven into the ground and connected to your receiver. Try several different grounding schemes to see which one provides the quietest reception. Don't be too surprised, however, to discover that you wind up having to still make body contact with the ground terminal for quietest reception!

The prototype receiver, with a whip antenna, was mounted on a piece of wood approximately 2″ × 60″. You can install the antenna at one end of the long rectangular piece of wood (it can be mounted on a bracket or just taped on) and the receiver can be mounted at the other end of the piece of wood, about one foot down from the antenna using duct tape, two-sided foam tape or a mounting bracket. A length of wire was run from the receiver's ground terminal down the wood and taped in place at several places. Now I had a handy walking stick that was easy to carry and use! The base can be stuck into any convenient hole, jammed into a rock pile, clamped onto a fence or vehicle, or even held by hand.

For the best chance of hearing whistlers and related phenomena, you will have to put some distance between the whistler receiver and the AC power grid. The bare minimum distance is about one quarter mile from any AC power line; the more distance you can put between the receiver and AC power lines, the better. You'll need to do some exploration to discover quiet sites near you. Once you arrive at a potential listening site, set up your receiver, put in a ground connection, and see what happens. What you will hear depends to some extent on the time of day, but you can always expect some sharp, crackling static. The intensity and volume of this static will depend on propagation conditions and where thunderstorms are in relation to your location. Chances are that you'll also hear some power line hum, but hopefully it won't be very loud. If your listening site is too noisy, you need to try another listening site.

If your ears have good high frequency response, you will probably hear a continuing sequence of one-second tones. These are from the OMEGA radio-navigation system. OMEGA is transmitted from 10 to 14 kHz, and OMEGA transmitters are very powerful. You may also hear anything that can produce an electrostatic discharge, particularly if the humidity is low. These miscellaneous noise sources can include wind noise, the buzz of insects flying near the antenna, dry leaves or grass moving, and even the electrostatic charges that build up on your clothing. Passing vehicles will often produce noise from

their electrical and ignition systems. And if you have a digital watch, keep it away from the antenna or you'll be listening to it instead of whistlers!

Anything you hear other than whistlers and related phenomena or OMEGA signals are likely to be the result of extraneous signals "overloading" your receiver. Burbling sounds (possibly mixed with OMEGA tones) are caused by military signals in the 15 to 30 kHz range, perhaps overloading the receiver. A "ticking" sound at a 10 Hz rate is from Loran-C radio-navigation signals at 100 kHz. In general, don't be surprised if you experience "overloading" from transmitters operating on any frequency if they are within sight of the location where you're using your whistler receiver! Table 14-3 illustrates some very low frequency stations around the world that broadcast low frequency signals, these signals are located at the upper edge of the VLF receiver's range.

Table 14-3
Very low frequency radio stations

Site	ID	Frequency kHz	Power KW
Cutler, ME	NAA	24.0	1000
Jim Creek, WA	NLK	24.8	250
Lualualei, HI	NPM	21.4	566
LaMoure, ND	NML4	25.2	
Aquada, Puerto Rico	NAU	40.75	100
Keflavik, Iceland	NRK	37.5	100
Niscemi, Italy		39.9	25
Harold E. Holt, Australia	NWC	19.8	1000
Rhauderfehn, Germany		18.5	500
Rosnay, France	HWU	15.1	400
St. Assie, France	FTA	16.8	23
Bombay, India		15.1	
Tavolara, Italy	ICV	20.27	43
Ebino Huyshu, Japan		23.4	
Noviken, Norway	JXN	16.4	45
Arkhanghelsk, Russia	UGE	19.4	150* input pwr
Batumi, Russia	UVA	14.6	100* input pwr
Kaliningrad, Russia	UGKZ	30.3	100* input pwr
Matotchkinchar, Russia	UFQE	18.1	100* input pwr
Vladivostok, Russia	UIK	15.0	100* input pwr
Anthorn, United Kingdom	GQD	.019	42

Time of listening

Most of the VLF signals, such as clicks, pops, and tweeks, can be heard almost at any time of the day or night. Whistler occurrence is a function of both thunderstorm occurrence and propagation conditions through the ionosphere. The rate is higher at night than during the day because of the diurnal variation in the D-region absorption, which is highest when the Sun is above the horizon. At night, the D-layer does not exist since it is ionized by solar ultraviolet rays. The whistler rate, or the number of whistlers heard per minute, has a marked dependence on sunspot activity, the rate increasing with the sunspot number. Whistlers are heard only when sufficient ionization exists along the path to guide the waves along the Earth's magnetic field. This ionization is assumed to be supplied from the Sun during a solar. To hear a whistler or other signals such as *dawn chorus*, you will no doubt have to get away from the power line and industrial noise. The listening time for these signals will be from near local midnight to early morning hours.

Observing and recording hints

Remember the best location to observe VLF signals is out in the countryside, in an area where few cars or trucks are likely to pass. You will want to get as far away from power lines as possible for optimum results. The lower the power line noise is, the stronger the VLF emission signals will be.

Some whistlers are related phenomena and so short-lived and impossible to predict that you might want to consider using a tape recorder to record the signals in the field unattended. Take a portable, battery-operated, cassette tape recorder with you and use a long audio cable to run to the tape recorder from the audio output of the VLF receiver. Any tape recorder with an external microphone input can be used. Best results are obtained with a recorder which has no ALC or automatic level control. The VLF receiver shown has an output jack which can be coupled to a recorder input for field recording. You may have to place the tape recorder away from the receiver to avoid picking up motor noise.

After returning home with your recording, if your receiver was out in the field, you can "play" your recording back into the sound card of your personal computer using an FFT or DSP sound recording program to analyze the recording. For those who are going to use their whistler receiver as a stationary receiver at home, you could use your PC to record on a continuous basis. Audio recording and FFT programs are available on the Internet for free. Good luck and happy exploring!

Additional notes

You can view spectrograms of whistlers as well as hear excellent recordings of many more whistlers than we used in generating the audio clip for this article, courtesy of NASA: http://image.gsfc.nasa.gov/poetry/inspire/advanced.html

A free Fast Fourier Transform (FFT) Waveform Analyzer program and an Audio Waveform Generator program can be downloaded for use on your own computer at:
FFT discussion:
http://www.dataq.com/applicat/articles/an11.htm

Some US Government agencies doing whistler research are:
NASA's Godard Space Flight center invites participation in a program called "INSPIRE" (an acronym for Interactive NASA Space Physics Ionosphere Radio Experiments) http://image.gsfc.nasa.gov/poetry/inspire

NOAA: http://ngdc.noaa.gov/stp/stp.html
http://www.ngdc.noaa.gov/stp/stp.htm
gopher://proton.sec.noaa.gov

Some of the university labs involved with whistlers are:
University of Alaska Geological Institute
University of Iowa: http://www-w.physics.uiowa.edu/~/jsp/polar/sounds/whistlers.html
Stanford University: http://www.stanford.edu/~vlf/Science/Science.html
[An excellent site, but I could only access it via the "Google" search engine]
UCLA: http://www.physics.edu/plasma-exp/Research/TransportWhistlers
Kyoto University, Kyoto, Japan

Chapter 15

Induction Loop Receiving System

Parts list

Parts Bin

Induction Loop Transmitter I

TX Mono amplifier or receiver – 5-10 watts output

R1 10 ohm 10 watt 10% resistor

L1 loop coil around room (see text)

Induction Loop Transmitter II

R1 10k ohm potentiometer chassis mount

R2 220 ohm ¼ watt resistor, 5% (red-red-brown)

R3 2.2k ohm ¼ watt, 5% resistor (red-red-gold)

R4 1 ohm ½ watt, 5% resistor (brown-black-gold)

R5 10 ohm 10 watt, 10% resistor

C1 10 µf, 25 volt electrolytic capacitor

C2,C5 100 nF, 35 volt poly capacitor

C3 100 µF, 35 volt electrolytic capacitor

C4 470 µF, 35 volt electrolytic capacitor

C6 2200 µF, 35 volt electrolytic capacitor

U1 TDA2002/3 6 to 10 watt audio amplifier module

HS heatsink for U1

L1 loop coil (see text)

Misc PC board, wire, hardware, chassis, etc.

Induction Loop Receiver

R1 1 megohm ¼ watt, 5% resistor

R2 10k ohm ¼ watt, 5% resistor

R3 1000 ohm potentiometer (tone)

R4 470 ohm ¼ watt, 5% resistor

R5 10k ohm potentiometer (volume)

R6 10 ohm, ¼ watt, 5% resistor

C1 39 nF, 35 volt mylar capacitor

C2 1 µF, 35 volt electrolytic capacitor

C3,C7 .1 µF, 35 volt disk capacitor

C4 4.7 nF, 35 volt poly capacitor

C5 10 µF, 35 volt electrolytic capacitor

C6 220 µF, 35 volt electrolytic capacitor

```
Q1    BC548 transistor

U1    LM386 audio
      amplifier IC

L1    800 turns on a
      9 × 70 mm ferrite rod
      (see text)

S1    SPST on-off switch

B1    9volt transistor
      radio battery

PH    8 ohm headphones

Misc  PC board, wire,
      1/8" phone jacks,
      battery clip, case
```

Induction loop technology has been around for many years, for use with the hearing impaired. Induction loop communication systems were originally designed to provide assistance in the school classrooms for the deaf, but today induction loop communications technology is now used for many other applications, such as museum displays, theater communications, mine and cave communication systems, and signal tracing. Induction loop systems have been installed in churches, public halls and auditoriums, schools, lecture halls, cinemas, service counter windows, drive-thru order and pick up windows, information kiosks, offices, airports, train stations, parks, tour and guide buses, automobiles, boats, riding academies, and homes.

Induction loop amplification operates on the basic principle of electronics called magnetic induction. When an electrical current is amplified and passes through a loop of wire, a magnetic field is generated around the wire. The field varies in direct proportion to the strength and frequency of the signal being transmitted. When another wire is placed close to the field of the first wire, an identical electrical current is induced within it. The second current can be amplified and converted into an exact duplicate of the original sound signal.

In technical language, a current is said to be "induced" in the second wire. Hence the term "induction." In an induction loop system, the induction loop coil in the transmitter sends an electromagnetic signal, which is "induced" to the receiver's induction loop coil in the receiving unit. An Induction Loop System consists of an amplifier and a loop, the amplifier

is connected to a sound source such as a TV, a radio, a public address system or a dedicated microphone. It then amplifies this sound signal and sends it out, in the form of an alternating current, through the loop. The loop itself consists of insulated wire, one turn of which is placed around the perimeter of the room in a simple loop system. When the alternating current from the amplifier flows through the loop, a magnetic field is created within the room. The magnetic field "induces" the signal into the Induction loop coil in the induction receiver unit.

In the portable induction receiver, the induction coil picks up the induced signal from the transmitting loop coil. The fluctuations in the magnetic field are converted into alternating currents once more. These are in turn amplified and converted by the induction loop receiver into sound. The magnetic field within the loop area is strong enough to allow the person with the hearing aid or Induction Loop Receiver to move around freely in the room and still receive the sound at a comfortable listening level.

Many modern day hearing aids incorporate a telecoil into the hearing device. So if a hearing aid user switches to the T or MT position on their hearing aid, the telecoil in the hearing aid picks up the induced signal from an induction loop transmitter and amplifies the signal for the hearing aid user to hear. This is a convenient feature for hearing aid users, in classrooms, churches, and museums with induction coil transmitters.

The listener can receive an audio signal without being physically or electrically connected to the loop, so in effect induction loop transmission is a form of wireless telephony. The listener receives the sound in one of two ways; either through the use of the T-coil or telephone switch on a hearing aid instrument or through an induction receiver that can work with earphone, which we will discuss in our project.

The basic induction loop system consists of a "transmitter" section and a "receiver" section, as shown in Figure 15-1. The "transmitter" section consists of a microphone or signal source, an audio amplifier to boost the signal, and a large loop or wire around the room or listening area. The "receiver" section consists of a smaller loop coil, an audio amplifier which boost the incoming audio and a pair of headphones.

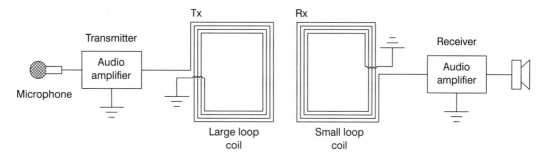

Figure 15-1 *Induction loop communications system block diagram*

"Transmitter" loop unit

You can easily build your own induction loop communication system. First we will begin by describing the "transmitter" section of the system, as shown in Figure 15-2. In its simplest form an induction loop "transmitter" can consist of a mono or stereo Hi-Fi receiver, a resistor and a large loop coil placed around the perimeter of a house, room, or your easy chair. If you have a stereo receiver with a speaker output A-B switch, you can select the (B) output for your induction loop system. You would place an 8 ohm, 10 watt resistor across the output of the left channel's output and then place an 8 ohm, 10 watt resistor in series with a large loop coil as shown. When you wanted to use the induction loop system, you would switch to the (B) speaker output; it's that simple.

If you wanted to construct your own amplifier specifically for an induction loop audio system, you could utilize the 6 to 10 watt audio amplifier shown in Figure 15-3. In order to drive an induction loop, you would need an amplifier with at least about 5 watts of power or more to have a workable system. The amplifier shown in this diagram would work just fine for driving a large loop coil. An audio input signal is fed into the audio input section at the point marked (A) on the schematic. Your audio source could be adjusted via the potentiometer at R1. Note, that you may need an additional pre-amp ahead of the amplifier if you use a low output microphone. You can feed a radio or TV signal directly into the potentiometer at R1 to drive the induction loop "transmitter." The audio amplifier module shown is an 8 watt National Semiconductor LM2002 module, but you could also use a higher power module such as their LM2005. The output of the audio amplifier is wired in series with an 8 to 10 ohm, 10 watt

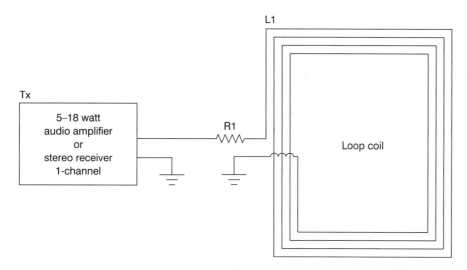

Figure 15-2 *Induction loop transmitter I*

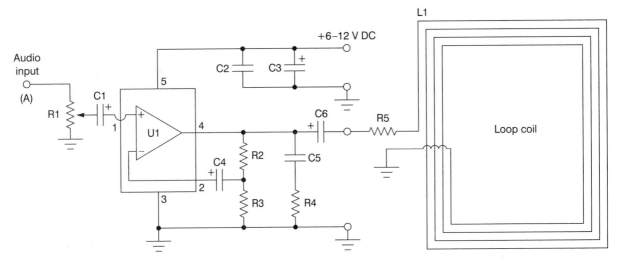

Figure 15-3 *Induction loop transmitter II*

resistor in order to drive the coil. This resistor also protects the output of the audio amplifier.

The inductive loop is a winding of wire through which an audio signal passes, which creates a field within which the audio signal can be detected by appropriate equipment, either a telecoil in a hearing aid or an induction loop receiver. The loop coil is basically a simple 4 or 5 turn loop placed around the room fed from an audio amplifier. The loop then converts the sound to a magnetic field that can be received by a second loop coil and amplified. Note that more turns in the transmission coil result in a stronger transmitted signal. The loop should have a DC impedance of at least 8 ohms, or you can add a resistor in series with the coil to reach 8 ohms.

Pick a path around the room on the ceiling, or under the carpet. A room measuring 15′ × 20′ will have a perimeter of 70′, which will require a minimum of four complete turns of 22 gauge wire, six turns of 20 gauge wire, or nine turns of 18 gauge wire. Pick a path around the room near the ceiling, and wind (or hang) your "coil," or loop, starting and ending at the amplifier. If the back of your amplifier says something like "8 ohms minimum," double the required wire length (load). Buy 120% of the wire you'll need to ensure that you can complete the loop. You should add a power resistor to ensure an appropriate load and reduce wire length, but the amplifier output will be divided between the loop and the resistor, proportionately, which means less radiated energy. This solution requires a lot of energy so try not to compromise. If required, get 4 ohms

at 10 watt minimum. There's another and simpler approach: I have a coil built into my favorite chair. It's only 18″ in diameter, but includes 120′ of 28 gauge wire (0.071 ohms per foot). That's about 25 turns of that relatively thin wire.

Another approach to constructing a large loop is to take a long length of 25 pair telephone cable 'Inside Wire' that's long enough to go around the perimeter of a room. Then get a 66 block (or some other kind of punch down block) and punch down the wires so they connect in a multiple turn loop. In this way you can add or subtract turns as you wish if you need that later. You could tack this "telephone" cable to the ceiling of the basement underneath the room you choose.

Connect either end of the loop coil wire to the plus (+) or red terminal and the other end to the minus (−) black terminal, on the audio amplifier's speaker terminals. Most amplifiers or stereo receivers have SPEAKER A and B switches; put your coil circuit on the unused output (B). Remember to place an 8 to 10 ohm, 10 watt resistor in series with the speaker output and the coil loop windings.

"Receiver" loop unit

The induction loop receiver, shown in the photo Figure 15-4 and the schematic diagram in Figure 15-5, is very sensitive and can be the basis for a monaural induction loop wireless headphone system. The induction coil receiver is used inside the area of transmission loop. The receiver coil at L1 consists of 800 turns of #26 ga. enamel wire wound on a 9 × 70 mm ferrite core.

Another approach to constructing a small loop coil for your inductive loop receiving coil is to obtain a "telephone pickup coil" if available or a suitable coil from some other device. The coil in the prototype was salvaged from a surplus 24 volt relay. Actually, two relays were needed since the first was destroyed in the attempt to remove the surrounding metal so that a single solenoid remained. Epoxy putty was used to secure the thin wires and the whole operation was a bit of a challenge. A reed relay coil will give reduced sensitivity but would be much easier to use. The experimentally inclined might try increasing the inductance of a reed relay by replacing the reed switch with soft iron. Avoid shielded inductors or inductors with iron pole pieces designed to concentrate the magnetic field in a small area or confine it completely (as in a relay or transformer) unless you can remove the iron. The resulting coil should be a simple solenoid like wire wrapped around a nail. Don't try to wind your own—it takes too many turns. Evaluate several coils simply by listening. Coils with too little inductance will sound "tinny" with poor low frequency response, and other coils will sound muffled, especially larger iron core coils. This prototype was tested with a large 100 mH air core coil with superb results but the 2″ diameter was just too big for this application. The induction receiver develops a 1–2 mV signal across its internal coil from the fluctuating magnetic field. Transistor Q1 provides pre-amplification to drive a headphone amplifier. The tone at R3 control varies the amount of gain in the high frequencies. Capacitors C1 and C2 block AC hum (C2 also blocks DC to the amplifier input). C3 functions as a low-pass filter to prevent oscillation. The output of the volume control at R5 is fed to the input of an LM 386 audio amplifier IC. In this configuration the amplifier has a gain of 200. Capacitor C7 and R6 form the output "boost" circuit which drives the output to the speaker through C6. The output of the LM385 at pin is used to drive an 8 ohm headphone. The induction loop receiver circuit can be powered from a 9 volt DC power supplies or batteries.

Are you ready to begin constructing the induction receiver circuit? First you will need to locate a clean, well lit work table or work bench. You will want a large table so you can spread out all the components, the diagrams and tools that you will need to build the project. Next you will need to locate a small pencil tipped 27 to 33 watt soldering iron and some 60/40 tin/lead rosin core solder for the project. Try and locate a small container of "Tip Tinner," a soldering iron tip cleaner/dresser which conditions the soldering iron tip; this is available from your local Radio Shack store. You should also locate a few small hand tools such as a pair of end-cutters, a pair of needle-nose pliers, a pair of tweezers, a small flat-blade and a Phillips screwdriver and a magnifying glass.

The induction loop receiver prototype was constructed on a small glass-epoxy circuit board which measured 2¼″ × 3″. A circuit board makes for a more reliable and professional circuit. Place all the diagrams in front of you on the work table along with all of the components for the induction loop receiver and we will begin building the project.

The induction loop receiver project contains about four resistors and two potentiometers. Locate the resistors using the color chart in Table 15-1, this will help you identify each of the resistors. Place the resistors into their respective locations on the circuit board, then solder them in place on the PC board. Locate your end-cutters and trim the excess component

Figure 15-4 *Induction receiver*

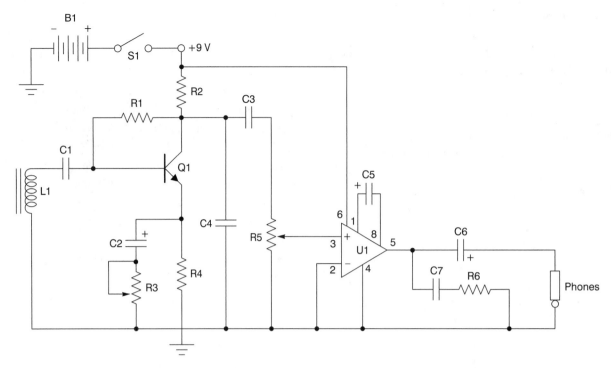

Figure 15-5 *Induction loop receiver schematic*

leads, by cutting them flush to the edge of the circuit board. Potentiometer R3 is a CP board mounted type, you can go ahead and install it on the board and then solder it in place; you can remove any excess leads if necessary.

Next we are going to move to installing the capacitors for the project. There a two distinct types of capacitors—small ceramic, mylar or polyester capacitors and larger electrolytic types. The small capacitors will most likely have a three-digit code stamped on them and not their actual value, since often the capacitor body is very small. Refer to the chart shown in Table 15-2, which illustrates the capacitor codes. You can search through the parts pile and try and identify the capacitors against the code chart and place the correct value into its respective location on the circuit board. After all of the small capacitors have been placed on the PC board, you can solder them in place. Remember to cut the excess leads from the components. Now, locate the larger electrolytic capacitors and place them in front of you and determine where they will go on the circuit board. Once you are certain where the capacitors should be placed on the board, you should take another look to make sure that you have installed the capacitors correctly with respect to its polarity. If an electrolytic capacitor is not

installed correctly the circuit will not work properly and may be damaged upon power-up. So you will want to install them correctly the first time through.

Note, there are a number of capacitors in this project, and in looking over the schematic, you will notice that a few of the capacitors will have polarity marking on them. These capacitors are electrolytic types and you must observe the polarity when installing them on the circuit board in order for the circuit to work properly. Electrolytic capacitors will have either a white or black band at one side of the capacitor body or they may be marked with a plus (+) or minus (−) marking on the capacitor body and this denotes the polarity. Refer to the schematic and parts layout diagrams when installing electrolytic capacitors in order to install them correctly with respect to their polarity markings.

Before we go ahead and install the semiconductors, take a brief look at the diagram depicted in Figure 15-6. The semiconductor pin-out diagrams will help you orient the semiconductors when building the induction receiver. The induction loop receiver also incorporates one transistor at Q1 which is connected to R1/R2 and R4 and C2. Transistors generally have three leads protruding from the bottom of the body. A Base lead,

Table 15-1

Resistor color code chart

Color Band	1st Digit	2nd Digit	Multiplier	Tolerance
Black	0	0	1	
Brown	1	1	10	1%
Red	2	2	100	2%
Orange	3	3	1,000 (K)	3%
Yellow	4	4	10,000	4%
Green	5	5	100,000	
Blue	6	6	1,000,000 (M)	
Violet	7	7	10,000,000	
Gray	8	8	100,000,000	
White	9	9	1,000,000,000	
Gold			0.1	5%
Silver			0.01	10%
No color				20%

a Collector lead and and an Emitter lead. Referring to the schematic you will note that the vertical line on the transistor symbol is the Base lead, while the Collector lead is connected to the junction of R1 and R2. The Emitter lead has the small arrow pointing downwards which connects to the junction of C2 and R4. Refer to the actual transistor pin-out diagram to see where each lead exits the transistor case. Generally small plastic TO-92 transistor case bodies will have all three leads in-a-row with the Emitter at one end, the Base lead in the center, and the Collector lead at the opposite end of the device.

The induction loop receiver has one integrated circuit at U1, this is an LM386 audio amplifier chip, which has eight leads in a dual in-line package. Prior to installing the IC, you should install a good quality IC socket. An IC socket is good insurance in the event of a possible circuit failure at some point in time. Since most people cannot successfully un-solder an integrated circuit without damaging the PC board, an IC socket makes good sense. Place the IC socket on the circuit board and solder it in place on the PC board. Locate the LM386 and take a look at the chip. You will note that the IC package will have either a cut-out, a notch or small indented circle at one end of the IC package. Pin 1 of the IC will be just to the left of the notch or cut-out.

Orient the IC correctly into the socket, while referring to the schematic and layout diagrams. Place the IC into the socket.

The receiver loop coil at L1 can be either the 800 turn coil wound on a plastic form, an old "high" impedance relay, or one of the black plastic suction cup-telephone pickup devices. Solder the coil in place between the circuit ground and the capacitor at C1. You could also elect to use a second 1/8" phone jack to allow the coil to be plugged into the circuit using a 6" to 8" length of coax cable.

The volume control used for the prototype was a chassis mounted miniature 10k ohm potentiometer. We used a mini 1/8" PC style phone jack on the circuit board at the output of the audio amplifier at C6; this will allow you to plug in a headphone for listening.

Take a short well deserved rest and when we return we will check the circuit board for possible "short" circuits and "cold" solder joints. Pickup the circuit board with the foil side of the circuit board facing upwards toward you. Look the circuit board over carefully to see if there are any "cold" solder joints. The solder joints should look clean, smooth and shiny. If any of the solder joints look dull, or "blobby," then you

Table 15-2
Capacitance code information

This table provides the value of alphanumeric coded ceramic, mylar and mica capacitors in general. They come in many sizes, shapes, values and ratings; many different manufacturers worldwide produce them and not all play by the same rules. Some capacitors actually have the numeric values stamped on them; however, many are color coded and some have alphanumeric codes. The capacitor's first and second significant number IDs are the first and second values, followed by the multiplier number code, followed by the percentage tolerance letter code. Usually the first two digits of the code represent the significant part of the value, while the third digit, called the multiplier, corresponds to the number of zeros to be added to the first two digits.

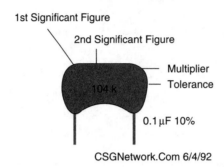

CSGNetwork.Com 6/4/92

Value	Type	Code	Value	Type	Code
1.5 pF	Ceramic		1,000 pF /.001 µF	Ceramic / Mylar	102
3.3 pF	Ceramic		1,500 pF /.0015 µF	Ceramic / Mylar	152
10 pF	Ceramic		2,000 pF /.002 µF	Ceramic / Mylar	202
15 pF	Ceramic		2,200 pF /.0022 µF	Ceramic / Mylar	222
20 pF	Ceramic		4,700 pF /.0047 µF	Ceramic / Mylar	472
30 pF	Ceramic		5,000 pF /.005 µF	Ceramic / Mylar	502
33 pF	Ceramic		5,600 pF /.0056 µF	Ceramic / Mylar	562
47 pF	Ceramic		6,800 pF /.0068 µF	Ceramic / Mylar	682
56 pF	Ceramic		.01	Ceramic / Mylar	103
68 pF	Ceramic		.015	Mylar	
75 pF	Ceramic		.02	Mylar	203
82 pF	Ceramic		.022	Mylar	223
91 pF	Ceramic		.033	Mylar	333
100 pF	Ceramic	101	.047	Mylar	473
120 pF	Ceramic	121	.05	Mylar	503
130 pF	Ceramic	131	.056	Mylar	563
150 pF	Ceramic	151	.068	Mylar	683
180 pF	Ceramic	181	.1	Mylar	104
220 pF	Ceramic	221	.2	Mylar	204
330 pF	Ceramic	331	.22	Mylar	224
470 pF	Ceramic	471	.33	Mylar	334
560 pF	Ceramic	561	.47	Mylar	474
680 pF	Ceramic	681	.56	Mylar	564
750 pF	Ceramic	751	1	Mylar	105
820 pF	Ceramic	821	2	Mylar	205

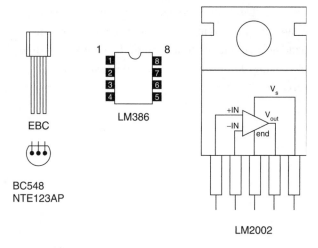

EBC

BC548
NTE123AP

1 · · 8
LM386

+IN · V$_s$
−IN · V$_{out}$
· end

LM2002

Figure 15-6 *Semiconductor pin-outs*

should remove the solder from the joint and re-solder the connection point over again until it looks good. Next we will inspect the PC board for any possible "short" circuits which may be the result of "stray" component leads which may have stuck to the foil side of the PC board. These "stray" components leads often "stick" to the board by solder residue. Solder blobs are often a cause of "short" circuits, so look the board over carefully for any solder blobs as well. Once the circuit board has been inspected, we can move on to installing the circuit board into a case or enclosure.

We mounted the induction loop receiver in a 3″ × 6″ by 1½″ metal chassis box. We placed one ⅛″ jack at the rear of the case for coil L1, then we placed the second ¼″ phone jack, power on-off switch and volume control on the front side of the chassis box. Locate a 9 volt battery clip and solder the black or minus (−) lead to the circuit ground. Next you will want to wire the red or positive (+) battery clip in series with the On-Off switch to the two power junction points on the PC board, one at R2 and the other at pin 6 of the LM386 IC. Once the circuit is inside the enclosure, you can begin testing the induction loop system. First you will want to make sure that the "transmitter" portion of the system is up and running. If you elected to use an existing stereo receiver or amplifier, make sure that the large coil is connected through a series resistor to the speaker terminals of the amplifier. Set the receiver or amplifier to MONO. If you elected to build your own audio amplifier to drive the large loop, again connect up the coil via an 8 ohm 10 watt series resistor. Once the

"transmitter" is ready, with your music or audio source connected, we can move on to testing the induction loop receiver.

Switch the induction receiver to the "On" position. When you first turn on the unit you will probably hear a lot of buzzing from the wiring in the room. Rotate the receiver in a horizontal plane to find a "null" where the hum is minimal. If you can get a reasonable null, you should be able to hear the source material from the "transmitter" loop or at least some hum coming from the earphone. If your "transmitter" loop is "broadcasting" then you should hear the source music or speech on the headphone of your induction loop receiver. If all goes well, both your transmitter and receiver section will be working and you will be hearing the source material in your remote headphones.

In the event that the induction receiver does not work when first power-up, you will have to remove the battery and carefully examine the circuit board for any errors. The most common cause for failure are improper placement of resistors, electrolytic capacitors and diodes installed backwards, and semiconductors such as transistors installed incorrectly. Have a knowledgeable electronics friend provide a second pair-of-eyes to help you examine the circuit for errors. It is very easy to miss a problem since, as the builder, you will continue to see the same circuit, the same way over and over. Once the error has been found and corrected, you can re-connect the battery and test the circuit once again.

As mentioned earlier, you can use the system for hard of hearing people, for museum displays and demonstrations, for theater personnel, late night TV listening, etc. You could wind a transmitter loop around your easy chair and use the system to "broadcast" TV sound to your earphones for late night listening. This is a great Boy Scout project for helping old folks who live in nursing homes to better enjoy TV, radio and music.

You can also use the induction loop receiver by itself without the transmitter loop, to trace power wires behind a wall or ceiling by listening for a sharp increase in hum as the coil passes near the wire. Make sure that current is flowing in the wires to be traced by turning on a lamp or other appliance. Other wires can be traced if they are carrying alternating current in the audio range or a signal generator can be connected to produce the current. Connect the generator to the wire to be traced and connect the generator's "ground" to the house wiring ground. Also ground the far end of the wire you are tracing so that current flows in the wire. This ground connection can also just be a temporary wire laying on the floor running from the generator ground to the far end of the wire you wish to trace. Have fun using your new induction loop receiver!

Chapter 16

Lightning Storm Monitor

Parts list

Lightning Storm Monitor

R1   47 k ohm ¼w,
5% resistor

R2,R4,R7   10 k ohm ¼w,
  5% resistor

R3   5.6 k ohm ¼w,
  5% resistor

R5   470 k ohm ¼w,
  5% resistor

R6,R10,R11   22 k ohm ¼w,
  5% resistor

R8   390 ohm ¼w,
  5% resistor

R9   4000 megohm or
  greater (see text)

R12   220 k ohm ¼w,
  5% resistor

R13   1 megohm ¼w,
  5% resistor

R14   100 k ohm ¼w,
  5% resistor

R15   10 ohm

VR1   500 k ohm
  potentiometer
  (chassis mount)

VR2,VR9   1 megohm
  potentiometer
  (trimmer)

VR3   3 k ohm
  potentiometer
  (trimmer)

VR4,VR5,VR6   10 k ohm
  potentiometer
  (chassis mount)

VR7   25 k ohm
  potentiometer (trimmer)

VR8   10 megohm
  potentiometer
  (chassis mount)

VR10   10 megohm
  potentiometer
  (trimmer)

C1,C9   2.2 pF, 50 volt
  mica capacitor

C2,C10   180 pF, 50 volt
  mica capacitor

C3,C4,C6,C8,C11,C12,C22,
  C23   10μF, 50 volt
  electrolytic capacitor

C5   .01μF, 50 volt disk
  capacitor

C7,C21   100 pF, 50 volt
  mica capacitor

C13   .22μF, 50 volt disk
  capacitor

C14,C15,C16,C19,C24
  .1μF, 50 volt disk
  capacitor

C17   .47μF, 50 volt disk
  capacitor

C18   220μF, 50 volt
  electrolytic capacitor

C20   470μF, 50 volt
  electrolytic capacitor

D1,D2   1N914 silicon
  diode

L1   coil optional -
  3000 turns on 2′ × 6″
  PVC form

Q1   J176 P-Channel JFET

U1   LF355 (JFET)

U2,U3,U5   LM741N op-amp

U4   LM380 audio amplifier

S1   SPDT switch (Hi/Lo)

S2   SPST switch (load)
ceramic high voltage
type

M1   500 ua meter
(RM1200 ohm)

M2   250-0-250 FS meter
(RM450 ohm)

J1   ⅛″ switches jack

J2   RCA jack - signal
output jack for data
collection

J3   SO-239 - Antenna-2
input jack

J4   Antenna-1 - RCA jack
for whip or outdoor
Ant.

CT   Capacitive Hat
(see text)

Misc   IC and transistor
sockets, wire,
PC board.

Multi-voltage Power Supply

R1   5k potentiometer -
(trimmer)

R2   240 ohm, ¼ watt,
5% resistor

C1,C4,C7   4700μF, 50 volt
electrolytic capacitor

C2,C5,C8   .1μF, 50 volt
ceramic disk capacitor

C6,C6,C9   10μF, 50 volt
electrolytic capacitor

BR1   4-diode bridge
rectifier - 3 amp,
100 volt

U1   LM317 - (+) voltage
adjustable regulator

U2   LM7805 - (+) voltage
regulator

U3   LM7905 - (−) voltage
regulator

T1   24-VCT transformer
2-amp-center tapped

S1   SPST toggle switch -
On-Off

F1   2-amp fast blow fuse

Misc   PC board, wire,
connectors, etc.

Intense electrical storms created by unstable weather conditions frequently occur all around the world. At any given point in time there is likely to be an electrical storm occurring somewhere in the world. Most people would really appreciate having an advance warning of an approaching electrical storm, especially hikers, boaters or backyard party goers. Why not construct your own lightning monitor to obtain an advance warning of an approaching storm before it arrives, thus giving you ample time to seek shelter from the elements. The lightning detector will give you ample warning of an approaching electrical storm before you see and hear the lightning. The sensitive lightning detector project will alert you to an oncoming storm over 50 miles away, giving you time to take cover or to go inside to safety.

During a lightning strike, is the Earth considered positive or negative, is a question often asked. In an electrical storm, the storm cloud is charged like a giant capacitor. The upper portion of the cloud is positive and the lower portion is negative. Like all capacitors, an electrical field gradient exists between the upper positive and lower negative regions. The strength or intensity of the electric field is directly related to the amount of charge build-up in the cloud. This cloud charging is created by a colliding water droplets.

As the collisions continue and the charges at the top and bottom of the cloud increase, the electric field becomes more intense—so intense, in fact, that the electrons at the Earth's surface are repelled deeper into the Earth by the strong negative charge at the lower portion of the cloud. This repulsion of electron causes the Earth's surface to acquire a strong, positive charge.

The strong electric field also causes the air around the cloud to break down and become ionized (a plasma).

A point is reached (usually when the gradient exceeds tens of thousands of volts per inch) where the ionized air begins to act like a conductor. At this point, the ground sends out feelers to the cloud, searching for a path of least resistance. Once that path is established, the cloud-to-earth capacitor discharges in a bright flash of lightning.

Because there is an enormous amount of current in a lightning strike, there's also an enormous amount of heat (in fact, a bolt of lightning is hotter than the surface of the sun). The air around the strike becomes super heated, hot enough in fact that the air immediately close to the strike actually explodes. The explosion creates a sound wave that we call thunder. A typical high energy lightning strike is shown in Figure 16-1.

Cloud-to-ground strikes are not the only form of lightning though. There are also ground-to-cloud (usually originating from a tail structure) and cloud-to-cloud strikes. These strikes are further defined into normal lightning, sheet lightning, heat lightning, ball lightning, red sprite, blue jet, and others that are lesser defined. For more information on lightning, see: http://science.howstuffworks.com/lightning.htm.

Satellites are more often used to follow lightning strikes around the world but haven't advanced to the point where they can accurately map local areas. There are two major types of sensors commonly used—magnetic direction finders and VHF Interferometery.

The National Lightning Detection (NLDN), which is operated by Global Atmospherics, Inc., (GAI) in

Tucson, AZ, is a network of more than 130 magnetic direction finders that covers the entire USA—more than twice the coverage of existing weather radar networks.

Each direction finder determines the location of a lightning discharge using triangulation and is capable of detecting cloud-to-ground lightning flashes at distances of up to 250 miles and more. Processed information is transmitted to the Network Control Center where it's displayed in the form of a grid map showing lightning across the USA: (www.lightningstorm.com/tux/jsp/gpg/lex1/mapdisplay_free.jsp).

Recently, NASA has improved the resolution of the system by adding acoustical measurements to the mix. Although the flash and resulting thunder occur at essentially the same time, light travels at 186,000 miles per second, whereas sound travels at the relative snail pace of one-fifth of a mile in the same time. Thus, the flash—if not obscured by clouds—is seen before the thunder is heard. By counting the seconds between the flash and the thunder and dividing by 5, an estimate of the distance to the strike (in miles) can be made.

In the NASA lightning sensor, a low frequency receiver detects the lightning strike. The leading edge of the electric-field pulse is used to start a timer and the leading edge of the thunder pulse is used to stop the timer. A microcontroller in each receiver transmits the time measured to a processing station, where the times are converted to distances that are used to compute the location of the lightning strike to within 12 inches. However, the NASA sensors have to be located within a 30 mile radius of the strike to be accurate.

If you are interested in monitoring static levels around you in stormy weather, and the ability to have some advance warning of an impending storm, then this circuit will be of great interest to you especially if you are involved in the radio hobby and have antennas high in the air. The Lightning Storm Monitor has proven to be very reliable and useful for advanced storm warning activities.

The Lightning Monitor circuit is shown in the schematic in Figure 16-2. The Lightning Monitor is actually two receivers in one, a spheric monitor and a static electricity monitor. The static monitor portion of the circuit begins at the outdoor Antenna-2, a capacitive hat type antenna which is fed with coax from the antenna to the receiver. The gate of the P JFET is tied

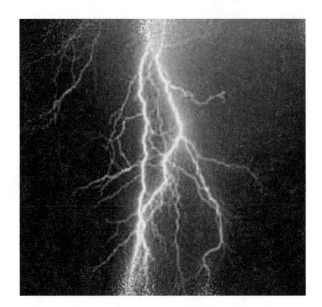

Figure 16-1 *Lightning*

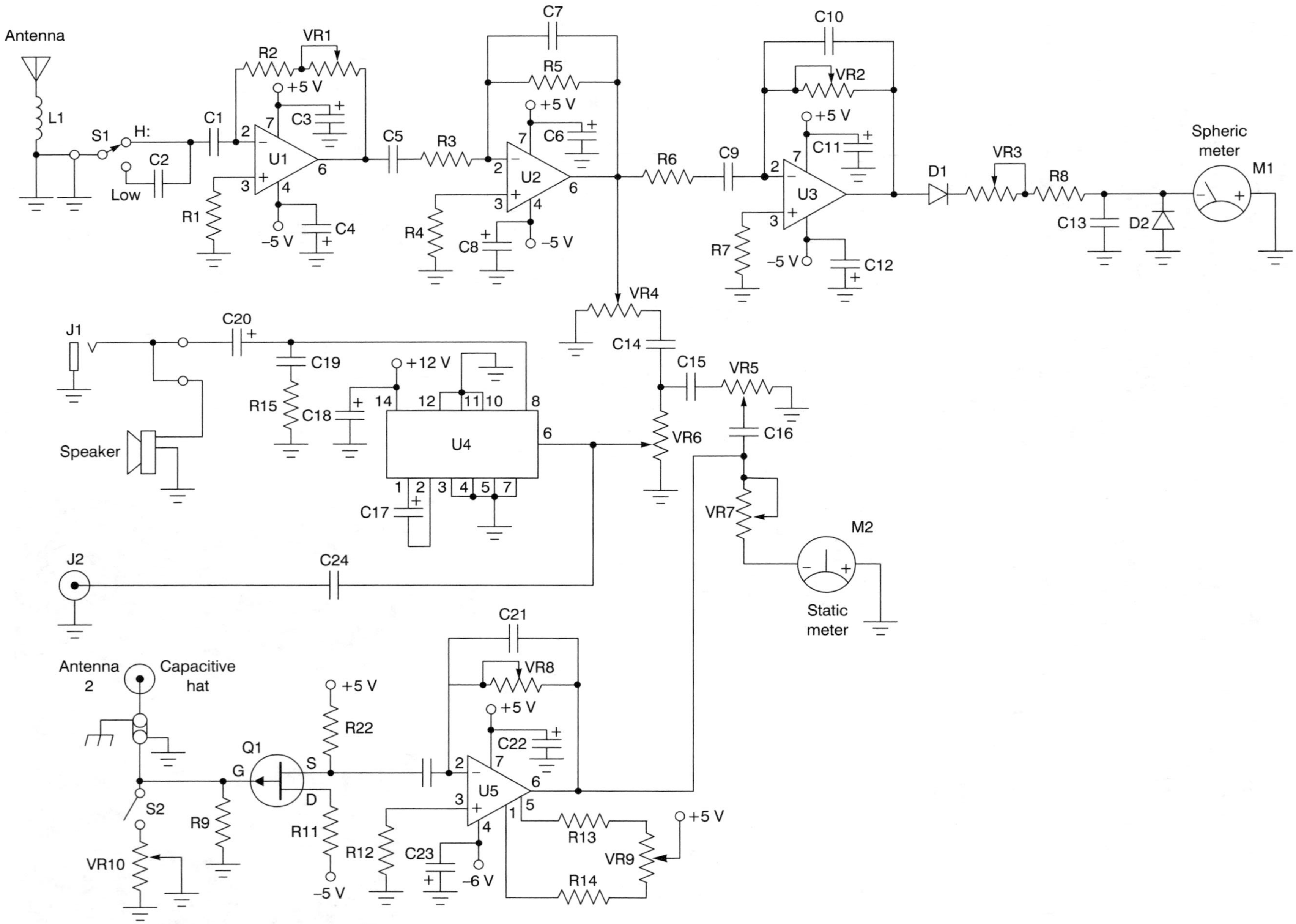

Figure 16-2 *Lightning monitor: Courtesy of Russell Clit*

to ground since a dual polarity supply is used, through a very high resistance. In the original prototype, I experimented with an 11,000 megohm resistor. These high value resistors are hard to find; you may notice that it is a moot point if there is too much leakage in your circuit. In that light, it is best to standoff the gate and use a high quality new coax to your outside antenna, usually a capacitive hat type. You must also come up with a mount design which keeps rain from the point on the mount where current could flow to ground, or you will notice that your meter becomes very insensitive as it starts raining.

The prototype used a 22″ shaft from an amateur radio antenna in a mobile antenna mount, with two nuts on top to hold the capacitive hat and a plastic umbrella to protect the mount from moisture. Likewise the coax connection must be well sealed from moisture, an SO-239 type connectors at the antenna and on the chassis for the meter inside the house. As far as the resistance load goes, if you cannot locate a high value for R9, you can fabricate a high value type if need be, for the strong field load. A 10 meg potentiometer which could be switched out of circuit was used in the prototype. For the switch at S2, you could use a ceramic high voltage type to minimize leakage. The FET type is not critical; in the prototype a J176 came from All Electronics, also the meter and the 10 meg potentiometer came from them as well.

Through experimentation it was found that controlling the offset of an op-amp gave a very sensitive way to control meter balance, far superior to trying to shift the meter itself. If you cannot locate a zero center meter you might ground the end, or put it on a trimmer tap where the ends of the pot are on the plus and minus of the supply, say a 5 or 10 K pot. If you can find it, a 250-0-250 μA meter is suggested. In the maximum gain mode you can notice field gradient changes in clear weather from morning to night, as well as detect storms more than a state away. One problem you will notice with this circuit, especially at high gain settings, is that you have to re-zero the meter often as the polarity changes overhead in stormy weather.

The higher the gain of the op-amp or the higher the input impedance of the FET gate, the more this becomes a problem, which is why a provision was made to lower the impedance of the gate circuit and reduce the op-amp gain in strong fields. Note the provision for audio to be taken from the op-amp, and mixed with separate levels for the static and RF signals, as well as a master volume.

In the Spheric, or RF section of the Lightning Monitor, the whip antenna at Antenna-1 was fed to a coil consisting of a 4′ foot long 6″ inch diameter PVC pipe, the coil consists of about 3000 turns of insulated wire on it. You will find that just a straight whip antenna works fine, so it was found that the large coil is not mandatory. The original thought for using the coil at the antenna was to have a lot of signal at low frequencies with a high Q inductor, to give more signal while trying to reduce the overall gain of the circuit, to minimize 60 cycle hum. On that note, long whips or especially long wires are not very desirable here, so stay with a 5′ to 6′ vertical whip antenna.

The input signal is amplified with a JFET input op-amp at U1, with selectable input range through a small value capacitor to allow high gain to be achieved with minimum 60 cycle hum. A LM741 op-amp gives gain to provide audio signal and then another LM741 is used to drive a 500 μA single ended meter for a reading of RF levels. You may find it easier to make the series resistance control for the meter to be a front mounted control as well as the gain control for the meter driver, i.e. the 741 op-amp, as it give much more flexibility under differing weather conditions. This meter is very helpful in determining the number of lightning strokes per unit time in severe weather.

The audio output from the Spheric monitor section is derived form an LM380 IC circuit; you will notice interaction between controls if you build it as shown. A buffer and mixer circuit might be helpful here, but in the original design we decided to keep the parts count low. Also, an equalizer circuit for the signal out is a very good addition, so you can tailor the output response and reduce unwanted signals such as line power noise. A signal output is provided so the output can be fed to a digital multi-meter with PC interface, if desired to log readings of the Lightning Monitor over time.

Building the lightning monitor

Prior to constructing the Lightning Monitor, you will want to locate a well lit and well ventilated work area. A large clean work bench or work table would be suitable for building the project. Grab a 27 to 33 watt pencil tipped soldering iron, a roll of 22 gauge 60/40 tin/lead rosin core solder. You should also have a small jar of "Tip Tinner," a soldering iron tip dresser which is available from your local Radio Shack store. You'll probably want to gather a few tools, such as a pair of end-cutters, a small pair of needle-nose pliers, a pair of tweezers, a magnifying glass, a small Phillips screwdriver as well as a small flat-blade screwdriver. Place your schematic diagram along with your circuit layout diagram and various charts and tables in front of you on your work table. Place your project components and circuit board in front of you and we can begin constructing the Lightning Monitor. A printed circuit is highly recommended for building this project rather than trying to use perf-board construction to permit more reliable operation from your circuit. Warm up your soldering iron and we will begin construction!

Let's begin building the monitor by first identifying the project resistors. Locate the resistors from the parts pile. Refer to the chart in Table 16-1, which illustrates the resistor color codes and their values. Resistor R1, for example, is a 47k ohm ¼ watt 5% resistor. Try and locate it from the parts pile. The first color band at one edge of the resistor body should be yellow, which is the number (4). The second color band should be violet or number (7) on the chart. The third color band is the multiplier and should be orange in color and

represents a multiplier value of 1000, so 47 times 1000 is 47,000 ohms or 47k ohms. The fourth color band on resistors represents the resistor's tolerance. A silver band denotes a 10% tolerance, a gold band represents a 5% tolerance resistor and no color band denotes a 20% tolerance resistor. Once you have found R1, you can try and locate resistor R2 and so on. You can place the resistors on the circuit board in groups of four or so in their respective locations on the circuit board. Once you have the first four resistors placed on the PC board, you can go ahead and solder them in place on the board. After the resistors have been soldered onto the circuit board, you can cut the excess component leads flush to the edge of the circuit board using your end-cutters. Now, you can locate the next batch of four resistors and place them in their respective locations on the circuit board, then go ahead and solder them in place. Remember to trim the excess component leads. Finish up installing the resistors and then we will move on to installing the two silicon diodes.

There are two silicon diodes in this project, and they are located in the meter range and display circuit. Diodes have two leads, a cathode lead and an anode lead. Take a look at the resistors and you will notice that there will be either a black or white color band at one edge of the diode's body. The color band denotes the cathode lead of the diode. It is important that you install

Table 16-1
Resistor color code chart

Color Band	1st Digit	2nd Digit	Multiplier	Tolerance
Black	0	0	1	
Brown	1	1	10	1%
Red	2	2	100	2%
Orange	3	3	1,000 (K)	3%
Yellow	4	4	10,000	4%
Green	5	5	100,000	
Blue	6	6	1,000,000(M)	
Violet	7	7	10,000,000	
Gray	8	8	100,000,000	
White	9	9	1,000,000,000	
Gold			0.1	5%
Silver			0.01	10%
No color				20%

the diode correctly with respect to this polarity. One you have installed the two diodes in their proper place, you can solder them in place on the circuit board. Don't forget to cut the extra component lead lengths.

Next, we will try and locate the capacitors for the Lightning Monitor project. This project uses two types of capacitors, small ceramic disk types and the larger electrolytic types which have polarity. The small capacitors come in many different styles and shapes and sometimes they are so small that there is not enough room to print their actual value on the component, so a three-digit code is used. Refer to the chart shown in Table 16-2 when installing the small capacitors. For example, a .1 µF capacitor would have a code value of (104). The Lightning Monitor also utilizes larger value electrolytic capacitors, these capacitors are usually large and have polarity marking on them, such as a black or white color band or a plus (+) or minus (−) marking on them. These polarity markings must be observed when installing the capacitors on the circuit board. Failure to observe the correct polarity could result in either a damaged component or a damaged circuit when first power-up.

Look through your component parts pile and let's try and identify some capacitors. Let's begin with some of the small types such as C1 and C2. Capacitor C1 may be marked 2.2 or 2n2, while Capacitor C2 will most likely be labeled 181. Try and locate the lower value capacitor from the parts stock and place about four components at a time on the PC board in their respective locations. After placing four or so of the first lot of capacitors, you can go ahead and solder them in place. Remember to trim the excess component leads from the board. Take the next grouping of four capacitors and install them on the PC board while referring to the schematic and parts layout diagrams. Go ahead and solder the next group of four capacitors and trim the excess lead lengths. Install the remaining smaller capacitors and then we will move on to installing the larger value electrolytic capacitors. As mentioned, electrolytic capacitors will have polarity markings on them and they must be installed correctly in order for the circuit to work properly. Take your time when identifying the electrolytic capacitors. Observe their polarity marking and then make sure you orient them correctly and in their proper location on the PC board. Refer to the schematic and parts layout diagram. Install a few electrolytic capacitors at one time, and them

solder them in place, follow up by removing the excess lead lengths. Install the remaining electrolytic capacitors and solder them in place. Use your end-cutters to remove the extra lead lengths by cutting them flush to the edge of the circuit board.

The Lightning Monitor uses a single FET transistor at Q1, which is connected to antenna-2. The P channel J-FET at Q1 is a J176 FET, which must be handled with care using an anti-static wrist band to avoid damage. This component is very sensitive to static electricity and can be easily damaged in handling. The three leads of the transistor Q1 are all in a single row with the Drain lead at one end and the Gate lead in the center and the Source lead at the opposite end of the plastic case of the transistor. Refer to Figure 16-3, which illustrates the semiconductor pin-outs.

Before we install the integrated circuits in this project, we would highly recommend using IC sockets. Integrated circuit sockets are good insurance in the event of a possible circuit failure at some possible distant point in time. It is much easier to simply unplug a defective IC, rather than trying to un-solder a 14-pin IC from the circuit board. Integrated circuit sockets are inexpensive and easy to locate. There a number of integrated circuits in this project, from 8-pin to 14-pin dual in-line types. In order to install the integrated circuits correctly, you must observe the correct orientation. Failure to install the integrated circuits correctly can cause serious damage to the IC and to the circuit itself, so you must get it right the first time! Each plastic IC package will have either a cut-out, a notch or small indented circle at one end of the plastic package. Looking just to the left of the notch or cut-out you will find pin 1 of the IC. You must align pin 1 of the IC to pin 1 of the IC socket when installing the chip on the PC board. Take a look at U1, this is an 8-pin dual in-line IC. Since we will not be using pin 1 on this IC, you will have to make sure that pin 2 or the minus (−) input is connected or aligns with C1 and R2, while pin 3 of U1 is connected to resistor R1 and so on. Now, go ahead and insert the integrated circuit into their respective sockets.

There are a number of potentiometers in this project, some of which are chassis mounted controls, while other potentiometers will be circuit board mounted types. Refer to the parts list below and Table 16-3. You can locate the circuit board trimmer type potentiometer and install them at their respective locations on the

Table 16-2
Capacitor code information

This table is designed to provide the value of alphanumeric coded ceramic, mylar and mica capacitors in general. They come in many sizes, shapes, values and ratings; many different manufacturers worldwide produce them and not all play by the same rules. Most capacitors actually have the numeric values stamped on them; however, some are color coded and some have alphanumeric codes. The capacitor's first and second significant number IDs are the first and second values, followed by the multiplier number code, followed by the percentage tolerance letter code. Usually the first two digits of the code represent the significant part of the value, while the third digit, called the multiplier, corresponds to the number of zeros to be added to the first two digits.

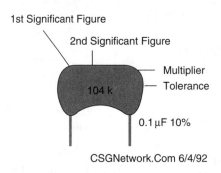

CSGNetwork.Com 6/4/92

Value	Type	Code	Value	Type	Code
1.5 pF	Ceramic		1,000 pF /.001 μF	Ceramic / Mylar	102
3.3 pF	Ceramic		1,500 pF /.0015 μF	Ceramic / Mylar	152
10 pF	Ceramic		2,000 pF /.002 μF	Ceramic / Mylar	202
15 pF	Ceramic		2,200 pF /.0022 μF	Ceramic / Mylar	222
20 pF	Ceramic		4,700 pF /.0047 μF	Ceramic / Mylar	472
30 pF	Ceramic		5,000 pF /.005 μF	Ceramic / Mylar	502
33 pF	Ceramic		5,600 pF /.0056 μF	Ceramic / Mylar	562
47 pF	Ceramic		6,800 pF /.0068 μF	Ceramic / Mylar	682
56 pF	Ceramic		.01	Ceramic / Mylar	103
68 pF	Ceramic		.015	Mylar	
75 pF	Ceramic		.02	Mylar	203
82 pF	Ceramic		.022	Mylar	223
91 pF	Ceramic		.033	Mylar	333
100 pF	Ceramic	101	.047	Mylar	473
120 pF	Ceramic	121	.05	Mylar	503
130 pF	Ceramic	131	.056	Mylar	563
150 pF	Ceramic	151	.068	Mylar	683
180 pF	Ceramic	181	.1	Mylar	104
220 pF	Ceramic	221	.2	Mylar	204
330 pF	Ceramic	331	.22	Mylar	224
470 pF	Ceramic	471	.33	Mylar	334
560 pF	Ceramic	561	.47	Mylar	474
680 pF	Ceramic	681	.56	Mylar	564
750 pF	Ceramic	751	1	Mylar	105
820 pF	Ceramic	821	2	Mylar	205

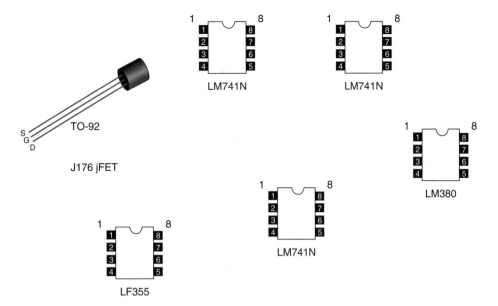

Figure 16-3 *Semiconductor pin-outs*

circuit board. Solder them in place and then trim the excess leads if necessary.

Let's take a well deserved break and when we return we will inspect the circuit board for possible "short" circuits and "cold" solder joints. Pick up the Lightning Monitor circuit board with the foil side of the board facing upwards towards you. First we will inspect the PC board for possible "cold" solder joints. Take a look at the solder joints, they should all look clean, shiny and bright. If you see any solder joints that appear dull, dark or "blobby," then you should remove the solder from the joint and re-solder the joint all over again until it looks good. Once you are confident that all looks well,

we can move on to inspecting the PC board for possible "short" circuits. There are two main causes for "short" circuits, small round solder blobs laying across a PC "trace" can cause a circuit to "short-out." The second cause of "short" circuits is a "stray" component lead. Often when cutting excess component leads, they will "stick" to the circuit board from rosin core solder residue. Make sure that there are no "stray" leads bridging between the circuit board traces. Once you are confident that there are no "shorts," we can move on to finishing the project.

Now, let's move on to mounting the larger components and hardware on the chassis box. The chassis mounted

Table 16-3

Potentiometer functions

Potentiometer	Function	Type
VR1	Spheric Gain Control	500k chassis mount
VR2	Meter Gain Control	1 megohm trimmer
VR3	Meter Range Control	3k trimmer
VR4	Spheric Volume Control	10k chassis mount
VR5	Static Volume Control	10k chassis mount
VR6	Main Volume Control	10k chassis mount
VR7	Static Range Control	25k trimmer
VR8	Static Gain Control	10M chassis mount
VR9	Static Meter Balance Control	1 megohm trimmer
VR10	Load Control	10 megohm trimmer

potentiometers were all mounted on the front of the metal chassis used to house the project, and wired to the circuit board with 20 ga. insulated wire. The two toggle switches, S1 and S2, were also mounted on the front panel of the chassis and were wired to the circuit board using 22 ga. insulated wire.

The Lightning Monitor utilizes two analog meters, one for each monitoring circuit. The meter for the Spheric circuit at M1, was a 500 µA meter, while the Static meter circuit uses a 250-0-250 µA center reading meter at M2. Both of these meters were found surplus and were mounted on the front panel of the metal chassis.

The speaker was mounted on the rear panel of the chassis and wired to the circuit board using two 6″ lengths or #22 ga. insulated wire. A ⅛″ switched mini audio panel jack for headphone use was mounted on the front panel. An RCA jack was provided for antenna-1; you can use a whip antenna or an outdoor whip antenna. An SO-239 antenna connector was mounted at the rear panel to allow the Capacitive Hat antenna-2 to connect to the circuit board, via a small piece of coax cable. The Capacitive Hat antenna is shown in Figure 16-4, it can be built with an antenna bracket mounted on a pole outdoors. An SO-239 jack was mounted on the small antenna bracket. A threaded shaft was silver soldered to the center terminal of the connector with a small rain guard on it. At the top of the threaded shaft a metal coffee can lid is held in place with two nuts. The antenna can be placed outdoor on a pole near the house with a length of coax cable leading from the antenna to the antenna jack on the Lightning Monitor.

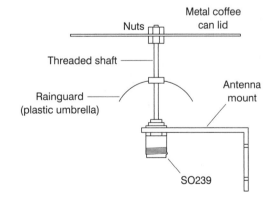

Figure 16-4 *Antenna-2 – Capacitive hat antenna*

If you wish to experiment, you can try an indoor 5′ whip without the coil and see how the circuit performs. You could also experiment with an outdoor whip antenna without the coil and you could build the coil and place it in series with an outdoor whip antenna if desired.

The Lightning Monitor needs a power supply which can provide a 12 volt source for the audio amplifier circuit as well as a plus and minus power supply for the op-amps. The diagram in Figure 16-5 illustrates a 12 volt and dual plus/minus 5 volt power supply which can be used to power the Lightning Monitor. The prototype Lightning Monitor incorporated the power supply in the same chassis as the actual monitor. You will need to locate a 12 volt, 250 mA center-tapped transformer ahead of the rectifier circuit. The rectifier circuit is fed directly to the regulator integrated circuits as shown. You will need to connect the dual voltage power supply to the Lightning Monitor. The plus voltage is connected to pins 7 on both the LM741 and the LF355 op-amp as well as to the J-FET at Q1. The negative voltage supply is connected to pin 4 of both the LM741 and the LF355 op-amp, as well as to the JFET at Q1. The 12 volt power supply is used to power the LM380 audio amplifier.

Once the power supply and Lightning Monitor have been completed, you connect the whip antenna-1 and the Capacitive Hat antenna-2 and prepare for testing the Lightning Monitor. To best test the circuit you will need to wait for a thunderstorm. However, you can do some testing of the circuit. First, apply power to the Lightning Monitor circuit, turn the Off-On switch to the "On" position. Make sure both antennas are connected. Turn up the main volume and the Static Volume controls past the ½ position and adjust VR7 to past ½. You will need to generate some static electricity by rubbing a lucite rod with some animal fur to create a static charge. Bring the lucite rod near the Capacitive Hat antenna and you should see the Static Meter at M2 swing into action. This will give you a good indication that the Static portion of the meter is working. You may be able to test the Spheric Meter portion of the circuit if you have an RF broadband noise generator.

If the Lightning Monitor doesn't seem to work at all or appears DEAD, then you will want to turn "Off" the circuit and remove the power and inspect the circuit board once again. While inspecting the circuit board

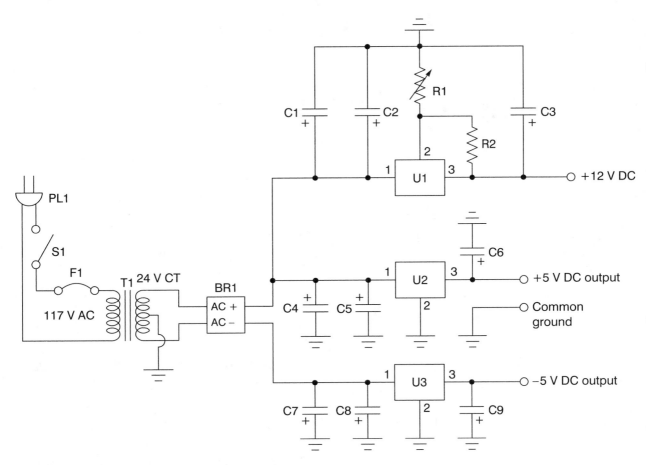

Figure 16-5 *Multi-voltage power supply*

you will be looking to make sure that the electrolytic capacitors have been installed correctly, to be sure that the diodes and the JFET have been installed correctly as well. Have a knowledgeable electronic hobbyist inspect the circuit, since it is easy to overlook something and not catch it the second time. Make sure that your power supplies are connected to the correct points and that the op-amps have been installed correctly. Once you are confident of your inspection and hopefully you found your error, you can re-apply power and re-test the circuit again. Good luck and be safe when experimenting with outdoor antennas, keeping them away from power lines.

Chapter 17

Ambient Power Receiver

Parts list

Parts Bin

Original Ambient Power Module I

C2,C3 .047 µF, 50 volt
    capacitor

C1,C4 40 µF, 50 volt
    electrolytic capacitor

D1,D2 1N34 diodes

D3,D4 1N34 diodes

ANT 50+ foot broadband
    monopole antenna

GND Earth ground stake

Misc PC board, wire,
    hardware, enclosure

Ambient Power Module II

D1,D2,D3,D4,D5,D6 1N60
    germanium diodes

D7,D8,D9,D10,D11,D12
    1N60 germanium diodes

C1,C2,C3,C4,C5,C6
    470 µF, 50 volt
    electrolytic
    capacitors

C7,C8,C9,C10,C11,C12
    470 µF, 50 volt
    electrolytic capacitors

L1 480 turns #22 wire
    on 3″ plastic tube

ANT 50+ foot broadband
    monopole antenna

GND earth ground stake

Misc printed circuit
    board, wire, hardware,
    case

In 1979, Joe Tate began experimenting with methods of turning radio energy in the air into usable electric power. He first developed a clock which drew its power from an antenna that was just a long piece of wire stretched out horizontally about 20 feet above the ground. The power supply for the clock worked something like an old-style crystal radio, except that it did not have a tuning circuit. Because of this, the "Crystal Clock" was able to absorb a wide spectrum of radio noise from the antenna and yield electric power. The power supply was able to deliver much more current than was developed in a crystal radio, although its output was just still just a few millivolts.

Joe began recording the power supply's output over a long period of time to determine its dependability. After all, the device relied completely on whatever stray signals happened to be in the air. Using an Atari computer, the output of the clock's power supply was measured continuously and recorded on floppy disk. This was done by feeding the unregulated voltage output directly into the computer's joystick port.

Joe began calling this power supply the "Ambient Power Module" (APM) because it extracted power from ambient background radio noise. This small circuit, when connected to antenna and ground, used the potential difference between air and ground to generate a small direct current continuously. Joe's original claims stated that the APM could produce up to 36 volt/9 watts. Joe applied for a US Patent and was later granted patent #4,628,299. As he studied the recorded data, mild fluctuations were noted in a daily cycle. The patterns were consistent over long periods of time, though they differed in different locations. Aside from that, the APM looked like a very dependable source of power. In the spring of 1984, he discovered an interesting correlation between voltage fluctuations from the APM and a local earthquake event; this point will be discussed later.

The APM operates, as a broadband receiver of energy in the lower portion of the radio spectrum, receiving

Figure 17-1 *Ambient power module*

most of its power from below 1 MHz. The basic circuit may be combined with a variety of voltage regulation schemes.

The original Ambient Power Module is shown in Figure 17-1 and the schematic diagram circuit in Figure 17-2 is used to convert the ambient RF energy to a direct voltage which can be used and handled by data processing equipment. This design makes use of doubler, splitter and rectifier circuit configurations. Designated as an ambient power module (APM), the circuit is connected to an antenna, preferably a 50 foot or greater broadband monopole antenna. The antenna is connected to the APM circuit via two capacitors C2 and C3, each being in series with the antenna input line for coupling and each having a value of 0.047 microfarad. Taking the left or negative side of the circuit first, it comprises two rectifiers (diodes) D3 and D4 (1N34 type) and a filter capacitor C4 (40 microfarads). Rectifier D2 is connected in parallel

to the signal path and rectifier D1 is connected in series, in the well-known voltage multiplier arrangement. Capacitor C1 is connected in parallel across the output of the APM to smooth the rectified output. The right or positive side of the circuit is similar, except for the polarity of the diodes.

In operation, an RF voltage is developed across the antenna; this voltage is voltage multiplied by the two rectifiers on each side of the circuit. The resultant voltage output is smoothed or filtered by capacitors C2 and C4. The circuit is very symmetrical with the plus side mirroring the minus side of the circuit. The output of the APM is ideal for low voltage, high impedance devices, like digital clocks, calculators, radios and smoke alarms, which are the most likely applications for this power source.

One possible application for the APM is to charge small NiCad batteries which provide effective voltage regulation as well as convenient electrical storage. Charging lead acid batteries is not practical because their internal leakage is too high for the APM to keep up with. Similarly, this system will not provide enough power for incandescent lights except in areas of very high radio noise. The APM could also be used to power small electronic devices with CMOS circuitry, like clocks and calculators. Smoke alarms and low voltage LEDs, and small emergency power/lighting circuits can also be powered by the APM. Use your imagination to devise an application for the APM.

Building the APM circuit

The builder has a choice of wiring techniques which may be used to construct the module. It may be hand wired onto a terminal strip, laid out on a bread board, experiment board, or printed circuit. We chose to build the original circuit on a small printed circuit board. Locate a clean well lit work table to build the APM. Find a small 27 to 33 watt pencil-tipped soldering iron, as well as some "Tip Tinner." "Tip Tinner" is a soldering iron tip cleaner/dressing compound which helps to clean the soldering tip after use and can be obtained at your local Radio Shack store. Try and locate a pair of end-cutters, a pair of small needle-nose pliers and a set of screwdrivers. Grab your schematic

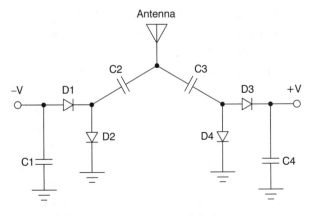

Figure 17-2 *Ambient power module 1*

diagram and look closely at it, before we begin construction.

The APM power receiver consists mainly of diodes and capacitors and you will recall that both of these type of components have polarity which must be observed for the circuit to function correctly. The diodes used in the APM circuit are germanium diodes and NOT silicon diodes which have a voltage drop. Diodes will have either a white or black band at one side of the diode's body. This colored band denotes the cathode end of the diode. When installing the diodes on the PC board make sure you observe the correct placement of all the diodes.

The APM circuit utilizes various capacitors, both electrolytic and non-electrolytic types. Electrolytic capacitors are often larger size capacitors with larger values. Electrolytic capacitors have polarity and may have either a black or white color bands with plus or minus marking at one end of the capacitor body. When installing the electrolytic capacitors make sure that you orient the capacitors correctly with respect to the circuit diagram. The chart in Table 17-1 may help you to identify the small value capacitors. In many instances, small value/size capacitors often use a three-digit code to represent a capacitor value.

Once all the components have been placed on the PC board, you can take a short break and when we return we will inspect the circuit board for possible "short" circuits and "cold" solder joints. Pick up the circuit board with the foil side facing upwards toward you. Look over the board carefully and inspect the solder connections. The solder joints should all look clean, smooth and shiny. If any of the solder joints look dull, dirty or "blobby" then remove the solder from the joint and re-solder the connection over again so that it looks good.

Next we will inspect the circuit board for any "short" circuit which could result from solder blobs or remaining component leads left from the building process. A component lead resting across a PC trace can result in a "short" circuit which could cause the circuit to fail.

Once you are satisfied that your circuit looks good, you can move on to testing the APM receiver. First you will need to attach a good antenna ground connection to the APM receiver module.

Antenna requirements

The antenna needs to be of sufficient size to supply the APM with enough RF current to cause conduction in the germanium diodes and charge the ground coupling capacitors. It has been found that a long horizontal antenna wire works best. A fifty foot monopole, or long-wire antenna will work well, but a small antenna should work, and of course you can experiment with your own antenna ideas. The antenna will work best when elevated to 20–30 feet. Lower elevations will work, but a longer wire may be necessary for best results.

In most locations, possible supporting structures already exist. The wire may be stretched between the top of a building and some nearby tree or telephone pole, see Figure 17-3. If live wires are present on the building or pole, care should be taken to keep your antenna and body well clear of these hazards. To mount the wire, standard commercial insulators may be used as well as homemade devices. Plastic pipe makes an excellent antenna insulator. Synthetic rope also works very well, and has the advantage of being secured simply by tying a knot. It is convenient to mount a pulley at some elevated point so the antenna wire may be pulled up to it using the rope which doubles as an insulator.

Grounding the APM

The APM requires a good ground circuit in order to operate efficiently. Usually a good ground can be established by connecting a wire to the water or gas pipes of a building. Solder or screw the wire to the APM ground terminal. In buildings with plastic pipes or joints, some other hookup must be used. A metal rod or pipe may be driven into the ground in a shady location where the earth usually is damper. Special copper coated steel rods are made for grounds, which have the advantage of good bonding to copper wire. A ground of this type is usually found within the electrical system of most buildings. Conduit is a convenient ground provided that the conduit is properly grounded. This may be checked with an ohmmeter by testing continuity between the conduit and system ground (ground rod). Just as with the antenna, keep the ground

Table 17-1

Capacitance code information

This table provides the value of alphanumeric coded ceramic, mylar and mica capacitors in general. They come in many sizes, shapes, values and ratings; many different manufacturers worldwide produce them and not all play by the same rules. Most capacitors actually have the numeric values stamped on them; however, some are color coded and some have alphanumeric codes. The capacitor's first and second significant number IDs are the first and second values, followed by the multiplier number code, followed by the percentage tolerance letter code. Usually the first two digits of the code represent the significant part of the value, while the third digit, called the multiplier, corresponds to the number of zeros to be added to the first two digits.

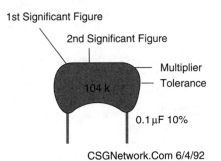

CSGNetwork.Com 6/4/92

Value	Type	Code	Value	Type	Code
1.5 pF	Ceramic		1,000 pF /.001 µF	Ceramic / Mylar	102
3.3 pF	Ceramic		1,500 pF /.0015 µF	Ceramic / Mylar	152
10 pF	Ceramic		2,000 pF /.002 µF	Ceramic / Mylar	202
15 pF	Ceramic		2,200 pF /.0022 µF	Ceramic / Mylar	222
20 pF	Ceramic		4,700 pF /.0047 µF	Ceramic / Mylar	472
30 pF	Ceramic		5,000 pF /.005 µF	Ceramic / Mylar	502
33 pF	Ceramic		5,600 pF /.0056 µF	Ceramic / Mylar	562
47 pF	Ceramic		6,800 pF /.0068 µF	Ceramic / Mylar	682
56 pF	Ceramic		.01	Ceramic / Mylar	103
68 pF	Ceramic		.015	Mylar	
75 pF	Ceramic		.02	Mylar	203
82 pF	Ceramic		.022	Mylar	223
91 pF	Ceramic		.033	Mylar	333
100 pF	Ceramic	101	.047	Mylar	473
120 pF	Ceramic	121	.05	Mylar	503
130 pF	Ceramic	131	.056	Mylar	563
150 pF	Ceramic	151	.068	Mylar	683
180 pF	Ceramic	181	.1	Mylar	104
220 pF	Ceramic	221	.2	Mylar	204
330 pF	Ceramic	331	.22	Mylar	224
470 pF	Ceramic	471	.33	Mylar	334
560 pF	Ceramic	561	.47	Mylar	474
680 pF	Ceramic	681	.56	Mylar	564
750 pF	Ceramic	751	1	Mylar	105
820 pF	Ceramic	821	2	Mylar	205

Figure 17-3 *Horizontal long-wire antenna*

wire away form the hot wires. The APM's ground wire may pass through conduit with other wires but should only be installed by qualified personnel. Grounding in extremely dry ground can be enhanced by burying some salt around the rod. The salt will increase the conductivity of the soil and also help retain water content around the ground rod. Some researchers have found that it is much better to get the largest surface area of metal in contact with the ground as possible, and have tried digging a hole and planting aluminum foil sheets in the ground.

Once you have tested the operation of the APM, you can disconnect the antenna and decide how you wish to enclose the APM circuit. You may want to experiment with the housing, since researchers have discovered that the germanium diodes are sensitive to sunlight and the voltage output is increased when the circuit is in sunlight. If you want to experiment along these lines, then you may wish to enclose the circuit in a clear plastic box. You can drill holes for two terminals posts for the plus and minus output connections, an RCA jack for the antenna input connection, a terminal for the ground connection, and your APM is complete and ready for operation.

The APM-II

The original APM is a great circuit for experimentation. It has been found that the germanium diodes are also sensitive to light, so if you house the APM in a clear plastic box and have it exposed to sunlight it will produce more output. Have fun experimenting with the APM receiver and you may find the perfect application for it, such as powering another receiver or night light, clock, emergency radio/lighting, etc.

Over time experimenters have improved upon the original APM design and claim increased output from the original design, so the reader is free to experiment and build both versions of the APM if desired. The second APM receiver, shown in Figure 17-4, illustrates the newer version of the APM with increased output voltage and current. The APM-II has quite a few more stages and has many more diodes and capacitors. You will note that the APM-II also uses germanium diodes which have more output than trying to use silicon diodes. Also note the higher capacitance values in the second design which were used to increase the output considerably. When constructing the APM-II be sure to pay careful attention to mounting the diode and capacitors; since there are many more components and they all have polarity marking on them, it is much more possible to make a mistake in orientation. Take your time when constructing the circuit. The APM-II also uses an input coil, which is said to improve the output. Some experimenters claim that the antenna can be eliminated and that the antenna connection can be connected to another separate ground, but this has not been verified. The APM-II is also sensitive to light which can increase the output if used outside. Some users claim that the power produced by the APM also seems to improve near bridges, ships and anything containing a great deal of metal.

Over time many scientists and researchers have reported bright flashes in the sky during strong earthquakes, as well as computer breakdowns during severe tremors. Scientists have long suspected that seismic activity is associated with a variety of electrical effects. Researchers have been taking a careful look at this link, with an eye toward using it to predict earthquakes.

Joseph Tate and William Daily of Lawrence Livermore National Laboratory in Livermore, CA. created a system of radio wave monitors distributed along California's San Andreas fault using the APM design. Tate and Daily recorded two kinds of changes in atmospheric radio waves prior to earthquakes that occurred between 1983 and 1986. Joe was able to witness and record the output of the APM during the earthquakes that followed, since he continuously monitored the output of the APM on a computer, so he could go back and look at the output on any given day.

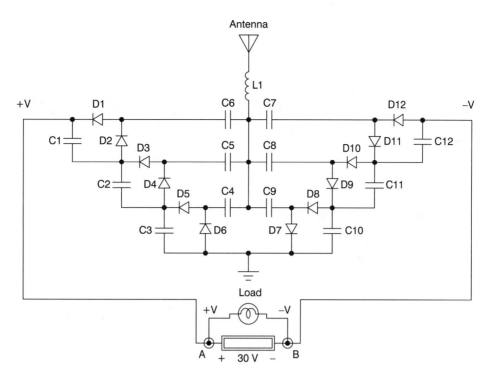

Figure 17-4 *Ambient power module II*

The most common change is a drop in the radio signals that normally pervade the air as a result of lightning and human sources such as car ignition systems and electric power grids. This reduction typically occurs one to six days before an earthquake and can last for many hours. For example, a magnitude 6.2 earthquake that shook Hollister, CA, in April 1984 was preceded six days earlier by a 24-hour drop in radio signals being monitored 30 miles from the quake's epicenter. Tate and Daily have found that the larger the earthquake, the longer the time between the radio wave depression and the quake.

Laboratory studies have shown that the electrical conductivity of rocks increases as they are stressed. Based on this and their electrical modeling of the ground, Tate and Daily think the increased conductivity of stressed rocks near the fault causes more radio waves to be absorbed by the ground rather than their traveling through the air. They also plan to test a possible link between radio wave drops and the emission of radon gas, which itself is thought to be a quake precursor. The radon may ionize the air, making it temporarily more absorptive than the detector antenna.

The researchers have also found, in addition to these drops, another pre-quake phenomenon in which short pulses of increased radio wave activity are emitted. For example, five days before the magnitude 6.5 earthquake hit palm Springs, CA, in July 1986, a station 15 miles from the epicenter detected a rise in radio signals. This sort of emission is consistent with laboratory work showing that cracking rocks release electromagnetic signals.

Tate says that in their first attempts at predicting earthquakes, they did not miss a single event, so he is optimistic about using this technique for short-term forecasting of San Andreas quakes.

If you live in an earthquake prone area such as California, you could use you APM not only for small low voltage emergency power/lighting system but you could connect the APM to an analog to digital converter card in a personal computer and record the fluctuations during earthquakes. The APM will provide many avenues for the builder and/or researcher. Disconnect your APM during thunderstorm activity to avoid damage to the APM and other household devices. Have fun and be safe.

Chapter 18

Earth Field Magnetometer Project

Parts list

Parts Bin

Earth Field Magnetometer

R1,R3,R4,R5 4.7 k ohm, $\frac{1}{4}$ w, 5% resistor

R2 100 ohms, $\frac{1}{4}$ w, 5% resistor

R6 10k ohm $\frac{1}{4}$ w, 5% resistor

C1,C2 15 pF, 35 volt capacitor

C3,C4 47 nF, 35 volt electrolytic capacitor

C5 10 µF, 35 volt capacitor

C6,C8 1 µF, 35 volt capacitor

C7 22 µF, 35 volt electrolytic capacitor

C9 10 µF, 35 volt Tantalum capacitor

D1,D2 1N914 silicon diode

D3 1N4001 silicon diode

X1 10 MHz crystal

U1 SCL006A - magnetometer output IC

U2 AD-557 - digital to analog converter chip

U3,U4 LM7805, 5 volt regulator

J1,J2 wire jumpers

S1 4-position rotary gain switch

S2 normally open pushbutton switch

S3 SPST toggle power switch

Logger HOBO H08-002-02 (ONSET)

Misc PC board, wire, IC sockets, chassis, etc.

Optional Field Magnetometer Power Supply

C1 470 µF, 50 volt electrolytic capacitor

C2 1 µf, 50 volt ceramic capacitor

C3 15 µf, 50 volt electrolytic capacitor

BR1 diode bridge rectifier 100v/2-amp

T1 24 volt center tapped transformer -.5A

U1 LM7809 IC regulator 9 volts

S1 SPST power switch

PL1 115 VAC power plug

MISC PC board, wire, hardware, etc.

1. Speake & Co. Ltd. (Elvicta Estate, Crickhowell, Powys. Wales, UK; http://www.speakesensors.com).

2. Fat Quarters Software
   (24774 Shoshonee
   Drive, Murieta, CA
   92562; Tel: 909-698-
   7950; Fax: 909-698-
   7913; Web: http://
   www.fatquarterssoftware
   .com)

3. ONSET Computers, PO
   Box 3450 Pocasset, MA
   02559-3450 (1-800-564-
   4377) http://www.
   onsetcomp.com

Magnetic fields are all around us. The Earth itself produces a magnetic field, which is why compasses work. Anytime an electrical current flows in a conductor, a magnetic field is generated. That is why transformers, inductors, and radio antennas work. There are several different devices that could be used to sense a magnetic field. One of the most familiar to electronics hobbyists is the Hall-effect device. However, in this chapter we'll take a look at a magnetic sensor that is as easy to use, but is more sensitive, more linear, and more temperature stable than typical Hall-effect devices. And just like Hall-effect devices, it can be used to make a variety of instruments, including magnetometers and gradiometers.

For those unfamiliar with them, magnetometers are used in a variety of applications in science and engineering. One high-tech magnetometer is used by Navy aircraft to locate submarines. Radio scientists use magnetometers to monitor solar activity. Earth scientists use magnetometers to study diurnal changes in the Earth's magnetic field. Archeologists use magnetometers to locate buried artifacts, while marine archeologists and treasure hunters use the devices to locate sunken wrecks and sunken treasure.

The Earth Field Magnetometer that we will be exploring uses a "flux-gate magnetic sensor." The device, in essence, is basically an over-driven magnetic-core transformer which the "transducible" event is the saturation of the magnetic material. These devices can be made very small and compact, yet will still provide reasonable accuracy.

In its most simple form the flux-gate magnetic sensor consists of a nickel-iron rod used as a core, wound,

with two coils. One coil is used as the excitation coil, while the other is used as the output or sensing coil. The excitation coil is driven with a square-wave signal with an amplitude high enough to saturate the core. The current in the output coil will increase in a linear manner so long as the core is not saturated. But when the saturation point is reached, the inductance of the coil drops and the current rises to a level limited only by the coil's other circuit resistances.

If the simple flux-gate sensor was in a magnetically pure environment, then the field produced by the excitation coil would be the end of the story. But there are magnetic fields all around us, and these either add to or subtract from the magnetic field in the core of the flux-gate sensor. Magnetic field lines along the axis of the core have the most effect on the total magnetic field inside the core. As a result of the external magnetic fields, the saturation condition occurs either earlier or later than would occur if we were only dealing with the magnetic field of interest. Whether the saturation occurs early or later depends on whether the external field opposes or reinforces the intended field.

Flux-gate magnetometer

The Earth Field Magnetometer project is illustrated in Figure 18-1 and the remotely located FM-3 flux-gate sensor is depicted in Figure 18-2.

The heart of the magnetometer is of course the flux-gate sensor. A compact low-cost line of flux-gate sensors are manufactured by Speake & Co. Ltd. and distributed in the United States by Fat Quarters Software. The FGM-3 sensor is the device used in

Figure 18-1 *Magnetometer*

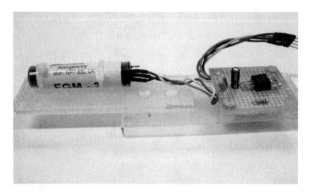

Figure 18-2 *FGM-3 Magnetometer sensor*

this project. It is a 62 mm-long by 16 mm-diameter (2.44 by 0.63 inch) device. Like all the devices in the line, it converts the magnetic field strength to a signal with a proportional frequency. The FGM-3 sensor has only three leads: Red: +5 VDC (power), Black: 0 volts (ground), and White: output signal (a square-wave whose frequency varies with the applied field). The output signal is a train of TTL compatible pulses with a period that ranges from 8 to 25 mS, or a frequency range of 40 to 125 kHz. The detection sensitivity of the FGM-3 device is +/– 0.5 oersted (+50 utesla). That range covers the Earth's magnetic field, making it possible to use the sensor in Earth-field magnetometers. Using two or three sensors in conjunction with each other provides functions such as compass orientation, three-dimensional orientation measurement systems, and three-dimensional gimbaled devices such as virtual-reality helmet display devices. It can also be used in applications such as ferrous metal detectors, underwater shipwreck finders, and in factories as conveyer-belt sensors or counters. There are a host of other applications where a small change in a magnetic field needs to be detected.

The FGM series also includes two other devices, the FGM-2 and the FGM-3h. The FGM-2 is an orthogonal sensor that has two FGM-1 devices on a circular platform at right angles to one another. That orthogonal arrangement permits easier implementation of orientation measurement, compass, and other applications. The FGM-3h is the same size and shape as the FGM-3, but is about 2.5 times more sensitive. Its output frequency changes approximately 2 to 3 Hz per gamma of field

change, with a dynamic range of +/– 0.15 oersted (about one-third the Earth's magnetic field strength).

The output signal in all the devices in the FGM series is +5 volt (TTL-compatible) pulse whose period is directly proportional to the applied magnetic field strength. This relationship makes the frequency of the output signal directly proportional to the magnetic field strength. The period varies typically from 8.5 μs to 25 μs or a frequency of about 120 kHz to 50 kHz. For the FGM-3 the linearity is about 5.5% over its +/– 0.5 oersted range.

The response pattern of the FGM-x series sensors is shown in Figure 18-3. It is a "figure-8" pattern that has major lobes (maxima) along the axis of the sensor, and nulls (minima) at right angles to the sensor axis. This pattern suggests that for any given situation there is a preferred direction for sensor alignment. The long axis of the sensor should be pointed towards the target source. When calibrating or aligning sensor circuits, it is common practice to align the sensor along the east-west direction in order to minimize the effects of the Earth's magnetic field.

The flux-gate sensor is coupled directly to a special interface chip, manufactured by Speake. This special SCL006A integrated circuit is shown in the main schematic diagram shown in Figure 18-4. It provides the circuitry needed to perform the Earth field magnetometry sensing. It integrates field fluctuations in one-second intervals, producing very sensitive output variations in response to small field variations. The magnetometer is of special interest to people doing radio-propagation studies, and those who want to monitor for solar flares. It also works as a laboratory magnetometer for various purposes. The SCL006A is housed in an 18-pin DIP IC package and is shown at U1. The FGM-3

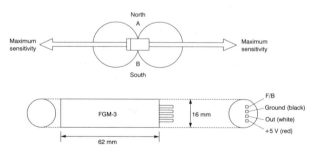

Figure 18-3 *FGM-3 Magnetometer sensor*

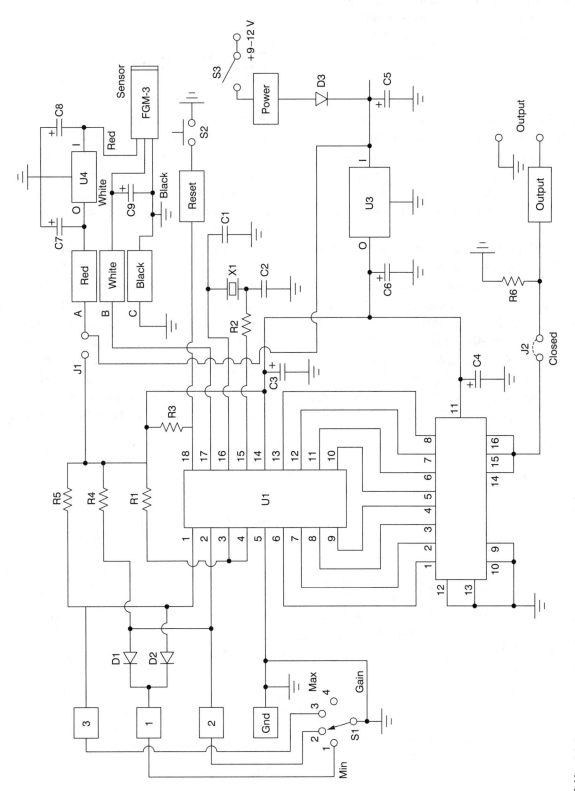

Figure 18-4 *Main magnetometer circuit*

flux-gate sensor is coupled to the input pin at pin 17 of U1. The reference oscillator of the SCL0006A is controlled by components X1, C1, C2 and R2. A sensitivity switch, S1, provides four positions, each with a different overall sensitivity range. Switch S2 is used to reset the SCL0006A, when applied for about two seconds.

The second op-amp at U2, is an Analog Devices AD557 D/A converter. The magnetometer circuit is designed so that it could be run from a 9 to 15 volt battery for use in the field. Power supplied from the battery is controlled by power switch S3 which supplies 9 volts to the regulator. For stationary research, the magnetometer should be powered from an 110 volt AC to 9–15 volt DC power supply. The regulator U3 drops the 9 volt input supply power down to 5-volts to power the magneto-meter circuit. The output signal is a DC voltage that can be monitored by a strip-chart or X-Y paper recorder, a voltmeter or the signal can be fed into a computer using an A-to-D converter. In the prototype we used the output to drive a portable mini data-logger such as the ONSET corporation HOBO H08-002-02. You will want to select the 2.5 volt model logger. You could also elect to use a digital multi-meter with RS-232 output

which can be fed to a computer's serial port for data-logging.

Construction

The Earth-field magnetometer project was built on two printed circuit boards. The main circuit board is shown in Figure 18-5, was constructed on a $3\frac{1}{2}'' \times 4\frac{1}{2}''$ PC board, while the remotely located FGM-3 sensor was paired with a second 5-volt regulator shown in Figure 18-6. Regulator U4 was mounted on a $2'' \times 3''$ PC board, which was attached to a sheet of flexible plastic using plastic screws. The FGM-3 sensor probe was then mounted on the $2'' \times 4\frac{1}{2}''$ plastic sheet. The plastic sheet was used to mount the probe, ahead of the regulator in order to give the flux-gate probe some isolation from surrounding metal. The flux-gate probe is mounted length-wise and parallel to the length of the PVC pipe. The FGM-3 sensor and remote regulator were then housed in a $2\frac{1}{2}''$ diameter $\times 12''$ long PVC pipe, which is separated by a length of three-conductor cable. Note that you could also elect to purchase a flux-magnetometer kit including PC board, the FGM-3

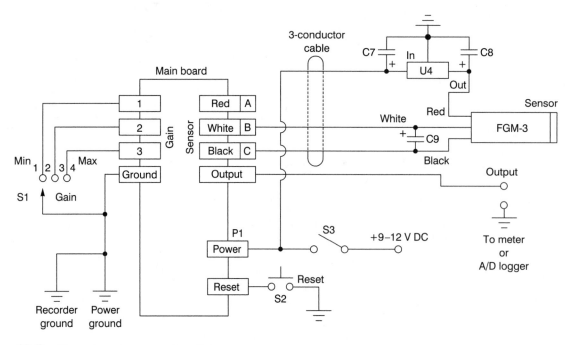

Figure 18-5 *Magnetometer connection diagram*

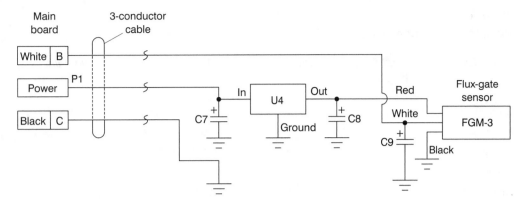

Figure 18-6 *Flux-gate sensor board*

sensor, ICs, and most other parts, but less the gain switch, can be from Fat Quarters Software for about $80.00.

Are you ready to begin constructing the flux-gate magnetometer? Before we begin you will want to secure a well lit and well ventilated work area, as well as a bench or large table for assembling your magnetometer project. Locate a small pencil-tipped 27 to 33 watt soldering iron, as well as a roll of 22 ga. 60/40 tin/lead rosin core solder. Go to your local Radio Shack store and purchase a small jar of "Tip Tinner"; this is a soldering iron tip cleaner/dresser compound, which works wonders. You will also need a few hand tools for the project. You will need a pair of small end-cutters, a pair of needle-nose pliers, a pair of tweezers, and a magnifying glass. A small flat-blade screwdriver and a small Phillips screwdriver should round out your tool requirements.

While you could build the magnetometer project on a perf-board or you could use point-point wiring it is recommended that you build the circuit on a printed circuit board for most reliable circuit operation and long life. Prepare your circuit board, and then we will be able to go ahead and build the magnetometer receiver project.

Locate all of the project diagram, such as the schematic, parts layout diagrams and the component pin-out sheets as well as the resistor and capacitor identification tables and we will be almost ready to begin. Place all of the magnetometer project components on the work bench in front of you, warm up your soldering iron and let's get going.

First, refer to the chart in Table 18-1, which illustrates the resistor color code. The resistor for this project will have three or four color bands on them. The first color band is closest to one end of the resistor body. The first color band represents the first digit of the resistor's value, the second color represents the second digit of the resistor value, the third color band represents the resistor's multiplier value, and the fourth band denotes the resistor's tolerance value. A silver fourth band denotes a 10% tolerance value, while a gold band represents a 5% tolerance value; no fourth color band denotes a resistor with 20% tolerance. Resistors R1, R3, R4 and R5, for example, are 4.7k ohm types. The first color band is yellow, which represents a four (4), while the second color band is violet or seven (7), and the third color band is red denoting a multiplier of (100), so you see (4) (7) times (100) equals 4700 or 4.7k ohms. Identify all of the 4.7k ohm resistors and place them on the circuit board at their correct location. Next, solder these resistors to the circuit board. Then follow-up by cutting the excess component leads flush to the edge of the circuit board with your end-cutters. Now locate R2 and R6 and install them in their proper locations, then solder them in place on the PC board. Remember to trim the excess resistor leads with your end-cutter.

Next we will move to identifying and installing the capacitors for the project. Capacitors are available in two major categories—electrolytic and non-electrolytic types. Electrolytic capacitors are generally both larger in size and larger in value than non-electrolytic types. Electrolytic capacitors have polarity and hence, they will have polarity marking on them to identify

Table 18-1

Resistor color code chart

Color Band	1st Digit	2nd Digit	Multiplier	Tolerance
Black	0	0	1	
Brown	1	1	10	1%
Red	2	2	100	2%
Orange	3	3	1,000 (K)	3%
Yellow	4	4	10,000	4%
Green	5	5	100,000	
Blue	6	6	1,000,000 (M)	
Violet	7	7	10,000,000	
Gray	8	8	100,000,000	
White	9	9	1,000,000,000	
Gold			0.1	5%
Silver			0.01	10%
No color				20%

their leads. You will usually find a black or white colored band or a plus (+) or minus (−) marking near one of the capacitor leads; this denotes the polarity. You will have to orient an electrolytic capacitor onto the circuit board with respect to the capacitor's polarity. When you mount electrolytic capacitors on the circuit board, you will have to refer to both the circuit schematic and the parts layout diagrams in order to install the capacitor properly. Failure to install electrolytic capacitors correctly may result in damage to the capacitor itself and perhaps damage to the circuit as well. Non-electrolytic capacitors will generally be smaller in value and physical size. Sometimes the non-electrolytic capacitors will have their value printed on them, and at other times the capacitor may be physically very small when a three-digit code is used to identify the capacitor. The chart in Table 18-2, helps to identify capacitors based on this code. Use the chart to help identify the capacitors. You will want to make sure that you install the correct value capacitor into the right location on the circuit board in order for the circuit to work correctly. For example, a capacitor marked with (102) denotes a value of .001 µF, while a capacitor with a marking of (103) denotes a capacitor value of .01 µF.

Look through the component pile and identify the capacitors and what type they are and then refer to the parts layout diagram and capacitor code chart before installing them on the circuit board. Go ahead and place a few of the capacitors at their respective locations on the PC board and solder them in place. Remember to trim the extra component lead length with your end-cutters.

The magnetometer main circuit board utilizes three silicon diodes. Diodes have polarity and they must be installed correctly for the circuit to operate correctly. Each diode will have either a black or white colored band at one end of the diode body. The colored band represents the diode's cathode lead, the opposite lead is the anode. Remember the diode's electronic symbol is an arrow or triangle pointing to a vertical line. The vertical line is the cathode of the diode. Refer to the schematic and parts layout diagram when mounting the diodes onto the circuit board. Place the diodes on the circuit board and solder them in place. Trim the excess leads as necessary.

The magnetometer project employs a 10 MHz crystal which is connected to pin 16 of U1. The crystal has no polarity so it can be placed in either direction on the board. Handle the crystal carefully so as to not break off the leads from the crystal body.

Before we begin placing the semiconductors on the PC board, you will want to refer to the semiconductor

Table 18-2

Capacitance code information

This table provides the value of alphanumeric coded ceramic, mylar and mica capacitors in general. They come in many sizes, shapes, values and ratings; many different manufacturers worldwide produce them and not all play by the same rules. Most capacitors actually have the numeric values stamped on them; however, some are color coded and some have alphanumeric codes. The capacitor's first and second significant number IDs are the first and second values, followed by the multiplier number code, followed by the percentage tolerance letter code. Usually the first two digits of the code represent the significant part of the value, while the third digit, called the multiplier, corresponds to the number of zeros to be added to the first two digits.

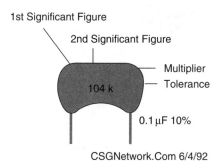

CSGNetwork.Com 6/4/92

Value	Type	Code	Value	Type	Code
1.5 pF	Ceramic		1,000 pF /.001 µF	Ceramic / Mylar	102
3.3 pF	Ceramic		1,500 pF /.0015 µF	Ceramic / Mylar	152
10 pF	Ceramic		2,000 pF /.002 µF	Ceramic / Mylar	202
15 pF	Ceramic		2,200 pF /.0022 µF	Ceramic / Mylar	222
20 pF	Ceramic		4,700 pF /.0047 µF	Ceramic / Mylar	472
30 pF	Ceramic		5,000 pF /.005 µF	Ceramic / Mylar	502
33 pF	Ceramic		5,600 pF /.0056 µF	Ceramic / Mylar	562
47 pF	Ceramic		6,800 pF /.0068 µF	Ceramic / Mylar	682
56 pF	Ceramic		.01	Ceramic / Mylar	103
68 pF	Ceramic		.015	Mylar	
75 pF	Ceramic		.02	Mylar	203
82 pF	Ceramic		.022	Mylar	223
91 pF	Ceramic		.033	Mylar	333
100 pF	Ceramic	101	.047	Mylar	473
120 pF	Ceramic	121	.05	Mylar	503
130 pF	Ceramic	131	.056	Mylar	563
150 pF	Ceramic	151	.068	Mylar	683
180 pF	Ceramic	181	.1	Mylar	104
220 pF	Ceramic	221	.2	Mylar	204
330 pF	Ceramic	331	.22	Mylar	224
470 pF	Ceramic	471	.33	Mylar	334
560 pF	Ceramic	561	.47	Mylar	474
680 pF	Ceramic	681	.56	Mylar	564
750 pF	Ceramic	751	1	Mylar	105
820 pF	Ceramic	821	2	Mylar	205

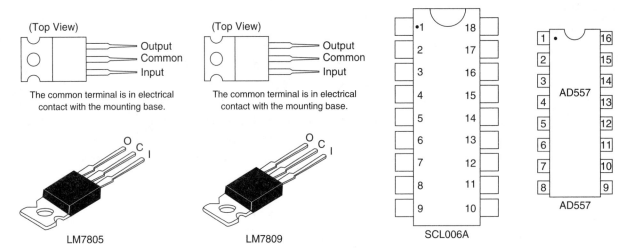

Figure 18-7 *Semiconductor pin-outs*

pin-out diagram shown in Figure 8-7, which will assist you in orienting the ICs. The magnetometer project utilizes two main integrated circuits which are rather expensive, so we recommend using IC sockets. As a good insurance policy locate two IC sockets and solder them in place for U1 and U2. If an IC fails at some later date 'down the road', you can simply unplug the defective one and replace it with a new one. Note that the IC sockets will usually have a notch or cut-out at one end of the plastic package. The lead just to the left of the notch is pin. You will have to make sure that pin one (1) of each socket is connected to the correct components; refer to the schematic and parts layout diagrams. Note pin one (1) of U1 goes to the anode of diodes D1 and D2, while pin one (1) of U2 is connected to pin six (6) of U1. When installing the ICs into their respective sockets, you will have to make sure pin one (1) of the IC is placed into pin one (1) of the socket to prevent damage to the integrated circuits.

It is a good time to take a short, well deserved rest and when we return we will examine the main circuit board for any possible "cold" solder joints or "short" circuits. Pick up the main circuit board with the foil side facing upwards towards you. Look over the main circuit board, we will be looking at the solder joints for possible "cold" solder joints. The solder joints should all look clean, bright and shiny, if you see any solder joints that look dull, dark or "blobby," you will need to un-solder the joint and remove the solder with a solder-sucker and then re-solder the connection, so that the new solder joint looks good. Next we will examine the

PC board for possible "short" circuits. Often "short" circuits can be caused by "stray" component leads or solder "blobs" which may have stuck to the underside of the PC board from solder residue. Look for solder blobs or solder bridges which may be "shorting" or bridging between circuit traces on the PC board. With the PC board free from any defect, we can continue on and prepare for connecting the sensor and power-up the circuit.

The main magnetometer board was mounted on standoffs in a metal chassis box, as seen in Figure 18-8. The on-off power switch and sensitivity rotary switch were mounted on the front panel of the chassis box, along with the meter binding posts. A 3-position screw terminal block or 3-circuit connector can be mounted on the side of the chassis box to allow the main circuit to connect to the remote sensor board.

Figure 18-8 *Magnetometer – inside view*

Now refer to the sensor power supply circuit diagram shown in Figure 18-6. This diagram illustrates the remote flux-gate sensor and power regulator PC board. The sensor and regulator were placed on a small circuit board, and in a water proof PVC cylinder so it can be buried below the frost-line at about 4 feet underground. The sensor should be placed well away from your house, buried level in the ground with the ends of the sensor facing East-West. The remote sensor is powered from the main circuit board. You will want to place the remote sensor and power supply in a non-metallic PVC case and waterproof the case so water cannot leak inside the case. The sensor will need to be facing the outside of the PVC cylinder away from the other electronics inside the package. Use no metal fasteners to secure the circuit to the PVC enclosure. You will have to make provisions to allow the three conductor to exit the opposite end of the PVC case away from the sensor end.

The main magnetometer circuit board was placed in a metal chassis which measures $6'' \times 3'' \times 2''$ as shown in the photo. The power switch, power LED, gain switch and reset switch along with the output binding posts, were mounted on the front of the chassis box. A four position terminal strip was mounted along the side of the chassis box to allow the remote sensor to be connected to the main circuit. A coaxial power jack was also mounted on the side of the chassis box to allow a 12 volt "wall wart" power supply to plug-in and power the circuit.

If you wish to construct your own power supply for the magnetometer project, refer to the diagram in Figure 18-9, which illustrates a 12 volt power supply from line-voltage. A 24 volt center tapped transformer feeds AC to the 4-diode bridge rectifier package at BR1. The DC output from the rectifier is filtered via C1 and C2 and then sent to a 9 volt IC regulator chip at U1. The output of the regulator chip is filtered at its output. This optional power supply will provide 9 volts which can be used to power the magnetometer circuit, if you choose not to use a 9 to 12 "wall wart" power supply.

The main circuit board was mounted on standoffs inside the bottom of the chassis box. Once you have mounted the main circuit board and the controls, you can locate a length of 20 to 24 gauge, three or four conductor cables and connect the sensor board to the main circuit board. Connect a digital voltmeter to the output terminal binding posts and set the meter to read 5 volts to 10 volts. Connect up your power supply to the main circuit board and we can now test the magnetometer to see if it is working. Turn the power switch to on, and set the meter range or gain switch to maximum sensitivity. With the sensor away from the main circuit board and chassis, bring a small magnet within rage of the sensor to say about 2 feet away. After about eight or ten seconds, you should see the effects of bringing the magnet near to the sensor. The reading will NOT be instantaneous, but you should see the meter read up to 2 volts or so. Move the magnet away and you should see the meter respond in 8 or 10 seconds. If all is well you will have a working portable flux-gate magnetometer, which can be used for many applications such as monitoring diurnal magnetic Earth variations or solar storms. You can even use the magnetometer as a UFO detector or driveway monitor.

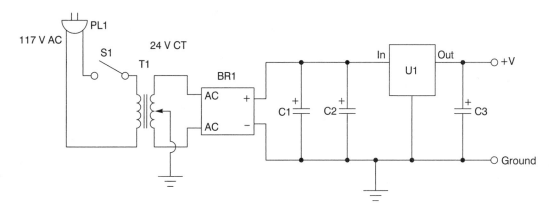

Figure 18-9 *Magnetometer power supply*

If your magnetometer did not "spring-to-life" then you will have to turn off the circuit and remove the power supply, and look the circuit over carefully to make sure you didn't install a component incorrectly. Grab your schematic, parts layout diagrams and charts and let's take a closer look at the circuit board. Remember we talked about certain components, like capacitors and diodes having polarity. Make sure that you installed the electrolytic capacitors and diodes with respect to their proper polarity; this is common cause for error. If need be, have another person, such as a parent or skilled technician or engineer or knowledgeable friend, look over the circuit for you. The same person looking over the same circuit numerous times often prevents you from finding your errors; you need a fresh look from another person to catch mistakes. Another cause for error is placing the integrated circuits in the wrong socket or in the wrong direction in the socket. After looking the board over carefully, you can re-apply the sensor and power supply connections and test the circuit once again, hopefully it will work fine this time around.

Once you know the magnetometer and sensor circuits are working properly, you can make provisions for mounting the sensor in a water-tight PVC case. The remote flux-gate sensor probe was mounted in a 2½″ × 12″ length of PVC pipe. Circuit board guides were epoxied to the inside of the PVC pipe to allow the circuit board to be slid into the PVC pipe. One 2½″ end cap was cleaned and cemented to the outside probe end of the PVC pipe. A ½″ hole was drilled and tapped in the center of the second PVC end cap. Try to use as little metal as possible when mounting the sensor in the PVC pipe, use brass or plastic screws where possible. A brass nipple fitting cemented in the tapped hole allows the three conductor cables to exit the probe assembly in the PVC pipe. Before cementing the final end cap to the PVC pipe you will want to test the Earth-Field magnetometer before burying the sensor probe assembly. Since the sensor is NOT an expensive temperature referenced or compensated type sensor it is subject to extreme temperature variations, so if you live in the northern climate zones, it will be necessary to mount the sensor underground, below the frost-line, to prevent temperature swings from affecting the reading of the meter. In most frost-prone areas, you will need to mount the water-proof sensor assembly about 4′ to 4½′ underground. The sensor should be mounted as far away from metal objects and traffic as possible to ensure the best results. The sensor package should be mounted so that it will face North-South as shown in Figure 18-3.

The output of the magnetometer can be connected to a logging digital multi-meter, so the imported data can be later analyzed by your personal computer. You could also use an inexpensive portable data-logger such as the HOBO series from Onset Computers, or you could use an analog to digital converter card in your PC to collect the data sample over time. If your meter is set to the most sensitive position, and everything is in working order, you should easily be able to monitor the Earth's diurnal variations twice a day. You should also be able to detect strong magnetic storms on the Sun, with your new magnetometer.

Chapter 19

Sudden Ionospheric Disturbance (SIDs) Receiver

Parts list

Parts Bin

SIDs Receiver

R1,R2,R3,R4 3.3k ohm
 ¼ w resistor

R5 10k ohm potentiometer
(chassis mount)

R6 1k ohm ¼ w resistor

R7 100k potentiometer
(PC mount)

R8 100 ohm ¼ w resistor

R9 10k ohm ¼ w resistor

R10 470k ohm ¼ w
resistor

R11 56k ohm ¼ w resistor

R12 22k ohm ¼ w resistor

R13 5k ohm
potentiometer
(PC mount)

C1 100pF, 35v ceramic
capacitor

C2 1500pF, 35v mylar
capacitor

C3 .001µF, 35v ceramic
capacitor

C4 1µF, 35v tantalum
capacitor

C5 10µF 35v tantalum
capacitor

C6,C7 10µF, 35v
electrolytic
capacitor

C8,C9 0.1µF, 35v
ceramic capacitor

C10,C11 1000µF, 35v
electrolytic capacitor

D1,D2 1N914 silicon
diode

D3,D4 1N34 germanium
diode

U1 LM353 op-amp
(National
Semiconductor)

U2 RC4136 op-amp (Texas
Instruments)

U3 79L09 - 5volt
regulator (- volts)

U4 78L09 - 5volt
regulator (+ volts)

T1 600 to 600 ohm
inter-stage/ matching
transformer

L1 Loop antenna
(see text)

S1 DPST toggle switch
(on-off)

F1 .50 amp fast-blow
fuse

BR1 bridge rectifier -
50v DC/2amp

PL1 117 VAC plug

Misc PC board, IC
sockets, wire, solder,
solder lugs, chassis
hardware, screws,
standoffs, etc.

A sudden ionospheric disturbance (SID) is an abnormally high plasma density in the ionosphere caused by an occasional sudden solar flare, which often interrupts or interferes with telecommunications systems. The SID results in a sudden increase in radio-wave absorption that is most severe in the upper medium-frequency (MF) and lower high-frequency (HF) ranges.

When a solar flare occurs on the Sun, a blast of ultraviolet and X-ray radiation hits the day-side of the Earth after 8 minutes. This high energy radiation is absorbed by atmospheric particles raising them to excited states and knocking electrons free in the process of photo-ionization. The low altitude ionospheric layers (D region and E region) immediately increase in density over the entire day-side. The Earth's ionosphere reacts to the intense X-ray and ultraviolet radiation released during a solar flare and often produces shortwave radio fadeout on the day-side of the Earth as the result of enhanced X-rays from a solar flare.

Shortwave radio waves (in the HF range) are absorbed by the increased particles in the low altitude ionosphere causing a complete blackout of radio communications. This is called a Short Wave Fadeout. These fadeouts last for a few minutes to a few hours and are most severe in the equatorial regions where the Sun is most directly overhead. The ionospheric disturbance enhances long wave (VLF) radio propagation. SIDs are observed and recorded by monitoring the signal strength of a distant VLF transmitter. As the fadeouts occur, reception of the station strength varies thus creating a fluctuating voltage output at the receiver which can be recorded and observed using a computer with a chart recorder program.

You can build and investigate the phenomena of SIDs with a special receiver and a low cost data-logger setup. You can not only observe when solar flares are occurring but you can collect and analyze the data and display it on your computer. The SIDs receiver is a great opportunity to observe first-hand when a solar event occurs, but you can use it to predict when radio blackouts will affect radio propagation, which is extremely useful for amateur radio operators.

VLF signal propagation

Why do VLF signals strengthen at night instead of getting weaker? If propagation is basically via the waveguide effect, why doesn't the signal drop DOWN at night when the waveguide disappears with the D layer? Is there some kind of reduced absorption at night? If so, where is it taking place and why? Also, what accounts for the big fluctuations in signal strength at night, apparently more or less at random?

The strength of the received signal depends on the effective reflection coefficient of the region from which the radio wave reflects in its multi-hop path between Earth and the ionosphere. In daytime the reflecting region is lower, the air density is higher, and the free electron density is controlled strongly by the solar radiation, etc. At night time the reflecting region is higher, the air density is lower, and the free electron density is controlled by variable ambient conditions as well as variable influences from electron "precipitation" from above, etc.

At noon the electron density is about 10 electrons/cm^3 at an altitude of 40 km, 100 electrons/cm^3 at an altitude of 60 km, 1000 electrons/cm^3 at an altitude of 80 km and 10,000 electrons/cm^3 at an altitude of 85 km.

At night these figures become 10 electrons electrons/cm^3 at 85 km, 100 electrons/cm^3 at 88 km and 1000 electrons/cm^3 at 95 km and then remains somewhat the same up to at least 140 km. At night the electron density in the lower part of the D region pretty much disappears. At 40 km the electron collision frequency is about 1,000,000,000 collisions per second whereas at 80 km the collision frequency drops to 1,000,000 collision per second. The reflection coefficient depends on (among other things) the number density of free electrons, the collision frequency, and the frequency of the radio signal. It is found by a mathematical integration throughout the entire D-region and of course the result depends on what time of the 24 hour day one performs the integration.

We can think of the E-Layer propagating the signal at night. Then the prominent sunrise pattern we see is a shift from E-Layer propagation back to D-Layer as the sun rises and forms the daytime D-Layer.

Figure 19-1 *Sudden ionospheric disturbance receiver*

The sunset pattern is the reverse. An interesting feature of waveguide mode propagation was that the signal was split into two components which can form an interference pattern.

You can build your own Sudden Ionospheric Disturbances (SIDs) receiver, shown in Figure 19-1, and begin your own investigation of solar flares and their effects on radio propagation. The SIDs receiver is a simple VLF receiver designed to be used with a loop antenna which can be placed either inside or outside. The receiver monitors the strong VLF signal from the US Navy NAA 24 kHz transmitter in Cutler Maine.

The Sudden Ionospheric Disturbance (SID) receiver is a modified Stokes' Gyrator tuned VLF receiver based on the circuit from Communications Quarterly Spring 1994 pp. 24–26. The main circuit diagram shown is Figure 19-2, it illustrates SIDs receiver which consists of two integrated circuits, a Texas Instruments RC4136 and a National LF353 operation amplifier. The SIDs receiver is fed from a loop antenna at L1. The antenna is coupled to the SIDs receiver via a miniature 600 to 600 ohm matching transformer at T1. The output from the secondary of T1 is fed to two protection diodes at D1 and D2 at the front end of the receiver. The output of the diode network is then coupled to a 100 pF capacitor at C1. Capacitor C2 is connected from the

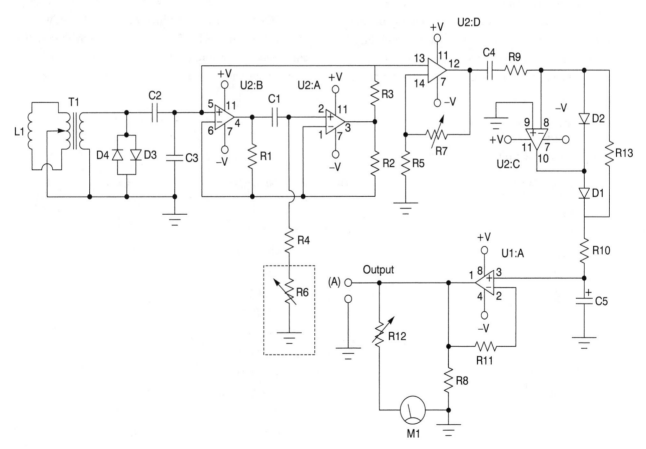

Figure 19-2 *Sudden ionospheric disturbance receiver*

output of C1 to ground at the input of the first op-amp stage U2:B. Integrated circuit sections U2:A and U2:B form an amplifier/filter and tuning. The "tuning" is controlled via R6, which is placed in a shielded enclosure to prevent circuit oscillation. The potentiometer is kept separate in a small shielded box formed by some scrap pieces of thin circuit board soldered together. The output from U2:A is next fed to IC:D and U2:C which forms an amplifier/integrator section. A final buffer amplifier section at U1 is used to drive the 0-1 mA meter at the output. The output of the SIDs receiver at (A) can be coupled a low cost data-logger in order to record signals over time.

The SIDs receiver was built on a circuit board circuit board, for most reliable operation, but other types of RF building techniques could be used. The circuit board is available from FAR Circuits—see Appendix. Before we begin building the circuit take a few minutes to locate a clean well lit work table or work bench. You will want to locate a small 27 to 33 watt pencil-tipped soldering iron, a roll of 60/40 tin/lead rosin core solder, and a small jar of "Tip Tinner," a soldering iron tip cleaner/dresser. "Tip Tinner" is available at a local Radio Shack store. You will also need some small hand tools, such as a small flat-blade screwdriver, a Phillips screwdriver, a pair of small needle-nose pliers, a pair of tweezers, a magnifying glass and a pair of end-cutters.

Once you have gathered all the tools, locate all the circuit diagrams, such as the schematics, the parts layout diagram, the resistor and capacitor code charts, etc. Place all the component parts, along with all your tools and diagrams, in front of you, warm up your solder iron and we will start building the SIDs receiver project.

First we will identify the resistors for the project. Refer to the chart shown in Table 19-1, which will help you to identify the resistor color codes. Resistors usually have three or four color bands on the body of the resistor which denotes the resistor value. The color bands begin at one end of the resistor body close to the edge. The first color band on resistor R1 is orange, this denotes the digit three (3). The second color band is also orange, which is the digit number three (3), and the third color band is the resistor's multiplier value which is red, and this suggests a multiplier value of 100. So the resistor value for R1 is (3) times (3) times (100) or 3.3k or 3300 ohms. The tolerance value of a resistor is noted in the four color band. A fourth color band which is silver is a 10% tolerance resistor, while a gold fourth band denotes a 5% tolerance value. If there is no fourth color band then the resistor will have a 20% tolerance value. Once you have identified resistor R1, you will note that resistors R1, R2, R3 and R4 all have the same color codes, therefore they all have the same value. Check your parts layout diagram and the schematic to

Table 19-1

Resistor color code chart

Color Band	1st Digit	2nd Digit	Multiplier	Tolerance
Black	0	0	1	
Brown	1	1	10	1%
Red	2	2	100	2%
Orange	3	3	1,000 (K)	3%
Yellow	4	4	10,000	4%
Green	5	5	100,000	
Blue	6	6	1,000,000 (M)	
Violet	7	7	10,000,000	
Gray	8	8	100,000,000	
White	9	9	1,000,000,000	
Gold			0.1	5%
Silver			0.01	10%
No color				20%

see where these resistors are to be placed on the circuit board. Once you verify where the resistors are placed, you can install all four resistors on the circuit board and solder them in place on the board. Now locate your end-cutter and trim the excess component leads. Cut the extra component lead lengths flush to the edge of the circuit board. Now go ahead and identify another group of three or four resistors and place them on the circuit board. Once you are sure of their respective locations you can solder the next grouping of resistors onto the circuit board, remember to trim the excess component leads from the board. Place all the remaining resistors on the circuit board and solder them all in place, trim the extra leads as necessary.

Now we will move on to installing the capacitors for the SIDs receiver project. Capacitors come in two major flavors, non-polarized and polarized types. First, we will talk about non-polarized capacitors. Usually non-polarized capacitors are smaller value types which could take form as ceramic disks, mica, tantalum or polyester types. Often the capacitors are small in size and frequently the capacitor values are printed as three-digit codes rather than their actual values, since this code system takes less space to print. We have included a chart in Table 19-2, which lists the actual capacitor values and their respective codes. For example, a capacitor with a value of .001 μF will have a code designation of 102, a .01 μF capacitor will have 103, and a 0.1 μF capacitor will have a code of 104 marked on it. The SIDs project has a number of non-polarized type capacitors which you will need to install. Take a look through the component pile and see if you can identify the small non-polar capacitors before attempting to install them on the PC board.

Polarized capacitors are called electrolytic capacitors, and they are often larger in physical size. Electrolytic capacitors will always have some sort of polarity marking on them. You will see either a black or white band or a plus (+) or minus (−) marking somewhere on the capacitor body. It is very important that you install these types of capacitors with respect to their polarity in order for the circuit to work properly and to not be damaged upon power-up. You will need to refer to the schematic and parts layout diagrams when installing electrolytic capacitors to make sure that you have installed them correctly. The plus (+) marking on the capacitor will point to the most positive portion of the circuit.

Let's go ahead and identify the non-polarized capacitors first. Check the capacitor code chart and make sure that you can identify each of the non-polarized capacitors. In groups of two or three capacitors install them on the PC board. When you have determined that you have installed the correct values into their respective location on the circuit board, you can solder the capacitors in place on the board. Remember to trim the excess capacitor leads from the circuit board. Once you have the first grouping of capacitors installed you can move on to installing another group of non-polarized capacitors, until you have installed all of the non-polarized capacitors.

Next, find the electrolytic capacitors from the parts pile and we will install them in their proper locations. Refer now to the schematic and parts layout diagrams so that you can orient the capacitors correctly when mounting them on the PC board. Install two or three electrolytic capacitors and then solder them in place on the PC board. Remember to cut the extra lead lengths flush to the PC board. Identify and install the remaining electrolytic capacitors on the board and solder them in place. Trim the component leads as necessary.

The SIDs receiver employs two silicon diodes at the "front-end" and two germanium diodes near the final output stage. Diodes have polarity so they must be installed in the proper orientation for the circuit to function properly. Diodes will generally have either a white or black band on the side of the diode's body. The colored band denotes the cathode of the diode. Remember that a symbol for a diode is a triangle pointing to a vertical line. The vertical line is the cathode of the diode. Make sure you install the proper diodes in their proper locations with respect to polarity: you will need to refer to the schematic and parts layout diagrams.

The SIDs receiver uses two integrated circuits at U1 and U2. Integrated circuit sockets are highly recommended as an insurance policy against any possible failure in the distant future. Before we go ahead and install the semiconductors, take a quick look at the semiconductor pin-out diagram shown in Figure 19-3. The pin-out diagram will assist you install the integrated circuits. IC sockets are inexpensive and will greatly be appreciated in the event of a circuit failure, since most people cannot un-solder a 14 or 16 pin integrated circuit

Table 19-2
Capacitance code information

This table provides the value of alphanumeric coded ceramic, mylar and mica capacitors in general. They come in many sizes, shapes, values and ratings; many different manufacturers worldwide produce them and not all play by the same rules. Most capacitors actually have the numeric values stamped on them; however, some are color coded and some have alphanumeric codes. The capacitor's first and second significant number IDs are the first and second values, followed by the multiplier number code, followed by the percentage tolerance letter code. Usually the first two digits of the code represent the significant part of the value, while the third digit, called the multiplier, corresponds to the number of zeros to be added to the first two digits.

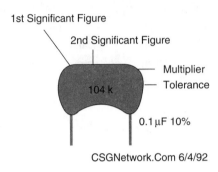

CSGNetwork.Com 6/4/92

Value	Type	Code	Value	Type	Code
1.5 pF	Ceramic		1,000 pF /.001 µF	Ceramic / Mylar	102
3.3 pF	Ceramic		1,500 pF /.0015 µF	Ceramic / Mylar	152
10 pF	Ceramic		2,000 pF /.002 µF	Ceramic / Mylar	202
15 pF	Ceramic		2,200 pF /.0022 µF	Ceramic / Mylar	222
20 pF	Ceramic		4,700 pF /.0047 µF	Ceramic / Mylar	472
30 pF	Ceramic		5,000 pF /.005 µF	Ceramic / Mylar	502
33 pF	Ceramic		5,600 pF /.0056 µF	Ceramic / Mylar	562
47 pF	Ceramic		6,800 pF /.0068 µF	Ceramic / Mylar	682
56 pF	Ceramic		.01	Ceramic / Mylar	103
68 pF	Ceramic		.015	Mylar	
75 pF	Ceramic		.02	Mylar	203
82 pF	Ceramic		.022	Mylar	223
91 pF	Ceramic		.033	Mylar	333
100 pF	Ceramic	101	.047	Mylar	473
120 pF	Ceramic	121	.05	Mylar	503
130 pF	Ceramic	131	.056	Mylar	563
150 pF	Ceramic	151	.068	Mylar	683
180 pF	Ceramic	181	.1	Mylar	104
220 pF	Ceramic	221	.2	Mylar	204
330 pF	Ceramic	331	.22	Mylar	224
470 pF	Ceramic	471	.33	Mylar	334
560 pF	Ceramic	561	.47	Mylar	474
680 pF	Ceramic	681	.56	Mylar	564
750 pF	Ceramic	751	1	Mylar	105
820 pF	Ceramic	821	2	Mylar	205

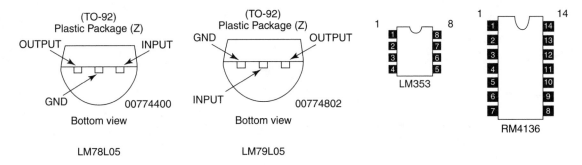

Figure 19-3 *Semiconductor pin-outs*

without damage to the circuit board. Integrated circuits must be installed correctly for the circuit to work properly. Looking at the IC, you will notice a small notch or cut-out or small indented circle at one end of the IC body. Pin one (1) of the IC will be just to the left of this cut-out or notch. The IC socket will often appear the same way with a notch or cut-out with pin one (1) to the left of the notch or cut-out. You must orient pin of the IC with pin 1 of the IC socket, and you must them be sure to orient pin of the IC socket to match up with the proper components on the circuit board. For example, pin one (1) of U1 is connected to the output of the circuit at the junction of R8 and R11. Note that pin 8 is the plus power supply input, while pin 4 is the ground on U1. Pin one (1) of U2 corresponds to the minus (–) INPUT #1 at U2:A, which is connected to the junction of R1 and R2.

A small metal compartment should be placed around the main tuning potentiometer R6 in order to shielded this control. This can be done with the use of some scrap circuit board material which can be soldered together to form a small box. You will have to use a soldering gun or a higher temperature soldering iron, in order to solder large areas of circuit board material.

The output of the SIDs receiver is fed to the 0-1 mA analog meter at M1. The output of the SIDs monitor can also be fed to the input of an analog to digital converter place in a personal computer, for recording over long lengths of time.

Since the SIDs receiver utilizes integrated circuits which require the use of a dual power supply providing both a plus and minus voltage to the circuit, and since the circuit is meant to be left on for long periods

of time the use of a dual voltage AC power supply is recommended. The diagram shown in Figure 19-4 illustrates a simple dual plus and minus voltage power supply which can be used to power the SIDs receiver. A 9 volt 500 mA center tapped transformer was used to drive a bridge rectifier, which provides both a plus and minus voltage that is sent to a 9 volt plus regulator. An LM78L09 regulator and the minus leg of the bridge is sent to an LM79L09 minus 9 volt regulator. The power supply could be built on perf-board or on a printed circuit board if desired. The power supply and SIDs receiver could be mounted alongside one another in the same chassis.

Take a short break and when we return we will inspect the circuit board for any possible "cold" solder joints or "short" circuits. Pick up the circuit board so that the foil side of the PC board faces upwards toward you. Look over the foil side of the board very carefully. The solder joints should all look clean, bright and shiny. If you see any solder joints that look dull, dark, dirty or "blobby" then you should remove the solder and re-solder the joint so that it looks clean, bright and shiny. Next we are going to inspect the PC board for any possible "short" circuits. Sometimes "cut" component leads will "stick" to the board from solder residue left on the board. These "stray" wires can often cause a "short" between the solder traces. Also look closely for any solder "blob" or "solder balls," which may have stuck to the underside of the PC board.

The SIDs receiver and power supply can both be installed in a metal chassis box enclosure. The prototype receiver was mounted in a sloping cabinet as in Figure 19-1. The power supply was mounted alongside

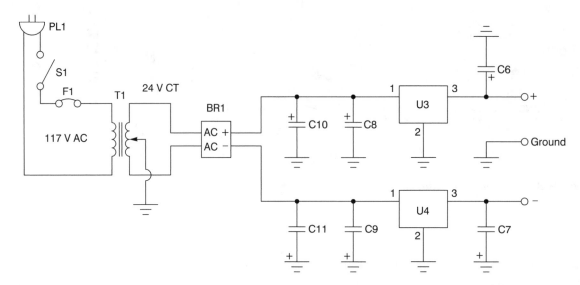

Figure 19-4 *Dual voltage power supply*

the SIDs receiver. The tuning control as well as the power on-off switch and meter were all mounted on the front panel of the enclosure. The power cord was led out of the rear panel through a stain-relief. A power fuse was installed inside a chassis mounted fuse holder on the rear of the panel. A dual binding post/banana chassis jack was mounted on the top of the chassis to connect the antenna loop to the SIDs receiver. An RCA jack was placed on the rear of the chassis to allow the receiver to connect to an analog to digital converter.

The SIDs receiver requires a rather large loop antenna, which can be placed inside your attic or placed outside away from large metal structures or aluminum siding, in a sheltered area. The coil at L1 is a diamond shaped loop antenna as shown in Figure 19-5. The loop consisting of 50 turns of solid #24 enameled or plastic coated wire, on a PVC pipe square or diamond form. The loop has an enclosed area of about 9 square feet. Locate four 2½″ diameter × 24″ long pieces of PVC pipe and form a square or diamond shape coil form as shown. A PVC elbow was placed on top and both sides of the loop form as shown, while a PVC "T" was used at the bottom of the loop form to allow winding of the coil around the four protrusions at each leg of the loop. The four hole center hub holds the loop form together. The main vertical PVC pipe was secured into a flange which was fastened to a wooden block at the base of the loop antenna. The lead-in from the

loop antenna should form a loose twisted pair. This balanced lead-in is routed as far away from metal as possible to the 600 to 600 ohm matching transformer at T1. This transformer arrangement reduces 60 Hz hum interference from entering the SIDs receiver.

The output of the SIDs receiver can be coupled to a solid state data-logger from the output of the receiver at point (A) at the receiver output terminals. In order to save and later view and correlate your recorded data you will need to acquire some form of data-logger. There are three good options for saving your data. The first option is the Onset Computer HOBO series of data-loggers. Onset offers a number of different models from 8-bit to 12-bit models which are reasonably priced from around $60.00. The data-loggers are powered by a small button battery which will last for a long time. Onset also offers low cost software for up/downloading information to the data-logger. Starting times and dates can be preset as well as voltage parameters and timing between samples. These are a great alternative. Another data-logger option is the new introductory "starter package" from Dataq DI-194RS 10-bit resolution PC data-logger kit for $24.95. This data-logger option is a real bargain, providing both hardware and software and it can get you started recording data in just a short time. The company offers many other data recorder options as well. Check their website for details http://www.dataq.com.

Figure 19-5 *Loop antenna*

In normal operation, the SIDs receiver would be permanently tuned to a powerful VLF station such as (NAA), the powerful US Navy 24 kHz VLF transmitter station in the state of Maine, and your data-logger or A/D converter would constantly monitor the signal from NAA to record propagation changes on your personal computer for study. Once the receiver has been completed, you can attach the antenna, power supply and data recorder. Connect the output of the data recorder to your personal computer and you are ready to begin recording. Now turn on the receiver and you should see a signal on your computer chart recorder display. If you remove the antenna from the circuit the output should go to zero or base-line. If your SIDs receiver does not appear to work, then you will need to disconnect the power supply and data recorder and antenna and inspect the circuit board for possible errors. The most common cause for circuit failure after building a new circuit is the improper installation of components, such as electrolytic capacitors, diodes and semiconductors such as transistors or integrated circuits. You may also want to check the placement of resistors on the circuit board. Ask a knowledgeable electronics enthusiast to help you inspect the circuit board for errors. Having completed the inspection, and hopefully an error was found, you can reconnect the antenna and power supply and then re-test the receiver.

SIDs research opportunities

You can join the foremost group involved with SIDs research. The AAVSO or American Association of Variable Start Observers SID Program consists of solar observers who monitor very low frequency (VLF) radio stations for sudden enhancements of their signals. Earth's ionosphere reacts to the intense X-ray and ultraviolet radiation released during a solar flare. The ionospheric disturbance enhances VLF radio propagation. By monitoring the signal strength of a distant VLF transmitter, sudden ionospheric disturbances (SIDs) are recorded and indicate a recent solar flare event. All SID monitoring stations are home built by the observers. The SID station operates unattended until the end of each month. Recordings are then analyzed for the beginning, end, and duration of SID events. A simple A/D converter design for specific use with the VLF receivers is available by contacting the chairman. SID observers submit strip-charts or computer plots to the SID coordinator for visual inspection at the end of each month. Many observers analyze their own stripcharts and computer plots. Analyzed results are submitted via e-mail to the SID Analyst for correlation with other observers' results. The final SID report combines individual observers' reports with the SID Coordinator visual analysis. SID event results are sent monthly to the National Geophysical Data Center (NGDC) for publication in the Solar-Geophysical Data Report where they are accessed by researchers worldwide. The reduced SIDs data and particularly interesting plots are reproduced in the monthly AAVSO Solar Bulletin mailed to all contributing members.

Chapter 20

Aurora Monitor Project

Parts list

Parts Bin

Aurora Monitor:
Sensor Head Unit

R1,R2   1.5 megohm $\frac{1}{4}$ watt,
  5% resistor

R3   5.6 megohm, $\frac{1}{4}$ watt,
  5% resistor

R4   100 k potentiometer
  PCB trimmer

R5   10 megohm $\frac{1}{4}$ watt,
  5% resistor

C1,C2,C3   .1 µF, 35 volt
  disk capacitor

D1   1N914 silicon
  diode

LM4250   low noise/high
  gain op-amp

L1   sensing coil, see
  text

B1,B2   1.5 volt "C" cell
  battery

J1   RCA input jack

Misc   PC board, coil
  form, wire, coax, etc.

Control/Display Unit

R6   18 k ohms, $\frac{1}{4}$ watt,
  5% resistor

R7   12 k ohms, $\frac{1}{4}$ watt,
  5% resistor

R8,R9,R12,R20
  10 k ohms, $\frac{1}{4}$ watt,
  5% resistor

R10   30 k ohms, $\frac{1}{4}$ watt,
  5% resistor

R11   50 k ohms, panel-
  mount potentiometer

R13   200 k ohms, $\frac{1}{4}$ watt,
  5% resistor

R14   3 megohm, $\frac{1}{4}$ watt,
  5% resistor

R15   1 megohm, PCB
  trimmer potentiometer

R16   1 megohm, $\frac{1}{4}$ watt,
  5% resistor

R17   1 megohm,
  panel-mount
  potentiometer

R18,R19   5 k ohm
  panel-mount
  potentiometer

R21   1 k ohm, $\frac{1}{4}$ watt,
  5% resistor

R22   100 ohms, $\frac{1}{2}$ W,
  5% resistor

R23   4.7 k ohms, $\frac{1}{4}$ watt,
  5% resistor

C4   0.005 µF, 25 volt
  ceramic disk capacitor

C5,C7,C9,C10   0.1 µF,
  25 volt ceramic disk
  capacitor

C6   0.02 µF, 25 volt
  ceramic disk capacitor

C8   4.7 µF, 25 volt
  aluminum electrolytic
  capacitor

C11   .010 µF, 25 volt
  disk capacitor
  (optional)

C12   1 µF, 35 volt

aluminum electrolytic
capacitor

D1,D2,D3,D4 1N914
silicon diodes

Q1 2N2907 PNP
transistor

Q2,Q3 2N394, NPN
transistor

U1 LM4250 low noise
op-amp (National
Semiconductor),

U2 LTC1062 (Linear
Technology)

U3 LM201A dual op-amp
(National
Semiconductor)

U4 LM747 op-amp
(National
Semiconductor)

S1 DPDT toggle power
switch

S2,S4 SPST toggle
switches

S3 3-position rotary
switch, break before
make

M1 100-micro-ampere
moving-coil panel-
mounted meter

SPKR1 8 ohm speaker,
2" dia.

B1,B2,B3,B4 1.5v (AA)
or (C) cells (+) plus
voltage supply

B5,B6,B7,B8 1.5v (AA)
or (C) cells (-)
minus voltage supply

J2,J3 RCA jack

Misc PC board, chassis,
wire, coax, hardware,
standoffs, RTV
compound,
washers, etc.

The Aurora Borealis is one of nature's most spectacular nighttime displays. Shimmering curtains of green, white, and even red light dance in the northern skies. Visible effects of charged particles from the Sun raining down on the Earth's ionosphere, northern lights or auroras, are visible in the northern night sky during high sunspot activity. The Aurora Australis, the southern hemisphere's counterpart of the Aurora Borealis, can be seen at night by looking toward the south pole.

These displays of undulating light are formed when flares from the Sun's surface (sunspots) launch showers of high-energy ionized particles and X-rays into space. Mostly electrons, the showers stream out from the sun and are attracted by the Earth's magnetosphere, an invisible magnetic field around the Earth.

Shaped like a pumpkin, the magnetosphere terminates at both magnetic poles but is many miles thick above the equator. Dimples at both poles form "sinks" that funnel the particles toward the poles where they ionize the gas in the ionosphere. Those collisions induce the gases to emit their characteristic light wavelengths—as in neon signs and fluorescent lamps.

The charged particle bombardment of the magnetosphere initially compresses it, temporarily increasing the strength of the Earth's geomagnetic field. The aurora monitor described here is sensitive enough to detect changes in the field caused by those "magnetic storms." Thus it can indirectly sense sunspots and predict the presence of auroras in the night sky.

The monitor also senses changes or anomalies in the magnetic field caused by large metal objects such as cars or trucks moving near the monitor. This permits it to act as an intrusion detection monitor able to detect the approach of vehicles at night in restricted areas. The monitor can also detect the presence of permanent magnets (such as those in speakers), and stray fields from AC-power lines.

Early warning of auroras will both permit you to observe them in the night sky or use them for boosting the range of your amateur radio transmissions. Auroras and their accompanying magnetic storms generally block or scramble the lower radio frequencies, but the higher frequencies can overcome this interference. Radio amateurs aim their antennas north during those storms, thus taking advantage of the phenomena to reach other hams on the opposite side of the Earth that

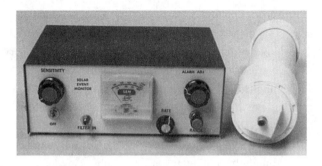

Figure 20-1   *Aurora monitor*

of one meter, which corresponds to one milligauss. (The Earth's magnetic field is about 0.5 ga.)

The control/display unit contains an active filter, additional amplification circuitry, a moving-coil ammeter, and an audio oscillator with speaker. Figure 20-1 shows the front-panel controls and indicators of the monitor: moving-coil ammeter, SENSITIVITY, RATE, and ALARM ADJUST potentiometers as well as power (OFF), filter-in and RANGE switches.

could not be contacted during periods of low sunspot activity.

In addition to scrambling low-frequency radio communications, the magnetic storms caused by auroras can induce large currents in power transmission lines. Those currents can cause overload, plunging large regions of the country into darkness. Auroras and related magnetic storms are quite common during the decreasing parts of the 11-year sunspot cycle such as the period we are now in.

The Aurora Monitor consists of two main components, a sensing head and the control/display unit which are connected by a coaxial cable. The sensing head contains a sensing coil, a DC nano amplifier capable of current amplification of 500, and a separate power pack, all enclosed in the tubular case shown on the right side of Figure 20-1. Figure 20-2 illustrates the interior of the Aurora Monitor prototype. The Aurora Monitor is sensitive to a pulse of one ampere at distance

The sensor head unit

The sensor head schematic circuit is illustrated in Figure 20-3. Gain is provided by U1, a National Semiconductor LM4250 programmable operational amplifier. It is protected from over voltage and transients by diode D1 and D2, and its overall gain is set by resistors R1 and R2. The output of U1 is driven to zero or balanced by network R3 and R4. Its output should remain at zero as long as no changes occur in the ambient magnetic field.

Bypass capacitors C2 and C3 are placed across the positive and negative power supply. Power for the sensing amplifier is obtained from two "C" cells. The sensor circuit draws very little current, so it can be left on at all times, thus eliminating the power switch.

The control/display unit

The control/display circuit schematic is depicted in the schematic in Figure 20-4. It provides an additional gain of 200 over that of the sensor circuit. The control/display circuit includes an adjustable low-pass notch filter, U2, a Linear Technology LTC 1062. By adjusting the clock frequency of the filter with resistor RI1 and the capacitors C4, C5 and C6, the filter cancels interference frequencies and noise in the 2 Hz to 10 kHz band. The notch filter can also screen out 60 Hz noise. Switch S2 inserts or removes the filter. U3, an LM201A general purpose op-amp, filters out the clock noise generated within the filter chip. The output of U3 is fed into the non-inverting input of U4-A, half of a dual 747 general purpose op-amp. The overall gain of U4-A is adjusted by resistor R16. Trimmer potentiometer R15

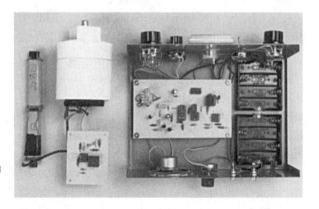

Figure 20-2   *Aurora monitor console and sensor*

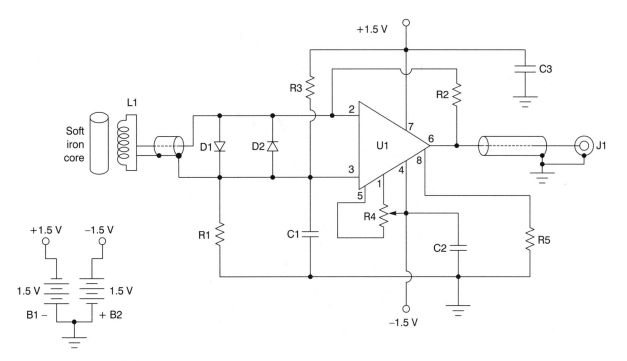

Figure 20-3 *Aurora monitor sensor head*

adjusts the offset or balance. PULSE SHAPE potentiometer R17 and capacitor C5 form an optional pulse-shaping network for coupling the Aurora Monitor to a chart recorder or an analog-to-digital conversion board of a personal computer.

The output of U4-A is coupled to a voltage follower at U4-B. The output of U4-B is divided into two channels. One channel is fed through 5000-ohm SENSITIVITY potentiometer R18, which adjusts the output level of the signal fed to the 100-microampere panel meter MI. Movement of the meter's needle shows changes in the local magnetic field. Potentiometer R18 also adjusts the output signal that can be fed to a chart recorder for data-logging.

The other channel is fed through ALARM ADJUST 5000-ohm potentiometer, R19, which sets the threshold or set-point for the reflex oscillator circuit that follows it. The oscillator consists of transistors Q1, Q2, and Q3 and associated components. Speaker SPKR1 gives an audible indication of changes in the local magnetic field. The network of diode D3 and aluminum electrolytic capacitor C8 performs additional filtering for the input signal to the reflex oscillator section.

Transistor Q1 controls the audible alarm by clamping the negative voltage returning through the ground path. When a magnetic event occurs, the speaker emits an audible alarm, and the meter gives a visual indication of a changing magnetic field. The adjustment of ALARM ADJUST potentiometer R24 can remove distortion from the sound of the speaker.

The author's prototype control/display unit is powered by eight AA cells: four cells provide positive voltage and four cells provide negative voltage. As an alternative, the monitor can be powered by rechargeable nickel-cadmium cells.

Construction

In order to begin constructing the Aurora Monitor, you will need to secure a clean will lit work table or work bench so you can spread out all the tools, project components, charts and diagrams. You will need to locate a small 27 to 33 watt pencil-tipped soldering iron, a small roll of 60/40 rosin core solder, a small jar of "Tip Tinner," soldering iron tip cleaner/dresser, from your local Radio Shack store. A few small hand tools

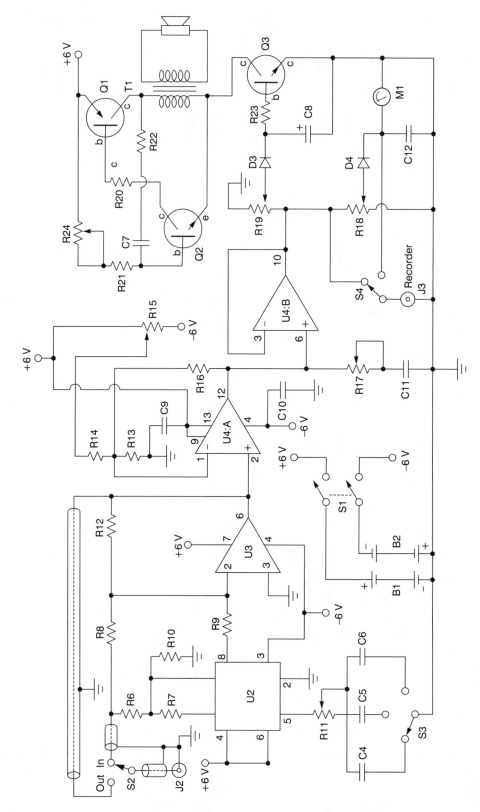

Figure 20-4 *Aurora monitor display/control unit*

are also required to construct the project. You should locate a small pair or end-cutters, a pair of needle-nose pliers, a pair of tweezers, a magnifying glass a small flat-blade screwdriver and a small Phillips screwdriver.

The Aurora Monitor project was constructed on two circuit boards, one for the sensor head unit and a second PC board for the control display unit. We will begin constructing the sensor head unit first, see the schematic in Figure 20-3. The sensor head board measured 2 1/16 by 2 9/16, it was housed along with the sensor head coil in a suitable PVC pipe housing.

Place the project components in front of you on the table, locate the resistor and capacitor identifier charts in Tables 20-1 and 20-2, heat up your soldering iron and we will begin. First refer to the resistor identifier chart, it illustrates the resistor color codes and their representative values. Most resistors will have three or four color codes on the body of the resistor. The color codes begin at one end of the resistor's body. The first color band denotes the first digit of the resistor value, while the second color band represents the second digit of the resistor code. The third color band on the resistor denotes the multiplier value of the resistor, and the fourth color code represents the tolerance of the resistor. A silver band notes a 10% tolerance value, while a gold band denotes a 5% resistor, and the absence of a fourth band implies a 20% tolerance for the resistor. For example, resistor R1 has a 1.5 megohm value. The first

color band would be brown or one (1), the second color band would be green or five (5), and the third color or multiplier would be green or (100,000). Go ahead and locate resistors R1 and R2 which are used in the sensor head unit and place them on the sensor head PC board in their respective locations. Solder them in place and then trim the excess component leads with your end-cutter. Cut the excess leads flush to the edge of the circuit board. Next locate the remaining resistors on the sensor head unit and populate the board. Solder the resistors to the board and trim the excess leads.

Next we are going to install the capacitors on the sensor head unit of the Aurora monitor. Capacitors are listed as electrolytic or non-electrolytic types. Non-electrolytic types are usually small capacitors, often they are so small that their actual value is not printed on the body of the capacitor but a three-digit code is used, see Table 20-2. For example, a .01 µF capacitor would be marked (103) while a .1 µF capacitor would be marked (104). Locate capacitors C1, C2 and C3 which are .1 µf values. Place them on the sensor head PC board and solder them in place on the board, remember to trim the excess component leads with your end-cutters. There are no electrolytic type capacitors on the sensor head unit but there are some on the control/display PC board. Electrolytic capacitors are usually larger in size and value, and usually they will have a polarity marking of some kind on the body

Table 20-1
Resistor color code chart

Color Band	1st Digit	2nd Digit	Multiplier	Tolerance
Black	0	0	1	
Brown	1	1	10	1%
Red	2	2	100	2%
Orange	3	3	1,000 (K)	3%
Yellow	4	4	10,000	4%
Green	5	5	100,000	
Blue	6	6	1,000,000 (M)	
Violet	7	7	10,000,000	
Gray	8	8	100,000,000	
White	9	9	1,000,000,000	
Gold			0.1	5%
Silver			0.01	10%
No color				20%

Table 20-2
Capacitor code identification information

This table is designed to provide the value of alphanumeric coded ceramic, mylar and mica capacitors in general. They come in many sizes, shapes, values and ratings; many different manufacturers worldwide produce them and not all play by the same rules. Most capacitors actually have the numeric values stamped on them; however, some are color coded and some have alphanumeric codes. The capacitor's first and second significant number IDs are the first and second values, followed by the multiplier number code, followed by the percentage tolerance letter code. Usually the first two digits of the code represent the significant part of the value, while the third digit, called the multiplier, corresponds to the number of zeros to be added to the first two digits.

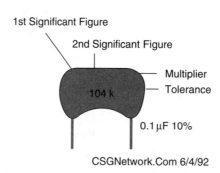

Value	Type	Code	Value	Type	Code
1.5 pF	Ceramic		1,000 pF /.001 µF	Ceramic / Mylar	102
3.3 pF	Ceramic		1,500 pF /.0015 µF	Ceramic / Mylar	152
10 pF	Ceramic		2,000 pF /.002 µF	Ceramic / Mylar	202
15 pF	Ceramic		2,200 pF /.0022 µF	Ceramic / Mylar	222
20 pF	Ceramic		4,700 pF /.0047 µF	Ceramic / Mylar	472
30 pF	Ceramic		5,000 pF /.005 µF	Ceramic / Mylar	502
33 pF	Ceramic		5,600 pF /.0056 µF	Ceramic / Mylar	562
47 pF	Ceramic		6,800 pF /.0068 µF	Ceramic / Mylar	682
56 pF	Ceramic		.01	Ceramic / Mylar	103
68 pF	Ceramic		.015	Mylar	
75 pF	Ceramic		.02	Mylar	203
82 pF	Ceramic		.022	Mylar	223
91 pF	Ceramic		.033	Mylar	333
100 pF	Ceramic	101	.047	Mylar	473
120 pF	Ceramic	121	.05	Mylar	503
130 pF	Ceramic	131	.056	Mylar	563
150 pF	Ceramic	151	.068	Mylar	683
180 pF	Ceramic	181	.1	Mylar	104
220 pF	Ceramic	221	.2	Mylar	204
330 pF	Ceramic	331	.22	Mylar	224
470 pF	Ceramic	471	.33	Mylar	334
560 pF	Ceramic	561	.47	Mylar	474
680 pF	Ceramic	681	.56	Mylar	564
750 pF	Ceramic	751	1	Mylar	105
820 pF	Ceramic	821	2	Mylar	205

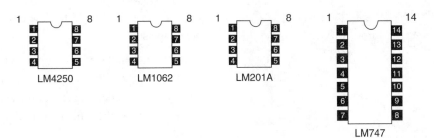

Figure 20-5 *Semiconductor pin-outs*

of the capacitor. You will find either a plus (+) or minus (−) marking near one of the leads on the capacitor body. Both the sensor head and control/display board utilize silicon diodes. Diodes also have polarity, so they must be installed correctly in order for the circuit to work properly. Look closely at a diode and you will see either a white or black color band near one end of the diode body. The colored band indicates the cathode end of the diode. Go ahead and install D1 on the sensor head board.

The sensor head and the display boards both have integrated circuits on them, so before we begin installing the ICs, take a quick look at the pin-out diagram depicted in Figure 20-5, which will help you orient the integrated circuits. Integrated circuits are highly recommended as an insurance against possible circuit failure at some later date. It is much easier to simply unplug an IC and place a new one in the socket rather than trying to un-solder a 14 or 16 pin IC from a circuit board. Integrated circuit sockets will usually have a notch or cut-out at one end of the plastic IC socket. To the immediate left of the notch or cut-out you will see pin one (1) of the socket. Pin (1) of U1 is connected to one end of potentiometer R4. The IC package itself will also have a notch or cut-out at the top end of the IC. When placing the IC into its respective socket make sure you align pin one (1) of the IC with pin one (1) of the socket. The integrated circuits must be inserted into the socket properly aligned or the circuit will not work and may damage the components. Go ahead and place U1 into its socket on the sensor head PC board.

Sensor coil assembly

The sensor coil at L1 detects changes in the local magnetic field. You will need to wind approximately 50,000 turns of 28 AWG magnet wire over a soft iron core ½″ in diameter and 10″. (The iron core concentrates the flux lines by offering a lower reluctance path than air.)

Wind the fine insulated magnet wire carefully on the iron core to avoid kinks and breakage. Tape the ends of the winding temporarily to the core and carefully solder hook-up wire at each end to form permanent terminals. The terminals can be secured to the core with room-temperature vulcanizing (RTV) adhesive to relieve any strains that might develop in the fine magnet wire.

In the prototype, the sensing circuit board, coil, and battery pack, consisting of two "C" cells were housed in a case made from standard diameter PVC water pipe cut to a length that will accommodate all of those elements as shown in Figures 20-6.

The covers for the sensing head housing are PCV end-caps that press-fit over the 2⅜″ outside diameter of the pipe. The upper cap is a simple cup, but the lower cap is a sleeve with a threaded insert at its end. Drill a hole in the square base of the threaded insert for jack J1 and fasten it with a ring nut. Then close the cover on the empty pipe and drill two pilot holes 180° apart in the sleeve for self-tapping screws to clamp the cap in position after the sensor head is assembled.

Cut about a 6″ length of RG-174/U coaxial cable, strip both ends and solder the inner conductor of one end to the jack terminal and its shield to the jack lug. Solder the inner conductor and shield of the other end to the sensor circuit board.

Attach the coil to the sensor circuit board with about a 6″ length of RG-174/U coaxial cable as shown. Connect the inner conductor to one terminal and the shield to the other and solder both in position. The prototype includes a twin "C" cell holder that, with cells in position, has a maximum width dimension of

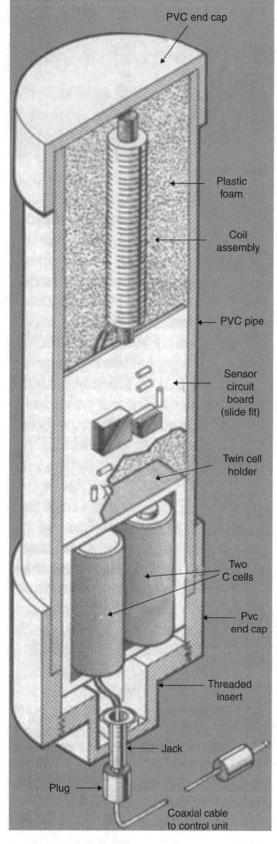

Figure 20-6 *Sensor*

less than 2″ permitting it to slide easily into the sensor head pipe section.

Cut three lengths of hook-up wire to extend the lengths of the positive, negative and ground leads of the battery pack. Solder one end of those wires to the holder leads and the other ends to the sensor circuit board.

Bond the upper cap to the PVC pipe section with PVC cement. Then position U1 in its socket, making certain that pin 1 is located correctly. Finally, insert the "C" cells in the holder. Wrap a layer of polyurethane foam around coil L1 to center it within the pipe section and insert both in the pipe section with the cover at the end. Then assemble the completed circuit board and battery pack to make sure that all of these parts will fit in the pipe section. Set the assembly aside until you are ready to perform the checkout steps.

Control/display unit

The control/display prototype circuit board measured $2\frac{7}{8}″ \times 5″$ to fit inside the instrument case selected. The schematic for the control/display board is shown in Figure 20-4. The dimensions of the circuit boards can be modified for packaging in the instrument and sensor head cases of your choice. Place the control/display circuit board in front of you, along with all its respective components, and we will begin constructing the control/display board.

First we will begin by placing the resistors on the PC board; locate R6, an 18k ohm resistor, and R7 a 12k ohm resistor, and place them on the board and solder them in place in their respective locations on the board. Next locate and install resistors R8, R9, R12, and R20; these are all 10k ohm resistors, solder them in place and cut the excess component leads with your end-cutter. Locate and install the remaining resistors, solder them in place and trim the excess component leads afterward.

Next, you can locate the capacitors C4, a .005 μF disk capacitor, and it may be marked (502). Install C4 on the PC board and solder it in place, remember to cut the excess lead lengths. Now, locate capacitors C5, C7, C9 and C10; these are all .1 μF capacitors and may be marked (104). Install these capacitors and solder them in place and then trim the extra leads. Now, locate C8, a 4.7 µF electrolytic capacitor. Capacitor C8 has polarity, so it must be installed with respect to these marking. Look at the schematic, and you will see that the plus (+) side of C8 is connected to the junction of resistor R23 and diode D3. Install C8, and then solder it place, trim the excess leads as necessary. Go ahead and install the remaining capacitors and solder them in place on the PC board.

The control/display section of the Aurora Monitor employs two silicon diodes, and they must be installed with respect to their polarity. The colored band denotes the cathode end of the diode. The cathode end of D3 is fed to R23, while the cathode end of D4 is sent to C12 and M1.

Transistors are generally three-lead devices, with a Base lead, a Collector lead, and an Emitter lead. The Base lead is depicted as a straight vertical line with two arrows pointing to it. The Emitter lead will have a small arrow pointing towards or away from it. When the arrow points toward the Base lead the transistor is a PNP type. If the arrow is pointing away from the Base lead then the transistor is an NPN type. Transistor Q1 is a PNP type, while Q2 and Q3 are NPN types. Identify the transistor leads and install them in their respective locations on the PC board, trim the leads as necessary.

The control/display portion of the Aurora Monitor utilizes three integrated circuits. Integrated circuits are highly recommended in the event of a circuit failure at a later date. Be sure to install the ICs correctly: pin one (1) of U2 is connected to R7, while pin one (1) of U3 is not used, pin two (2) is jointed with R9 and R8 and R12. Integrated circuit U4 has pin one (1) connected to the junction of R13 and R14. Failure to install the integrated circuits correctly may result in damage to the IC or to other components as well.

Finally you will need to connect the small audio transformer to the circuit. The higher impedance (1K) end of the transformer is connected to the junction of Q1 and R22, and the other 1k lead is connected to the collector of Q3. The secondary or 8 ohm winding of T1 is sent to the small 8 ohm speaker. The transformer can be mounted on the circuit board or on the bottom of the chassis box. The 100 uA panel meter is wired between the Emitter of Q3 and ground.

The control/display assembly is housed in a standard 7″ × 5″ × 3″ aluminum electronics instrument case with a removable cover. Drill the holes in the front panel of the case for power ON switch S1, FILTER IN switch S2, RANGE switch S3 and (if used) optional recorder switch S4. Also drill the holes for SENSITIVITY, ALARM ADJUST, and RATE potentiometers (R18, R19, and R11 respectively). Cut out the hole for mounting meter M1. Note: If you want to use the monitor strictly as a security monitoring system the meter can be omitted.

Drill the holes in the back panel for jacks J2 and J3, and PULSE SHAPE potentiometer R17. Drill a series of holes in a circular pattern for the speaker SPKR1. Drill holes in the bottom of the case for mounting the control/display circuit board and two quad "C" cell holders. Remember the circuit uses both a plus (+) 6 volt supply and a minus (−) 6 volt supply. We chose to power the circuit from batteries, but you could use a dual 6 volt power supply to power the control/display unit.

Assemble the panel-mounted switches, jacks, potentiometers, meter and speaker to the front and back panels of the case as shown in Figure 20-2. (You might prefer to bond the face of the speaker to the inside of the back panel with RTV adhesive rather than bolts and nuts.)

Determine the length of speaker wires needed to permit assembly of the circuit board in the instrument case, cut the wires to length and solder them to the board. Cut and solder all leads from panel-mounted components to the circuit board, allowing sufficient lengths to permit unimpeded circuit board assembly, yet not allowing too much slack.

Cut about a 6″ length of RG-174/U coaxial cable to connect INPUT jack J2 and the FILTER IN Switch S2 and two other lengths about 6″ long to connect S2 to the FILTER-IN and FILTER-OUT pads on the control/display board. Strip all cable ends and solder and trim all connections.

Assemble the control/display board to the base of the case with screws and ½″ insulating standoffs, lock washers and nuts. Bolt the two quad AA cell holders to the base of the case. Position the ICs in their sockets on the control-display board, making sure that all pin 1's are in their correct positions. Insert the eight AA cells in the spring-loaded holders, observing the correct polarity.

Cut a length of RG-174/U coaxial cable to the length that will suit your installation. Strip the wire ends and attach phono plugs PL1 and PL2 to cable ends and solder or crimp them in position.

Before applying power to the Aurora Monitor, you may want to inspect the two circuit boards for "cold" solder joints and "short" circuits. Look carefully at the foil side of the sensor head unit and examine the solder joints. The solder joints should all look clean, bright and shiny. If any of the solder joints look dark, dull, or "blobby" then remove the solder from the joint and re-solder the connection. Next examine the control/display board for "cold" solder joints. Now, we will inspect the two boards for possible "short" circuits. "Short" circuits are usually caused by "stray" component leads which get stuck to the board from the sticky residue left from rosin core solder or from small solder balls which may "short" across the circuit traces. Examine both circuit boards for "stray" leads and solder balls and remove them if you see any. Now, you are ready to apply power to the circuits in order to "test" the circuits.

Test and checkout

Test the sensing circuit first. Connect a general purpose oscilloscope or multi-meter to the output of U1. Position a permanent magnet near coil L1 and the oscilloscope display should show a pronounced pulse. If a multi-meter is used, its readout should jump.

Next, move L1 away from the magnet and the reading on the multi-meter should fall to zero. If the reading does not go to zero, adjust trimmer potentiometer R4 in the sensor circuit. When the sensing head is adjusted and working correctly, close the lower cap and fasten it with two non-magnetic stainless steel self-tapping screws to complete the assembly. Connect one end of the plug-terminated coaxial cable to the sensor head jack 1 and the other end to the control/ display unit jack J2.

Turn on the power toggle switch of the control/display unit and turn SENSITIVITY potentiometer R18 full clockwise. The meter should remain at zero. Rotate ALARM ADJUST potentiometer R19 clockwise until the alarm just begins to sound.

Then turn it back slightly so that no sound is heard to obtain the maximum sensitivity setting. Next, turn potentiometer R19 full clockwise. There should be no sound from the speaker.

Position a small magnet or piece of metal next to coil L1 in the sensor head and the unit should now be activated: the speaker should emit sound, and the meter should read full scale. If everything checks out, you can now start observing magnetic field disturbances or anomalies.

In a quiet magnetic environment it might be necessary to adjust potentiometer R4 in the sensor head to the threshold of the meter movement. This fine adjustment eliminates any small dead zones in sensitivity. Test the instrument's ability to detect the Earth's magnetic field by rotating the sensing head with short, quick, snapping motions in a counter-clockwise direction. The meter movement should jump off scale.

As rotation is continued, a direction will be found where the meter will have its lowest response. This nulling point is the north-south direction. Any objects containing permanent magnets such as speakers or meters that are brought into close proximity to the Aurora Monitor's sensing head coil L1 will affect the accuracy of the instrument's readings.

In the event that the Aurora Monitor does not work upon power-up, you will need to disconnect the power supply and sensor head unit and inspect the circuit board for possible errors which might have occurred while building the Aurora Monitor. The most common cause for error when constructing the circuit is the incorrect installation of the electrolytic capacitors, the diodes and semiconductors such as transistors and integrated circuits. You may have installed one of these components backwards, so you will need to carefully inspect the circuit board. A second pair of eyes could be very helpful in locating potential errors of this sort. Have a knowledgeable electronic enthusiast help you inspect the circuit for errors. Another cause for error is mis-locating resistors. You may have placed the wrong value of resistor in a particular location, this would cause the circuit to malfunction. After carefully inspecting the Aurora Monitor circuit boards, you may quickly find your error and correct it. Once the error has been located, you reconnect the sensor head unit, and then re-apply power and then you can test the Aurora Monitor once again.

The Aurora Monitor can be connected to a chart recorder or it can pass signals to a personal computer with an analog-to-digital converter board. The recorder or PC can collect data for the study of magnetic fields, magnetic storms, and sunspot activity over long periods of time for further analysis. A set of high-impedance (greater than 1 kilohm) headphones can be plugged into jack J3 if you wish to "hear" the changes in magnetic fields.

If you want a permanent installation, mount the sensing head assembly so that it is directed away from any large metal obstructions, oriented on a north-south axis, and pointed slightly upwards. It's a good idea to fasten it to a heavy wooden post to prevent wind-induced vibrations.

Chapter 21

Ultra-Low Frequency (ULF) Receiver

Parts list

Parts Bin

ULF Receiver

R1,R4 50 kohm trim
potentiometer

R2 50 kohm - 20 turn pot

R3 100 kohm trim pot

R5,R6,R10 10 k ohm, $\frac{1}{4}$w,
5% resistor

R7,R8,R9,R12,R21 33 k
ohm, $\frac{1}{4}$w, 5% resistor

R11,R16 1 k ohm, $\frac{1}{4}$w,
5% resistor

R13,R20 22 k ohm, $\frac{1}{4}$w,
5% resistor

R14,R15,R18,R19 47 k
ohm, $\frac{1}{4}$w, 5% resistor

R17 50 k ohm pot

R22 2.2k ohm, $\frac{1}{4}$w,
5% resistor

C1,C2,C3,C4,C5,C6 .1 µF,
35 volt disk capacitor

C11,C12,C13,C14 .1 µF,
35 volt disk capacitor

C8,C9 1 µF tantalum,
35 volt electrolytic
capacitor

C10 .01 µF, 35 volt
disk capacitor

C7 .05 µF, 35 volt disk
capacitor

D1 Zener diode

U1,U2 LTC1063 IC

U3,U4 TL074 op-amp IC

U5 INA114 Differential
amplifier IC

J1,J2,J3 3-terminal
screw terminal strips

S1 SPST power switch

F1 1-amp fast-blow fuse

Misc PC board, chassis,
wire, hardware, power
cord, etc.

**ULF Receiver Power
Supply**

C1,C4 2200 µF, 35 volt
electrolytic
capacitor

C2,C5 .33 µF, 35 volt
ceramic capacitor

C3,C6 .1 µF, 35 volt
ceramic capacitor

U1 LM7805 - fixed
positive voltage
regulator

U2 LM7905 - fixed
negative voltage
regulator

D1,D2,D3,D4 50 volt,
2 amp silicon
diode

T1 110v-12 v center
tapped transformer -
2amp

Misc circuit board,
wire, terminals, plug,
fuse, etc.

If you are interested in monitoring the ultra-low frequency spectrum from .001 Hz to 30 Hz, the ULF receiver will allow you to do some serious research into earthquake monitor and Earth field studies. This project is intended for curious individuals who want to participate in the research and experimentation of the study of extremely low frequency signals which travel through the interior of our Earth. This project will allow you to construct a ULF receive station to be used to collect data for later analysis and compare it with others who belong to the ELFRAD group and those researchers involved with the Public Seismic Network.

The ELFRAD low frequency group coordinates a research group which studies the low frequency spectrum and reports to one another the research that individual members do and makes available the information for comparison through their central website. In order to achieve their goal, it is necessary to have most of the equipment and construction parameters standardized as much as possible, so the ELFRAD Group will furnish software and support.

A complete standardized ELFRAD monitoring station consists of a ULF receiver, an antenna array, and interface card and a personal computer, and in this project we will discuss how to put a receive station together in your home workshop.

The Ultra-Low Frequency Receiver allows the desired ultra-low frequency signals to be recovered from the Earth. The signals are directed to a conditioning circuit which not only protects the receiver equipment from transients caused by lightning and other sources but attenuates the ambient 60 or 50 Hz frequency caused by commercial power, with the receiver's power-line notch line filter. The signals then pass through a circuit which removes any ground loop problems which may occur at your computer interface. The low frequencies of interest are then passed through a series of filters to remove most of the noise and unwanted signals above 25 Hz. This desired band of signals is then coupled to the analog-to-digital interface in your personal computer to be recorded on the hard drive. The interface used in the prototype accepted up to eight channels of data samples and was able to collect 200 samples per second, record them and then display them on the computer screen in either real time or compressed time format.

Figure 21-1 *ULF receiver*

At any time it is possible to take any period of data previously recorded, generate a file and display the data for analysis. You are also able to post filter the data, with band pass, notch, low pass, and high pass capability.

The Ultra Low Frequency Receiver project is shown in Figure 21-1. This ULF receiver circuit was designed as an all purpose ULF and ELF receiver with a frequency range from DC to 30 hertz. The signals from both leads of the non-polarized antenna array are connected to J3, as seen in the schematic diagram in Figure 21-2 and 21-3, a two sheet diagram The antenna input signal travels to pins 2 and 3 on U5, which is a high impedance differential instrumentation amplifier. This chip's gain is controlled by a single variable resistor at R3.

The combined signals then travel from pin 6 of U5 on to U3:A which is a buffer amplifier, followed by a 2 pole low pass filter at U3:B and U3:C and signal amplifier at U3:D. From pin 14 on U3, the partially filtered signal travels on to integrated circuit U2 and U1 which are 5 pole Butterworth low pass filters in cascade. The LTC1063 ICs at U1 and U2 are clock controlled and the frequency is set at 3000 hertz by R9. This configures the two LTC1063 chips to have a cut-off frequency of about 30 hertz. The clock ratio is 1/100.

The integrated circuit at U4 consists of a X5 gain amplifier at U4:D, a buffer at U4:C and another 2 pole analog low pass filter at U4:B and U4:A. Trim potentiometer R17 is the final gain control for the desired output voltage at U4:A. IC section U4B is a 2 pole low pass filter which removes any residual clocking pulses from the signal which may have been caused by U1 and U2. The amplified output signal at J1 can be routed to your analog-to-digital converter card in your PC. Power is supplied to the ULF receiver at the terminals of J2. The output from the ULF receiver can be routed to an analog-to-digital

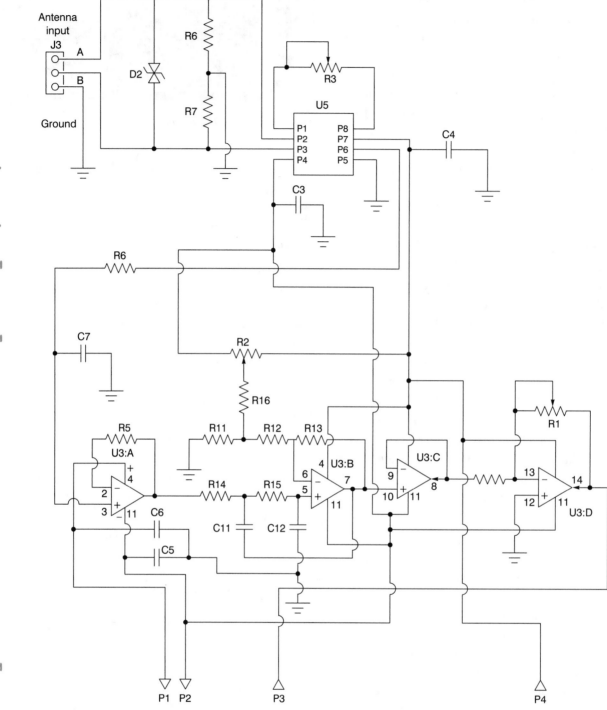

Figure 21-2 *ULF receiver schematic – Part I*

converter card in a personal computer for recording and displaying the receiver's output in real time or for later viewing.

The ULF receiver circuit is powered from a dual plus/minus 5 volt power supply, shown in the schematic at Figure 21-4. The power supply will furnish plus and minus 5 volts DC for all of the op-amps in the receiver.

A 110 to 12 volt center tapped transformer is fed directly to a four diode bridge rectifier. The resultant DC signal is then coupled to two 2200 µF filter capacitors at C1 and C4. The plus voltage output from C1 is then fed to the input of an LM7805. The output of U1 is then sent to a .1 µF capacitor at the positive output terminals. The minus voltage output from C4 is sent to a

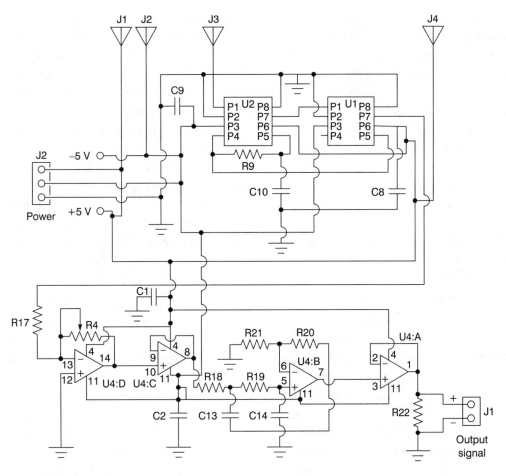

Figure 21-3 *ULF receiver schematic – Part II*

negative voltage regulator at U2, an LM7905. The
output from U2 is then sent to capacitor C6, at the
minus voltage output terminals.

Let's get started building the ULF receiver! Before
we begin building the ULF receiver, you will need to
locate a clean well lit and well ventilated work area.
A large table or workbench would be a suitable work
surface for your project. Next you will want to locate a
small 27 to 33 watt pencil-tipped soldering iron, as well
as a roll of #22 ga. 60/40 tin/lead solder and a small jar

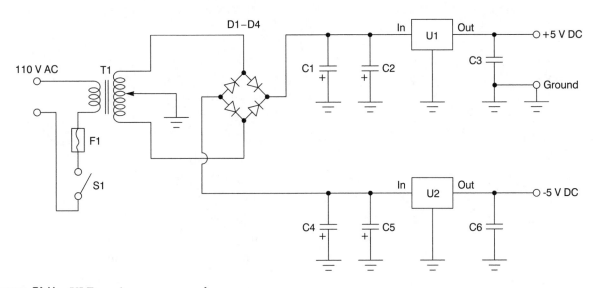

Figure 21-4 *ULF receiver power supply*

of "Tip Tinner," a soldering iron tip cleaner/dresser from your local Radio Shack store. You will also want to secure a few hand tools for the project, such as a pair of small end-cutters, a pair of tweezers and a pair of needle-nose pliers. Locate a small Phillips and a small flat-blade screwdriver, as well as a magnifying glass to round out your tool list. Grab the schematic, parts layout diagram as well as the resistor and capacitor identifier charts and we will begin our project. Place all the project components on the table in front of you. The ultra-low frequency receiver is an RF or radio frequency project and it is best constructed on a printed circuit board with large ground plane areas covering the board for the best RF grounding techniques. Once you have all the parts and PC board in front of you, heat up the soldering iron and we'll get started!

First, find your resistor identifier chart in Table 21-1, which will help you select the resistors from the parts pile. Resistors used in this project are mostly small ¼ watt carbon composition type resistors, which have colored bands along the resistor body. The first color band should be closest to one end of the resistor body. This first color band represents the first digit of the resistor value. For example, resistor R5, R6 and R10 has four color bands, the first one is a brown band followed by a black band followed by an orange band. The fourth band is gold. The first band is brown which denotes a

digit one (1), the second band is black which represents a digit five (0) and the third band is orange, which represents a multiplier of zero (000), so the resistor value is 10,000 or 10k ohms with a tolerance value of 5%. Identify the remaining resistors for the project and we can begin "populating" the PC board. Place a few resistors on the board at one time, so as not to confuse the process. Make sure that you place the correct resistor into the correct PC location before soldering it in place. Once you solder a few resistors in place on the PC board, you can use your end-cutter to trim the excess component leads. Cut the excess leads flush to the edge of the circuit board. Then place a few more resistors on the PC board and solder them on to the board. Remember to trim the component leads as necessary.

Next we will locate the capacitors for the ULF receiver. Capacitors are listed as electrolytic and non-electrolytic types. The non-electrolytic types are generally smaller in value and size as compared with the electrolytic types. Non-electrolytic capacitors in fact can be so small that their actual value cannot be printed on them, so a special chart was devised as shown in Table 21-2. The chart illustrates the three-digit codes which are often used to represent capacitors. For example a .001 µF capacitor will have (102), while a .01 µF capacitor will have (103) marked on it to represent its true value. Use the chart to identify the

Table 21-1
Resistor color code chart

Color Band	1st Digit	2nd Digit	Multiplier	Tolerance
Black	0	0	1	
Brown	1	1	10	1%
Red	2	2	100	2%
Orange	3	3	1,000 (K)	3%
Yellow	4	4	10,000	4%
Green	5	5	100,000	
Blue	6	6	1,000,000 (M)	
Violet	7	7	10,000,000	
Gray	8	8	100,000,000	
White	9	9	1,000,000,000	
Gold			0.1	5%
Silver			0.01	10%
No color				20%

small capacitors in the project. Non-electrolytic have no polarity markings on them so they can be installed in either direction on the PC board.

Electrolytic capacitors are usually larger in size and value and they will have a white or black band on the side of the capacitor body or a plus (+) or minus (−) marking on the body of the capacitor near the leads. These markings are polarity markings and that indicate the direction in which the capacitor must be mounted on the PC board. Failure to observe polarity when installing the capacitor may result in damage to the capacitor or to the circuit itself, so pay particular attention to capacitor polarity when placing the capacitors on the board.

Let's go ahead and place some of the non-electrolytic capacitors on the PC board. Identify a few small capacitors at a time and place them on the PC board and solder them in place. Trim the component leads after. Go ahead and install the remaining non-electrolytic capacitors after choosing your desired band. Install the capacitors and solder them in place, remember to cut the extra leads flush to the edge of the circuit board. Now, you can go ahead and install the electrolytic capacitors onto the circuit board. Go ahead and solder them in place and trim the leads as necessary.

The ULF receiver utilizes five integrated circuits and two regulators integrated circuits. Locate the semiconductor pin-out diagram illustrated in Figure 21-5, which will help you install the ICs. When constructing the project it is best to use IC sockets as an insurance against a possible circuit failure down-the-road; it's much easier to unplug an IC rather than trying to un-solder an IC from the PC board. IC sockets will have a notch or cut-out at one end of the plastic socket. Pin one (1) of the IC socket will be just to the left of the notch or cut-out. Note, pin 1 U1 connects to pin 7 of U2, while pin 1 of U2 connects to the output of U3:D. Now take a look at U3, pin 1 of U3 is connected to R14, while pin 1 of U4 is fed to the output through R22. Finally, pin 1 of U5 is connected to potentiometer R3. Now place each of the integrated circuits into their respective sockets.

A protection Zener diode is placed across the antenna input pins between resistors R6 and R7. Note that all diodes have polarity, including Zener diodes. Diodes will have either a black or white colored band at one

end of the resistor body. The lead closest to the colored band is the cathode lead of the diode.

The main ULF schematic diagram is broken into two diagrams, since the schematic is rather large. You have most likely noted that along the bottom of diagram (A), you will see designations of P1, P2, P3 and P4. These are not connectors as such but merely to help separate the two diagrams for illustration purposes. The receiver was constructed on a single printed circuit board. In the original prototype, the antenna input shown at J3 was a screw terminal strip, but this could be changed to an SO-238 antenna jack mounted on the chassis enclosure. The power connection was made at a second screw terminal jack at J2, which had three terminal posts for plus/minus and ground connections. The output of the ULF receiver was connected to a third screw terminal jack at J3, this could be changed to an RCA jack if desired.

The ULF receiver contains a number of variable resistors or potentiometers in the circuit. Potentiometer R1, R3 are PC board trimmer types, while R2, which is a 20 turn precision chassis mounted potentiometer, and R4 is a conventional chassis mounted linear taper potentiometer.

The ULF receiver is now almost completed, so take a short well deserved break, and when we return we will inspect the circuit board for any possible "cold" solder joints and "short" circuits before we apply power to the circuit. Pick up the PC board with the foil side of the board facing upwards toward you. Examine the foil side of the board carefully, take a look at the solder joints. All of the solder joints should look clean, bright, shiny and smooth; if any of the solder joints look dark, dull or "blobby" then you should remove the solder from the joint with a solder-sucker and then completely re-solder the joint all over again so that the solder joints all look good. Next we will examine the PC board for any possible "short" circuits, which may be caused from "stray" component leads which were cut from the board. Another possible cause of short circuits are solder blobs which may have stuck to the foil side of the board. Rosin core solder often leaves a sticky residue which can cause leads and solder blobs to stick to the board and possibly "bridge" across circuit traces.

Next we will need to build the dual voltage power supply circuit board. The power supply was built on the circuit board as well. The transformer was mounted off

Table 21-2
Capacitance code information

This table provides the value of alphanumeric coded ceramic, mylar and mica capacitors in general. They come in many sizes, shapes, values and ratings; many different manufacturers worldwide produce them and not all play by the same rules. Some capacitors actually have the numeric values stamped on them; however, many are color coded and some have alphanumeric codes. The capacitor's first and second significant number IDs are the first and second values, followed by the multiplier number code, followed by the percentage tolerance letter code. Usually the first two digits of the code represent the significant part of the value, while the third digit, called the multiplier, corresponds to the number of zeros to be added to the first two digits.

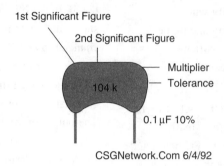

Value	Type	Code	Value	Type	Code
1.5 pF	Ceramic		1,000 pF / .001 µF	Ceramic / Mylar	102
3.3 pF	Ceramic		1,500 pF / .0015 µF	Ceramic / Mylar	152
10 pF	Ceramic		2,000 pF / .002 µF	Ceramic / Mylar	202
15 pF	Ceramic		2,200 pF / .0022 µF	Ceramic / Mylar	222
20 pF	Ceramic		4,700 pF / .0047 µF	Ceramic / Mylar	472
30 pF	Ceramic		5,000 pF / .005 µF	Ceramic / Mylar	502
33 pF	Ceramic		5,600 pF / .0056 µF	Ceramic / Mylar	562
47 pF	Ceramic		6,800 pF / .0068 µF	Ceramic / Mylar	682
56 pF	Ceramic		.01	Ceramic / Mylar	103
68 pF	Ceramic		.015	Mylar	
75 pF	Ceramic		.02	Mylar	203
82 pF	Ceramic		.022	Mylar	223
91 pF	Ceramic		.033	Mylar	333
100 pF	Ceramic	101	.047	Mylar	473
120 pF	Ceramic	121	.05	Mylar	503
130 pF	Ceramic	131	.056	Mylar	563
150 pF	Ceramic	151	.068	Mylar	683
180 pF	Ceramic	181	.1	Mylar	104
220 pF	Ceramic	221	.2	Mylar	204
330 pF	Ceramic	331	.22	Mylar	224
470 pF	Ceramic	471	.33	Mylar	334
560 pF	Ceramic	561	.47	Mylar	474
680 pF	Ceramic	681	.56	Mylar	564
750 pF	Ceramic	751	1	Mylar	105
820 pF	Ceramic	821	2	Mylar	205

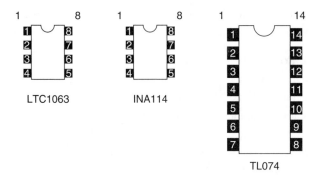

LTC1063   INA114   TL074

Figure 21-5 *Semiconductor pin-outs*

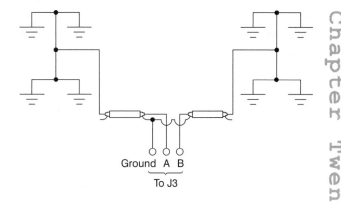

Ground A B
To J3

Figure 21-6 *ULF receiver antenna circuit*

the circuit board and on the bottom of the chassis. The diode bridge was placed on the circuit board, paying careful attention to the polarity of the diodes. You could elect to individual diodes or one of those four diode bridge rectifiers all in one package. If you choose the second option again you must observe the polarity of the plastic package. The AC input section will be marked with two wavy lines, the AC side has no polarity. The DC side of the bridge will have a plus (+) and a minus (−) marking. When installing the six capacitors be sure to observe polarity since four of them are electrolytic types. Note that there are two power supplies and the polarity placement orientation is different for the two power supply sections. The LM7805 is the plus voltage regulator, while the LM7905 regulator is the minus voltage regulator; don't mix them up, and take your time to install them at the correct locations to avoid damage to the circuit. A fuse was placed in series with the power switch to the transformer input leads. A power line-cord with a 110 volt plug was used to supply power to the transformer. Once the power supply circuit has been completed, you can carefully inspect the circuit board for any "cold" solder joints and "short" circuits before applying power for the first time.

The Ultra-Low Frequency Receiver requires an antenna "site" for proper operation. This is not the usual conventional type of antenna that you encounter with most receivers, since this is a special type of receiver project. The array is designed to be built completely underground. Space for construction is necessary and may be anywhere from 100 to 1000 linear feet, see Figure 21-6.

Due to the fact that we will be dealing with extremely long wavelengths on the order of miles rather than feet,

you are going to need a little space. As long as you have a strip of property at least one 100 or more feet in length you will be able to detect the signals. There are several items of interest in determining your location. You will be building two arrays widely separated from each other. The further apart the better. Also if you have a choice, a damp location is better than dry rocky soil.

The first step in constructing your antenna/detector is to determine the magnetic orientation. Take an ordinary pocket compass and locate magnetic North. The line of sight between the two points should be aligned with North-South. The next step is to obtain some copper clad ground rods at least eight feet in length, put on some gloves and start driving one into the ground. We found that a standard metal fence post driver purchased from the local farm store was ideal for the job. It helps also to take a pair of post diggers and dig out a hole about 18″ deep. Then drive the rod down into the ground below the surface of the hole leaving enough room to make a connection.

Next measure out a distance as far as possible, at least 30 meters (98 ft) in a North or South direction and repeat the procedure above, driving a copper clad eight foot ground rod into the soil. Then connect the two widely spaced rods together using a length of wire. Obtain an ohmmeter, and open the wire conductor and measure the resistance. Make a note of the reading, then reverse the ohmmeter leads and measure again. Then take this reading, average it with the other reading and you will have a close indication of the DC resistance between the two probes (ground rods). Very likely, unless you have excellent soil conductivity in your area, the reading will be several thousand ohms. The point is to lower this reading as much as possible.

The lower the better. A workable resistance is 50 to 500 ohms but the ideal resistance would be around 5 to 10 ohms. In most cases, however, this would be unobtainable unless you used many ground rods or copper strapping at each end.

Keep adding rods at each probe site, spaced 20 feet apart from one another connected together with #6 copper bare wire, until the resistance gets down to between 25 and 500 ohms. Keep in mind two items. The further apart the probe sites, the lower the resistance, and the more rods driven into the ground, the lower the resistance. We have found that on the average, five grounds at each end of the site are sufficient to obtain this reading.

Another option rather than using ground rods, in rocky or extremely hard soil, you may use a ditch trencher and bury #6 or #4 copper wire at least 24 inches deep. The copper wire should be placed in an "X" fashion with a length of at least 25 feet. The lead cable will then be connected to the center point of the "X."

Now obtain a length of CATV lead in wire such as RG-11 or RG-59, from your local electronic parts store. A good source to try is the local Cable TV company. Sometimes they have lengths of scrap lying around their warehouse and will be happy to get rid of it. Connect the center conductor only (not the outside shield) to each antenna array using copper clamps. Take each piece of co-ax to the location of your receiver. Then install a suitable connector to the two ends. These two cables will be the input leads to the receiver. The shield should be connected to earth ground using a single copper clad ground rod. Do not use the power ground and make sure the ground connection is at least 20 feet from power ground.

With the ULF receiver, power supply and antenna constructed and ready to go, we just have one more detail left to complete our ELFRAD ULF monitoring station. The most expensive components will be the computer and analog-to-digital interface device. A PC type of computer is recommended using Windows 98, NT, Win2000, or XP software will work nicely. Since most of everyone owns a computer or laptop these days, the only thing left to be purchased is the A/D interface. The software for the A/D converter is available from the supplier of the A/D interface listed below. You will need at least a two channel A/D board, since it is recommended that you also record a short wave radio tuned to WWV, which keeps computer synched to within plus or minus 2 milliseconds. Also if desired, a GPS type of time standard is available. The A/D converter is available at http://seismicnet.com/ serialatod.html.

For more information on Earthquake monitoring you may finds these sites helpful.

ELFRAD Research Group

The low frequency radio research group, which provides support to interested researchers.
Helps researchers in correlating and comparing data.

http://www.elfrad.com/

Public Seismic Network

A seismic research group dedicated to earthquake monitoring.
Helps researchers in correlating and comparing data.

http://psn.quake.net/

Chapter 22

Jupiter Radio Telescope Receiver

Parts list

Jupiter Radio Telescope

R1   68 ohm

R2   294 ohm

R3   17.4 ohm

R4   294 ohm

R5   100 ohm

R6   2.2 k ohm

R7   10 k ohm linear
    potentiometer

R8   2.2 k ohm

R9, R19   100 k ohm

R10   220 ohm

R11   1.5 k ohm

R12,R20,R21,R27   1 k ohm

R13,R18   27 k ohm

R14   100 k ohm

R15   10 k ohm
    potentiometer/
    switch

R16   10 k ohm

R17   1.5 k ohm

R22,R23   2 ohm

R24   1 ohm

R25   220 ohm

R26   47 ohm

R28,R29,R30,R31,R32
    10 ohm

C1   39 pF, 35v disk
    ceramic

C2   4-40 pF, variable
    capacitor

C3   56 pF, 35v disk
    ceramic

C4   22 pF, 35v disk
    ceramic

C5,C8,C11,C14   .01 mF,
    35v dipped ceramic

C6   4-40 pF, variable
    capacitor

C7   not used

C9,C12,C13   47 pF, 35v
    disk ceramic

C10   270 pF, 35v disk
    ceramic

C15   10 pF, disk
    ceramic

C16,C24,C25   10 mF, 35
    vdc, electrolytic

C17,C18,C21,C23,C26C29
    .1 mF, 35v dipped
    ceramic

C19   1 mF, 35v metal
    polyester

C20,C22   0.068 mF, 35v
    5% metal film

C27   10 mF, 35 vdc,
    tantalum, stripe,

C28   220 pF, 35v disk
    ceramic

C30,C31,C33   10 mF, 35
    vdc, electrolytic

C32   330 mF, 35 vdc,
    electrolytic

C34,C35,C36   0.1 mF, 35v
    dipped ceramic

C37   10 pF, 35v disk
    ceramic

C38 10 mF, 35 vdc,
    electrolytic

C39 100 mF, 35 vdc,
    electrolytic

C40,C41,C42,C43 0.1 mF,
    35v dipped
    ceramic

C44 10 mF, 35 vdc,
    electrolytic

D1 1N4001

D2,D3 1N914

LED Red LED

VD1 MV209, varactor
    diode

ZD1 1N753, 6.2v, Zener
    diode, 400 mw

ZD2 1N5231, 5.1v, Zener
    diode, 500mw

L1 0.47 mH, (gold,
    yellow, violet,
    silver)

L2 1 mH, (brown, gold,
    black, silver)

L3 3.9 mH, (orange,
    gold, white, gold)

L4,L5 1.5 mH,
    adjustable inductor

L6,L7 82 mH, fixed
    inductor

Q1 J-310, junction
    field effect (JFET)

Q2 2N-3904, bipolar,
    NPN

Q3 2N-3906, bipolar,
    PNP

U1 SA602AN,
    mixer/oscillator IC

U2,U3 LM387 or NTE824 -
    audio pre-amplifier IC

OSC1 20 MHz crystal
    oscillator module

J1 Power jack, 2.1 mm

J2 F female chassis
    connector

J3,J4 3.5 mm stereo
    jack, open ckt

Misc PC Board,
    enclosure, knobs,
    solder lugs, wire

Jupiter is the fifth planet from the Sun and is the largest one in the solar system. If Jupiter were hollow, more than one thousand Earths could fit inside. It also contains more matter than all of the other planets combined. It has a mass of 1.9×1027 kg and is 142,800 kilometers (88,736 miles) across the equator. Jupiter possesses 28 known satellites, four of which—Callisto, Europa, Ganymede and Io—were observed by Galileo as long ago as 1610. Another 12 satellites have been recently discovered and given provisional designators until they are officially confirmed and named. There is a ring system, but it is very faint and is totally invisible from the Earth. (The rings were discovered in 1979 by Voyager 1.) The atmosphere is very deep, perhaps comprising the whole planet, and is somewhat like the Sun. It is composed mainly of hydrogen and helium, with small amounts of methane, ammonia, water vapor and other compounds.

Colorful latitudinal bands, atmospheric clouds and storms illustrate Jupiter's dynamic weather systems, see Figure 22-1. The cloud patterns change within hours or days. The Great Red Spot is a complex storm moving in a counter-clockwise direction. At the outer edge, material appears to rotate in four to six days; near the center, motions are small and nearly random in direction. An array of other smaller storms and eddies can be found throughout the banded clouds.

Auroral emissions, similar to Earth's northern lights, were observed in the polar regions of Jupiter. The auroral emissions appear to be related to material from Io that spirals along magnetic field lines to fall into Jupiter's atmosphere. Cloud-top lightning bolts, similar to super-bolts in Earth's high atmosphere, were also observed.

The Jupiter Radio Telescope is a special shortwave receiver which will pick up radio signals from the planet Jupiter and also from the Sun, and it is a very exciting project, see Figure 22-2. This project will allow

Figure 22-1   *Jupiter*

you to do amateur research into radio astronomy in your own back yard.

Radio signals from Jupiter are very weak—they produce less than a millionth of a volt at the antenna terminals of the receiver. These weak radio frequency (RF) signals must be amplified by the receiver and converted to audio signals of sufficient strength to drive headphones or a loudspeaker. The receiver also serves as a narrow filter, tuned to a specific frequency to hear Jupiter while at the same time blocking out strong Earth based radio stations on other frequencies. The receiver and its accompanying antenna are designed to operate over a narrow range of shortwave frequencies centered on 20.1 MHz (megahertz). This frequency range is optimum for hearing Jupiter signals.

The block diagram shown in Figure 22-3 illustrates the various components of the Jupiter Radio Telescope. First shown the antenna intercepts weak electromagnetic waves which have traveled some 500 million miles from Jupiter to the Earth. When these electromagnetic waves strike the wire antenna, a tiny RF voltage is developed at the antenna terminals. Signals from the antenna are delivered to the antenna terminals of the receiver by a coaxial transmission line. Next the signals are delivered to the RF Bandpass Filter and Preamplifier, they are filtered to reject strong out-of-band interference and are then amplified using a junction field effect transistor (JFET). This transistor and its associated circuitry provide additional filtering and amplify incoming signals by a factor of 10. The receiver input circuit is designed to efficiently transfer power from the antenna

Figure 22-2   *Radio Jupiter receiver project*

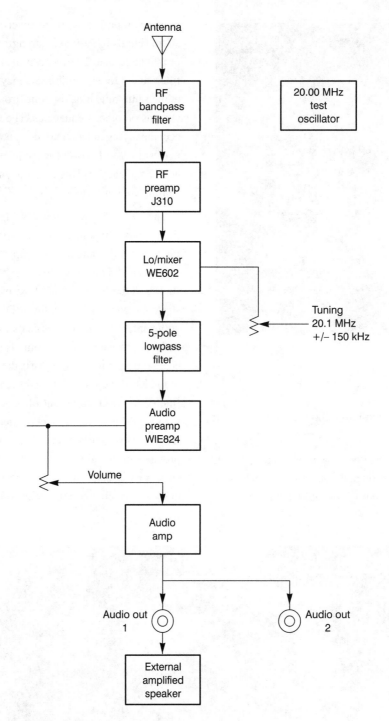

Figure 22-3 *Jupiter radio block diagram*

to the receiver while developing a minimum of noise within the receiver itself.

Next the signals move to the local oscillator (LO) and mixer to perform the important task of converting the desired radio frequency signals down to the range of audio frequencies. The local oscillator generates a sinusoidal voltage wave form at a frequency in the vicinity of 20.1 MHz. The exact frequency is set by the front panel tuning control. Both the amplified RF signal from the antenna and the LO frequency are fed into the mixer. The mixer develops a new signal which is the arithmetic difference between the LO and the incoming signal frequency. Suppose the desired signal is at 20.101 MHz and the LO is tuned to 20.100 MHz.

The difference in frequency is therefore 20.101–20.100 =.001 MHz, which is the audio frequency of 1 kilohertz. If a signal were at 20.110 MHz, it would be converted to an audio frequency of 10 kHz. Since the RF signal is converted directly to audio, the radio is known as a direct conversion receiver.

To eliminate interfering stations at nearby frequencies, a low pass filter is next used, which is like a window, a few kilohertz wide through which Jupiter signals can enter. When listening for Jupiter or the Sun, the radio will be tuned to find a "clear channel." Since frequencies more than a few kilohertz away from the center frequency may contain interfering signals, these higher frequencies must be eliminated. This is the purpose of the low pass filter following the mixer. It passes low (audio) frequencies up to about 3.5 kHz and attenuates higher frequencies.

Finally the low pass filter feeds an audio amplifier stage which takes the very weak audio signal from the mixer and amplifies it enough to drive headphones directly, or to drive an external amplified speaker assembly.

Circuit diagram

We have already seen a block diagram of the Jupiter radio receiver which shows the radio as a group of functional blocks connected together. While this type of diagram does not show individual components like resistors and capacitors, it is useful in understanding signal flow and the various functions performed within the radio. The next level of detail is the schematic diagram. A schematic is used to represent the wiring connections between all of the components which make up a circuit. The schematic diagram uses symbols for each of the different components rather than pictures of what the components actually look like.

A schematic diagram of the complete receiver is seen in Figure 22-4. On this schematic, the part types are numbered sequentially. For example, inductors are denoted L1 through L7, and resistors are denoted R1 through R31. Signal flow as shown in the schematic is as follows. The signal from the antenna connector (J2) is coupled to a resonant circuit (bandpass filter L1, C2, C3) and then to the J-310 transistor (Q1), where it

is amplified. The output of the J-310 goes through another resonant filter (L3, C6) before being applied to the resonant input circuit (L4, C9, C10) of the SA602 integrated circuit (U1), which serves as the local oscillator and mixer. The center frequency of the local oscillator is set by inductor L5 and adjusted by the tuning control R7. The audio output from U1 passes through the lowpass audio filter (L6, L7, C20, C21, and C22). The audio signal is next amplified by U2 (an NTE824) before going to the volume control R15. The final audio amplifier stages comprise U3 (another NTE824), and the output transistors Q2 (2N-3904) and Q3 (2N-3906). After the receiver has been assembled, the variable capacitors C2 and C6 and variable inductors L4 and L5 will be adjusted to tune the receiver for operation at 20.1 MHz.

PC board assembly

Before we begin building the circuit, take a few minutes to locate a clean well lit work table or work bench. You will want to locate a small 27 to 33 watt pencil-tipped soldering iron, a roll of 60/40 tin/lead rosin core solder, and a small jar of "Tip Tinner," a soldering iron tip cleaner/dresser. "Tip Tinner" is available at a local Radio Shack store. You will also need some small hand tools, such as a small flat-blade screwdriver, a Phillips screwdriver, a pair of small needle-nose pliers, a pair of tweezers, a magnifying glass and pair of end-cutters.

Once you have gathered all the tools, locate all the circuit diagrams, such as the schematics, the parts layout diagram, the resistor and capacitor code charts, etc. Place all the component parts along with all your tools and diagrams in front of you, warm up your solder iron and we will start building the Radio Jupiter receiver project.

First we will identify the resistors for the project. Refer to the chart shown in Table 22-1, which will help you to identify the resistor color codes. Resistors usually have three or four color bands on the body of the resistor which denotes the resistor value. The color bands begin at one end of the resistor body close to the edge. The first color band on resistor R1 is blue, this denotes the digit six (6). The second color band is

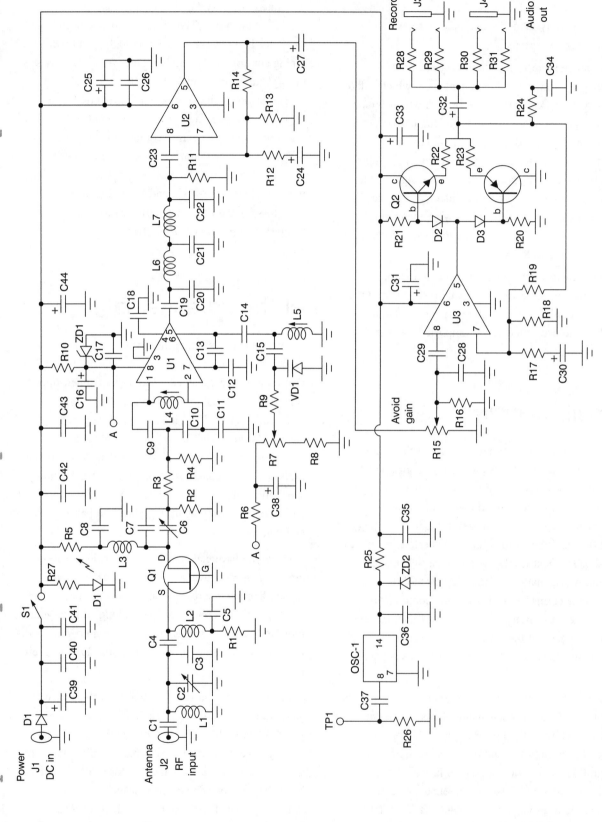

Figure 22-4 *Radio Jupiter receiver schematic*

Table 22-1

Resistor color code chart

Color Band	1st Digit	2nd Digit	Multiplier	Tolerance
Black	0	0	1	
Brown	1	1	10	1%
Red	2	2	100	2%
Orange	3	3	1,000 (K)	3%
Yellow	4	4	10,000	4%
Green	5	5	100,000	
Blue	6	6	1,000,000 (M)	
Violet	7	7	10,000,000	
Gray	8	8	100,000,000	
White	9	9	1,000,000,000	
Gold			0.1	5%
Silver			0.01	10%
No color				20%

gray, which is the digit number eight (8) and the third color band is the resistor's multiplier value which is black and this suggests a multiplier value of 1. So the resistor value for R1 is (6) times (8) times (1) or 68 ohms. The tolerance value of a resistor is noted in the fourth color band. A fourth color band which is silver is a 10% tolerance resistor, while a gold fourth band denotes a 5% tolerance value. If there is no fourth color band then the resistor will have a 20% tolerance value. Next, locate R2, R3 and R4, then check your parts layout diagram and the schematic to see where these resistors are to be placed on the circuit board. Once you verify where the resistors are placed, you can install these resistors on the circuit board and solder them in place on the board. Now locate your end-cutter and trim the excess component leads. Cut the extra component lead lengths flush to the edge of the circuit board. Now go ahead and identify another group of three or four resistors and place them on the circuit board. Once you are sure of their respective locations you can solder the next grouping of resistors onto the circuit board, remember to trim the excess component leads from the board. Place all the remaining resistors on the circuit board and solder them all in place, trim the extra leads as necessary.

Now we will move on to installing the capacitors for the Radio Jupiter receiver project. Capacitors come in two major flavors, non-polarized and polarized types. First, we will talk about non-polarized capacitors. Usually non-polarized capacitors are smaller value types which could take form as ceramic disks, mica, tantalum or polyester types. Often the capacitors are small in size and frequently the capacitor values are printed as three-digit codes rather than their actual values, since this code system takes less space to print. We have included a chart in Table 22-2, which lists the actual capacitor values and their respective codes. For example, a capacitor with a value of .001 µF will have a code designation of 102, a .01 µF capacitor will have 103 and a 0.1 µF capacitor will have a code of 104 marked on it. The Radio Jupiter project has a number of non-polarized type capacitors which you will need to install. Take a look through the component pile and see if you can identify the small non-polar capacitors before attempting to install them on the PC board.

Polarized capacitors are called electrolytic capacitors, they are often larger in physical size. Electrolytic capacitors will always have some sort of polarity marking on them. You will see either a black or white band or a plus (+) or minus (−) marking somewhere on the capacitor body. It is very important that you install these types of capacitors with respect to their polarity in order for the circuit to work properly and to not be damaged upon power-up. You will need to refer to the

Table 22-2
Capacitance code information

This table provides the value of alphanumeric coded ceramic, mylar and mica capacitors in general. They come in many sizes, shapes, values and ratings; many different manufacturers worldwide produce them and not all play by the same rules. Some capacitors actually have the numeric values stamped on them; however, many are color coded and some have alphanumeric codes. The capacitor's first and second significant number IDs are the first and second values, followed by the multiplier number code, followed by the percentage tolerance letter code. Usually the first two digits of the code represent the significant part of the value, while the third digit, called the multiplier, corresponds to the number of zeros to be added to the first two digits.

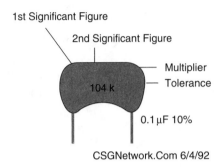

CSGNetwork.Com 6/4/92

Value	Type	Code	Value	Type	Code
1.5 pF	Ceramic		1,000 pF /.001 µF	Ceramic / Mylar	102
3.3 pF	Ceramic		1,500 pF /.0015 µF	Ceramic / Mylar	152
10 pF	Ceramic		2,000 pF /.002 µF	Ceramic / Mylar	202
15 pF	Ceramic		2,200 pF /.0022 µF	Ceramic / Mylar	222
20 pF	Ceramic		4,700 pF /.0047 µF	Ceramic / Mylar	472
30 pF	Ceramic		5,000 pF /.005 µF	Ceramic / Mylar	502
33 pF	Ceramic		5,600 pF /.0056 µF	Ceramic / Mylar	562
47 pF	Ceramic		6,800 pF /.0068 µF	Ceramic / Mylar	682
56 pF	Ceramic		.01	Ceramic / Mylar	103
68 pF	Ceramic		.015	Mylar	
75 pF	Ceramic		.02	Mylar	203
82 pF	Ceramic		.022	Mylar	223
91 pF	Ceramic		.033	Mylar	333
100 pF	Ceramic	101	.047	Mylar	473
120 pF	Ceramic	121	.05	Mylar	503
130 pF	Ceramic	131	.056	Mylar	563
150 pF	Ceramic	151	.068	Mylar	683
180 pF	Ceramic	181	.1	Mylar	104
220 pF	Ceramic	221	.2	Mylar	204
330 pF	Ceramic	331	.22	Mylar	224
470 pF	Ceramic	471	.33	Mylar	334
560 pF	Ceramic	561	.47	Mylar	474
680 pF	Ceramic	681	.56	Mylar	564
750 pF	Ceramic	751	1	Mylar	105
820 pF	Ceramic	821	2	Mylar	205

schematic and parts layout diagrams when installing electrolytic capacitors to make sure that you have installed them correctly. The plus (+) marking on the capacitor will point to the most positive portion of the circuit.

Let's go ahead and identify the non-polarized capacitors first. Check the capacitor code chart and make sure that you can identify each of the non-polarized capacitors. In groups of two or three capacitors install them on the PC board. When you have determined that you have the installed the correct values into their respective location on the circuit board, you can solder the capacitors in place on the board. Remember to trim the excess capacitor leads from the circuit board. Once you have the first grouping of capacitors installed you can move on to installing another group of non-polarized capacitors, until you have installed all of the non-polarized capacitors.

Next, find the electrolytic capacitors from the parts pile and we will install them in their proper locations. Refer now to the schematic and parts layout diagrams so that you can orient the capacitors correctly when mounting them on the PC board. Install two or three electrolytic capacitors and then solder them in place on the PC board. Remember to cut the extra lead lengths flush to the PC board. Identify and install the remaining electrolytic capacitors on the board and solder them in place. Trim the component leads as necessary.

In order to avoid damaging the receiver, it is important to install the capacitors, diodes as well as the semiconductors with respect to proper polarity marking on each of the components. Electrolytic capacitors have either a plus or minus marking on them and must be installed correctly by observing the polarity of the capacitor with respect to the PC board.

Next, we will install the semiconductors. Locate the semiconductor pin-out diagram shown in Figure 22-5, which will help you install the semiconductor devices properly. There are a number of diodes in the Radio Jupiter project, including Zener and varactor diodes and they all have polarity. In order for the circuit to work properly, these devices must be installed with respect to their polarity. Diodes will generally have a white or black colored band at one edge of the diode body. The band usually denotes the cathode end of the diode. Transistors generally have three leads—a Base, Collector and Emitter; pay particular attention to these leads when installing the transistors. Integrated circuits always have some type of markings which indicate their orientation. You will find either a plastic cut-out on the top of the integrated circuit package or an indented circle next to pin 1 of the IC. Be sure to use these markings when installing the integrated circuits. It is a good idea to incorporate integrated circuit sockets as a low cost insurance policy, in the event of a circuit failure, if it ever happens. It is much easier to remove an old dead IC, and simply insert a new one rather than trying to undo many pins and risk damage to the circuit board.

There are a number of coils in this project, and although there is no polarity to the coils, you will need

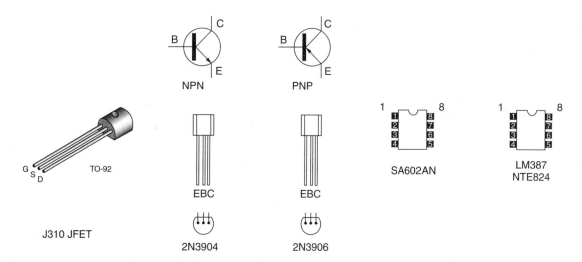

Figure 22-5 *Semiconductor pin-outs*

to pay attention to the values of the coils as well as which ones are adjustable and which are fixed before installing them on the PC board. Note, on the schematic there are two points marked (A), one is at R6 and the other at C16, both of these points are connected together. Note, there is a test point marked (TP1) at the OSC-1 module which can be used to ensure that the oscillator module is working correctly.

Power is brought to the circuit via the coaxial power input jack at J1. The antenna connection is shown at J2. This connector can be either a BNC or an "F" type chassis mounted connector. The receiver circuit has two audio output jacks both at J3 and J4. Two circuits of mini ⅛″ stereo jacks were used in the prototype.

Once the printed circuit has been assembled, you will need to recheck the solder joints to make sure they are smooth and that there are no cold solder joints. Also observe the PC circuit lines to ensure that there are no bridges or "shorts" formed by cut component leads. Once the circuit has been checked, you are ready to install the circuit board in the chassis.

A metal enclosure was chosen, since this is a sensitive RF receiver and you will want to keep outside interference from affecting the circuit operation, see Figure 22-2. You will have to wire the four jacks J1 through J4 to the circuit board. We elected to mount the power and antenna jacks on the rear panel of the chassis box. One audio jack was mounted on the front panel, while the second audio jack was mounted on the rear chassis panel. The power switch S1 was wired to the PC and mounted on the front panel of the receiver along with the power indicating LED. The tuning control R7 and volume control R15 were both mounted on the front panel of the chassis box.

Once you are satisfied that your circuit board is wired correctly. You are ready to apply power to the circuit and begin testing the circuit for proper operation.

Testing and alignment

The receiver requires 12 volts DC (vdc) which may be obtained from a well regulated power supply or from a battery. Current drain is approximately 60 milliamps (ma).

The power cable supplied with the kit has a female power plug on one end and stripped leads on the other. Notice that the power cable has a black stripe, or tracer, along one of the wires. This is the wire that is connected to the center conductor of the plug and must be connected to the (+) side of the power source. The Radio Shack RS 23-007 (Eveready) 12 volt battery or equivalent is suitable.

Next, turn the Jupiter receiver's power switch to OFF. Connect either headphones or an amplified speaker (Radio Shack 277-1008C or equivalent) to the receiver audio output (J3 or J4). These jacks accept 3.5 mm (⅛″) monaural or stereo plugs.

If you are using a Radio Shack amplified speaker, turn it ON and adjust the volume control *on the speaker* up about ⅛ turn. If you are using headphones, hold them several inches from your ear as there may be a loud whistle due to the internal test oscillator. Turn the receiver ON. The LED should light. Set the volume control to the 12 o'clock position. Allow the receiver to "warm-up" for several minutes.

Set the TUNING control to the 10 o'clock position. *Carefully* adjust inductor L5 with the white tuning stick until a loud low frequency tone is heard in the speaker (set volume control as desired). *Caution*: Do not screw down the inductor slugs too far, as the ferrite material could crack. By adjusting L5 to hear the tone, you are tuning the receiver to 20.00 MHz. The signal which you hear is generated in OSC1, a crystal controlled test oscillator built into the receiver. Once L5 has been set, DO NOT readjust it during the remainder of the alignment procedure. (When the receiver tunes 20.00 MHz with the nob set to the 10 o'clock position it will tune 20.1 MHz with the knob centered on the 12 o'clock position.) The following steps involve adjusting variable capacitors (C2 and C6) and a variable inductor (L4) to obtain the maximum signal strength (loudest tone) at the audio output. For some, it is difficult to discern slight changes in the strength of an audio tone simply by ear. For this reason three different methods are described, each using a form of test instrument. In the event that no test equipment is available, then a fourth method—simply relying on the ear—is possible. In all cases, adjust the receiver tuning knob so that the audio tone is in the range of about 500 to 2000 Hz. Use procedure A as a method to use in tuning up your receiver. Procedures B-D are optional tuning methods.

If no test equipment is available, then simply tune by ear for the loudest audio signal. Listen to the tone and carefully adjust the tuning knob to keep the pitch constant. If the pitch changes during the alignment, it indicates that the receiver has drifted off frequency. As you make adjustments, the signal will get louder. Reduce the receiver volume control as necessary to keep the tone from sounding distorted or clipped.

To test the receiver on the air, simply connect the antenna. For best performance, use a 50 ohm antenna designed to operate in the frequency range of 19.9–20.2 MHz. At certain times of the day, you should be able to hear WWV or WWVH on 20.000 MHz. These are standard time and frequency stations, located in Colorado and Hawaii, which broadcast the time of day as well as other information related to propagation and solar-terrestrial conditions. The Jupiter Radio Telescope direct conversion receiver design does not allow clear reception of amplitude modulated (AM) stations like WWV, so the voice will probably be garbled, unless you tune very precisely. The receiver does work well on single sideband (SSB) signals and code (CW).

In the event that the Jupiter receiver does not appear to work, you will need to disconnect the power supply and antenna and examine the circuit board for errors. The most common cause for building errors are the incorrect installation of the electrolytic capacitors and diodes which may have been installed backwards, since the components have polarity which must be observed. Another possible cause for errors are the incorrect placement of resistors. Often color codes are read wrong and the wrong resistor may be installed at a particular location. Finally one of the most common causes for errors is the incorrect installation of transistors and integrated circuits. Make sure that you can identify the manufacturer's pin-out diagrams and pay particular attention to if the pin-outs are for top or bottom views. Have a knowledgeable electronics enthusiast help you scan the circuit for possible mistakes; a second pair of eyes always helps to find errors you may have missed as the builder. Once you have located and fixed the problem, you can apply power and re-test the circuit for proper operation.

The jupiter radio telescope antenna

The antenna intercepts weak electromagnetic waves that have traveled some 500 million miles from Jupiter to the Earth or 93 million miles from the Sun. When these electromagnetic waves strike the wire antenna, a tiny radio frequency (RF) voltage is developed at the antenna terminals. A basic dipole antenna is shown in Figure 22-6. The recommended antenna for the Jupiter Radio Telescope is the dual dipole shown in Figure 22-7. Signals from each single dipole antenna are brought together with a power combiner via two pieces of coaxial cable. The output of the power combiner is delivered to the receiver by another section of coaxial transmission line. The antenna system requires a fair-sized area for setup: minimum requirements are a 25 × 35 ft. flat area that has soil suitable for putting stakes into the ground. As the antenna system is sensitive to noise it is best not to set it up near any high tension power lines or close to buildings. Also for safety reasons, please keep the antenna away from power lines during construction and operation. The best locations are in rural settings where the interference is minor. Since many of the observations occur at night it is wise to practice setting up the antenna during the day to make sure the site is safe and easily accessible.

If you want to locate your receiver indoors and your antenna is outside or on the rooftop, and your antenna lead will not reach, you must use a longer

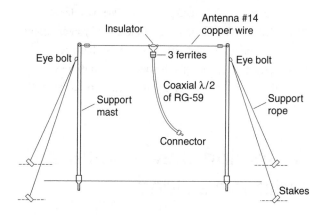

Figure 22-6 *Basic dipole antenna diagram. Courtesy Wes Greenman, Charles Higgins*

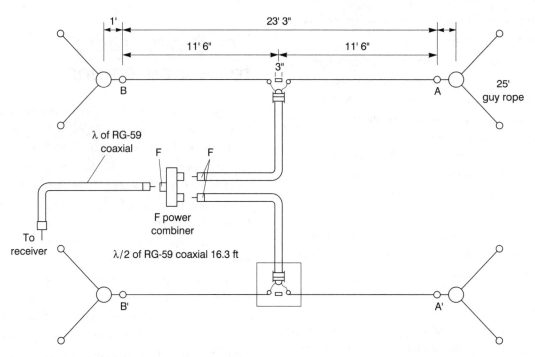

Figure 22-7 *Antenna combiner diagram. Courtesy Wes Greenman, Charles Higgins*

coax cable. Do not use just any length of coax for the combiner assembly. The cable going from the power combiner harness assembly to the receiver should be a multiple half wavelength long. The cable supplied with the kit is one wavelength—that is 9.85 meters long (taking into account the 66% velocity factor of the RG-59/U cable). The maximum recommended cable length is 5 wavelengths. There are many different manufacturers and qualities of coaxial cable. The 75 ohm cable supplied with the kit is manufactured by Belden and has a solid center conductor and velocity factor of 66%. Radio Shack does not carry RG-59/U cable but they do have RG- 6 and the higher grade RG-6QS (quad shield), which is also 75 ohm cable. Both of these cables have a velocity factor of 78%. One wavelength at 20.1 MHz in RG-6 cable is 11.64 meters. If you are going to put in a longer feedline we recommend that you completely replace the existing one wavelength piece—rather than splicing another length of cable onto the end.

Observing tips—jupiter's radio emissions

Most Jovian radio storms, received with the Jupiter radio telescope, will likely not be very strong. You will need to maximize your chances of receiving the Jovian storms by picking the best opportunities, selecting a good observing site, and using good observing techniques.

During the daytime, the Earth's Ionosphere blocks Jupiter's radio emissions, so therefore you will have to confine your investigations to night time listening. Radio noise from electrical sources, such as power lines, fluorescent lights, computers, and motors, can mask the signals from Jovian storms.

If possible, try to select an observing sight away from buildings and power lines, also try to use portable power supplies such as a battery power supply. Note that Jovian radio storms may be short in duration, so persistence and luck are needed to capture Jupiter's radio signals.

Night time observing from a temporary field setup can be dangerous, so PLAN CAREFULLY!! Get permission in advance to use the site, set up before dark, and use caution when walking around the site at night. The Jupiter Radio Telescope has two audio outputs, one channel can be used for ordinary listening and the second output jack can be used for recording. You can elect to use a portable audio tape recorder, however, note that nearly all common portable tape recorders use an automatic recording level control circuit (ALC). The ALC can mask the variations in the Jovian L bursts.

Use a tape recorder that allows you to SWITCH OFF the ALC. Keep the ALC turned off.

Another alternative for recording Jovian storms is the use of a computer or laptop to record the signals from your Jupiter radio telescope. Using a laptop computer to record the Jovian storms has the distinct advantage that you can analyze the audio recording with special spectral software which can help you make sense of the complex signals received.

Note you could also elect to use an audio tape recorder to record the Jovian signals in the field and later when you return home play the tape recording into your laptop or personal desktop computers. There are many audio analyzing software programs out there on the Internet, many are free. Many programs allow FTT and frequency division analysis.

Chapter 23

Weather Satellite Receiver

Parts list

Weather Satellite Receiver

R1 22 ohm, $\frac{1}{4}$ watt, 5% resistor

R2 150 k ohm, $\frac{1}{4}$ watt, 5% resistor

R3 110 k ohm, $\frac{1}{4}$ watt, 5% resistor

R4,R10,R11 22 k ohm, $\frac{1}{4}$ watt, 5% resistor

R5 47 k ohm, $\frac{1}{4}$ watt, 5% resistor

R6 1.8 k ohm, $\frac{1}{4}$ watt, 5% resistor

R7 360 ohm, $\frac{1}{4}$ watt, 5% resistor

R8 100 ohm, $\frac{1}{4}$ watt, 5% resistor

R9 470k ohm, $\frac{1}{4}$ watt, 5% resistor

R12,R13,R16,R27 1 k ohm, $\frac{1}{4}$ watt, 5% resistor

R14 240 ohm, $\frac{1}{4}$ watt, 5% resistor

R15 1.2 k ohm, $\frac{1}{4}$ watt, 5% resistor

R17 5.6 k ohm, $\frac{1}{4}$ watt, 5% resistor

R18 39 k ohm, $\frac{1}{4}$ watt, 5% resistor

R19,R21 10 k ohm, $\frac{1}{4}$ watt, 5% resistor

R20,R31 2.2 k ohm, $\frac{1}{4}$ watt, 5% resistor

R22 100 k ohm, $\frac{1}{4}$ watt, 5% resistor

R23 220 ohm, $\frac{1}{4}$ watt, 5% resistor

R24 390 k ohm, $\frac{1}{4}$ watt, 5% resistor

R25 10 ohm, $\frac{1}{4}$ watt, 5% resistor

R26 1.5 k ohm, $\frac{1}{4}$ watt, 5% resistor

R28 4.7 k ohm, $\frac{1}{4}$ watt, 5% resistor

R29 300 ohm, $\frac{1}{4}$ watt, 5% resistor

R30 47 ohm, 14watt, 5% resistor

VR1 50 k linear pot, 16mm PC board mount

VR2 25 k linear pot, 16mm PC board mount

VR3 50k log pot, 16mm PC board mount

VR4,VR5 50 k 10-turn trimpots, PC board mount

C1,C2,C3,C5 2.2 nF, 16v, disk ceramic capacitor

C6,C7,C8,C12 2.2 nF, 16v, disk ceramic capacitor

C4,C10 3-10 pF trimcaps

C9 1 nF, 16v, disk ceramic capacitor

C11,C34,C35 10 μF, 16v, tantalum capacitor

C13,C27 10 pF, 16v, NPO ceramic capacitor

C14,C19,C21,C23,C25 10 nF, 16v, mono ceramic capacitor

C15,C16 15 pF, 16v, NPO ceramic capacitor

C17 10 μF, 35v Tantalum capacitor

C18a,C18b,C20,C22 100 nF, 16v, mono ceramic capacitor

C24,C26,C28,C32 100 nF, 16v, mono ceramic capacitor

C29 390 pF, 16v, NPO ceramic capacitor

C30 4,7 nF, 16v, metal polyester capacitor

C31 1 nF, 16v, metal polyester capacitor

C33 470 nF, 16v, metal polyester capacitor

C36 330 μF, 16v, electrolytic capacitor

C37 47 nF, 16v, metal polyester capacitor

C38 22 nF,16v, metal polyester capacitor

C39 2200 μF, 25v, electrolytic capacitor

C40 470 μF, 35v, electrolytic capacitor

RFC1 4.3 μH slug tuned coil - see Table 23-3

RFC2 10 μH slug tuned coil - prewound

RFC3 68 μH slug tuned coil - prewound

L1 see Table 23-3

L2 see Table 23-3

L3 see Table 23-3

L4 see Table 23-3

D2 5.1V 400 mW Zener diode

D3,D4 3 mm red LED

D1 ZMV833ATA varicap

X1,X2 5.5 MHz ceramic filters, Murata SFTRD5M50AF00-B0

Q1 BF998 dual-gate MOSFET

Q2 PN100 NPN transistor

U1 SA605D mixer/IF amplifier/FM detector

U2 TL072 dual op-amp

U3 LM386 audio amp

U4 78L05 +5V regulator

J1,J2 BNC type jacks

J3 $\frac{1}{8}$, 3.5 mm stereo headphone jack

J4 1 2.5 mm concentric power socket, PC-mount

S1 SPDT miniature toggle switch

Misc PC board (double-sided, not plated through)

TO-220 heatsink,

1 coil former, 4.83 mm OD with F16 ferrite slug

1 F29 ferrite bead (for RFC1)

1 short length of 0.8 mm tinned copper wire for L1

1 length of 0.8 mm enameled copper wire for L2, L3

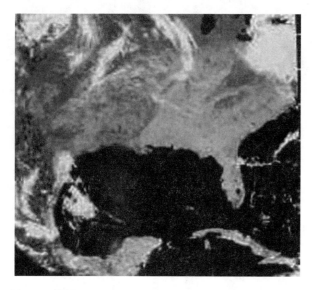

Figure 23-1 *Satellite view – USA*

The days of guessing the weather by looking at the clouds overhead have just ended. Now you can look at the clouds from above! This satellite receiver/decoding project will allow you to receive pictures from satellites 600 km overhead, see Figure 23-1. A typical NOAA satellite can cover nearly one-sixteenth of the Earth in a single pass! Living in New York, for example, we can clearly capture images from mid-Hudson bay (where there was still ice in late spring), all the way down past Cuba, as well as spanning from Wisconsin to far out in the Atlantic Ocean. The clarity of the image was enough to see the individual Finger Lakes (in New York), and shadows on the underside of thunderstorms. With the weather satellite receiver and a personal computer, you can easily receive and decode weather satellite pictures from overhead satellites.

The US National Oceanographic and Atmospheric Administration (NOAA) operates both GEO and LEO polar satellites. Here we concentrate primarily on the polar orbiters. The NOAA polar satellites orbit at 850 km and pass within view of all areas on Earth at least twice day. The satellite carries a number of instruments including cameras for both visible and infrared light.

The cameras scan back and forth at right angles to the ground path, like a broom sweeping side-to-side as you walk forward, taking picture strips that cover an area 3000 km wide. The satellite thus makes a continuous picture as if it were a tape reeling out from an endless roll. The image, however, is not recorded on the satellite. Each image strip is immediately broadcast to

the ground at a frequency just above 137 MHz. The satellite will be in range for up to 12 minutes as the satellite passes from horizon to horizon.

The broadcast uses the APT (Automatic Picture Transmission) analog format for the imagery. (A digital format High Resolution Picture Transmission (HRPT) signal is also transmitted but it is more difficult to receive and decode and we did not try to receive it in this project.) Currently, there are six operational NOAA polar orbiters but usually only two or three have the APT activated at a given time. (In this report, only images from NOAA-14 and NOAA-15 were received.) See NOAA satellite status.

The Russian METEOR series of satellites also broadcast with the APT format. In addition, the remote sensing SICH and OKEAN radar satellites occasionally broadcast images but only over Russia (scanners in nearby countries have picked up some of these images).

The APT system

The APT satellites give about 12 good images a day, wherever you are in the world. The resolution is 4 km per pixel and there are visible and infrared sensors. This is a simple system where the satellite is in low Earth orbit and very easily received at 137 MHz. The two frequencies in main use are 136.50 for NOAA 12 and 15 and 137.62 MHz for NOAA 14. A simple crossed dipole antenna can be used, or for marine use a stainless steel quadrifilar helix antenna. These simple omni-directional antennas do not need moving or tracking to receive the satellite. The satellite is so strong that when is about 20° above the horizon perfect results will be obtained. The APT signal is easily recognizable by its modulated frequency (FM) and carrier tonality ranging between 1500 and 2500 Hz. New satellites have a 2400 Hz carrier like this NOAA 14 APT signal.

In APT mode, satellites transmit at the rate of 120 LPM (lines per minute) alternating two channels, one for the visible image, the other for the infrared image. This is through the decoding software that you select one or another channel.

Russian weather satellites only transmit infrared images at 120 LPM. The sound of their transmission is thus slightly different from the other weather satellites like on this record of a RESURS 01-N4 signal with an IOC of 382. On their side, METEOR 3-05 uses a higher carrier closer to 2500 Hz with an IOC of 382 as well.

The WX satellite receiver project will allow you to receive weather satellite transmissions on the VHF band, where most of the polar-orbiting satellites are located. You will recognize these transmissions on the news when you see the time lapse of the clouds darting across the countryside. The weather man in this case has taken multiple images on the computer, aligned and pieced them together, and then run through one image after the other. It is possible to do this same thing with this kit and the proper software.

The way in which a weather satellite works is fairly simple. Just think of your office fax machine as an example. The satellites circle the Earth going north to south back to north again almost directly over the poles, which is why they call it a polar orbit. This means that the satellite will cover every location on the Earth at least twice per day. With a good antenna, and partly because of overlap of consecutive orbits, you can conceivably receive the same satellite up to six times a day! Notice though that the image received from polar orbits will be upside down on every other pass.

The satellite retrieves the data in a linear fashion, one line at a time using a *scanning radiometer*. The scanning radiometer transmits the equivalent of a single television horizontal line as the satellite circles the earth. The system uses a series of optics and a motor driven rotating mirror system to receive a very narrow line of the image of the Earth. Each line is received at a right angle to the satellite's orbital track, so as the satellite circles the Earth, a line is received from west to east or east to west depending on the orbit of the satellite. The total image is received from north to south or south to north, depending on the orbit also, and this motion is what relays the equivalent of the vertical scan in a television. You can continue receiving this satellite as long as it is within the line of sight.

Since all of the receivable satellites are similar, we will describe the ones you will most commonly receive. The NOAA/TIROS satellites, during the first half of the transmission, send visible light data to the receiver at the same time they are taking in the view. Meanwhile during the same part of the scan, they are recording the infrared view. During the second half of the scan, while the sensors are facing away from the Earth, it sends the infrared data. The user then sees the data as two images side by side, on the left the visible light data is seen, and on the right, infrared data is seen. In between the images are synchronization pulses that help computers to align the individual lines precisely.

These particular satellites continuously transmit an FM signal modulated with a 2400 Hz tone. This tone is very precise in frequency so the image seen is aligned properly. The 2400 Hz tone is AM modulated with the intensity of the current view of the Earth. The brighter or colder the point on the Earth, the higher in amplitude the 2400 Hz signal is.

The receiver demodulates the FM signal and retrieves the 2400 Hz tone. The detector board in the computer will then find the peak amplitude of each wave of the 2400 Hz tone, and each peak, upper and lower, now represents a single pixel on the screen. For the NOAA/TIROS satellites, each horizontal line represents 2400 pixels, since the incoming frequency is 2400 Hz, and the scanning radiometer rotates twice per second. The full 12 minute pass of a NOAA satellite requires approximately 3.5 MB (3.5 million 8 bit pixels) of storage! This is much more data (pixels) than can be seen on a super VGA screen at any one time.

Many people mistakenly think that they can easily utilize a conventional VHF scanner as an APT satellite picture receiver. Most scanners and communications receivers usually only provide a choice of two bandwidth settings for VHF FM reception: "narrow" and "wide." The narrow setting gives a bandwidth of ±15 kHz or less, which is fine for NBFM reception, and the wide bandwidth setting usually gives a bandwidth of about ±100 kHz, so this is the setting that must be used. Unfortunately, this "wide" bandwidth is actually too wide for weather satellite signals and, as a result, the demodulated audio level is relatively low. At the same time, the wider reception bandwidth allows more noise through, so the signal-to-noise ratio can become quite poor. Only a few scanners have the correct bandwidth and these currently include the AOR5000 and Icom IC-PCR1000. The second problem with all scanners,

Figure 23-2 *Weather satellite receiver*

including the ones mentioned, is that a weather satellite system needs a very good performance receiver, one that provides high sensitivity, good signal to noise and high immunity to other adjacent transmissions. So, you may get a scanner to work, but it will not provide the optimum results.

Ideally, in order to obtain the best results you really need to have a receiver with a ±30 kHz bandwidth.

This type of specialized VHF receiver is available but fairly pricey. Hence the motivation for developing the low-cost weather satellite receiver described here.

The weather satellite receiver project in this chapter features good sensitivity, at about 0.7V for 12 dB of quieting. At the same time, the effective bandwidth is approximately ±35 kHz, which is quite suitable for weather satellite reception. The satellite receiver is built into a very compact plastic instrument box, as shown in Figure 23-2. All of the circuitry is mounted on a double-sided PC board, so it's quite easy to build. It has switch tuning between two preset frequency channels, for ease of use. There are RF Gain, Audio Muting and Audio Gain controls and the receiver can drive a small monitor speaker or headphones, as well as providing a line level signal to feed into your PC for recording and decoding.

The heart of the receiver is a Philips SA605D IC high-performance low-power FM mixer and IF system on a single chip. As you can see from the block diagram of Figure 23-3, it contains a local oscillator transistor and balanced mixer, plus a high-gain IF amplifier and IF limiter, a received signal strength (RSSI) detector, an

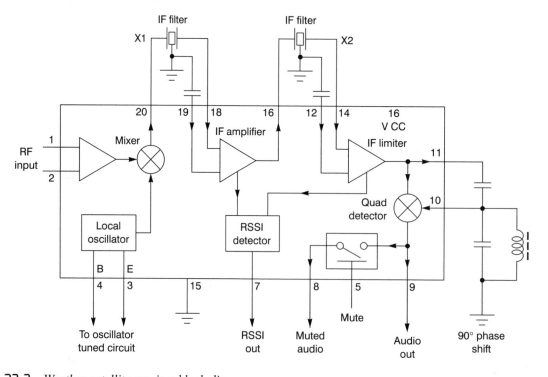

Figure 23-3 *Weather satellite receiver block diagram*

FM quadrature detector and finally an audio muting circuit. The local oscillator transistor can operate at frequencies up to about 500 MHz in an LC circuit, or up to 150 MHz with a suitable crystal. The mixer can operate up to 500 MHz as well, while the IF amplifier and limiter can operate up to about 25 MHz with a combined gain of about 90 dB.

The main receiver diagram illustrated in Figure 23-4 depicts the complete weather satellite receiver circuit details. In this receiver project, we're using the SA605D in a fairly conventional single-conversion superhet configuration, with the IF amplifier and limiter working at 5.5 MHz. This allows us to take advantage of high selectivity 5.5 MHz TV sound IF ceramic filters to provide most of our bandwidth shaping. The two filters in question are X1 and X2, which are both Murata SFT5.5MA devices.

As shown in the receiver schematic, X1 is connected between the mixer output and the IF amplifier input, while X2 is connected between the IF amplifier output and the limiter input. The resistors connected to the filter inputs and outputs are mainly for impedance matching, while the 10 nF capacitors are for DC blocking. The 90° phase shift required for U1's quadrature FM detector is provided by coil L4 and its parallel 390 pF capacitor, which are tuned to 5.5 MHz.

The local oscillator transistor inside U11 is connected in a Colpitts circuit. This includes coil L3, together with the two 15 pF capacitors (which provide the emitter tap) and a 10 pF capacitor in series with the Varicap diode D1.

The Varicap diode D1 is the receiver's tuning capacitor. Its tuning voltage for each of the two channels is set by 10-turn trimpots VR4 and VR5, with switch S1 selecting between them. We can tune the receiver simply by changing the local oscillator frequency because we only need to tune over a relatively small range (i.e. 137.3–137.85 MHz maximum), which is within the selectivity curve of the "front end" tuned circuits.

The satellite receiver's front end uses a BF998 dual-gate MOSFET (Q1) connected in a standard cascode amplifier configuration. The incoming VHF signals are fed into a tap (for impedance matching) on antenna coil L1, which is tuned to about 137.55 MHz using trimmer capacitor VC1. The signal from the top of this tuned

circuit is then fed directly to gate 1 of Q1, while gate 2 is bypassed to ground but also fed with an adjustable DC voltage via VR1 for RF gain control. The amplified VHF signal on Q1's drain is fed to pin 1 of U1 via a 1 nF coupling capacitor. Additional RF selectivity is provided by coil L2 and trimmer capacitor VC2, which are again tuned to about 137.55 MHz. The 100 ohm resistor and 10 μH RF choke form an untuned high-impedance load for Q1.

Notice that as well as being coupled to the tap on L1 via a 2.2 nF capacitor, the antenna input is also connected to the +12V supply line via RFC1 and a series 22 ohm resistor. As you may have guessed, these components are there to provide "phantom" DC power for the masthead pre-amplifier. At the output end of U1, we take the demodulated APT signals from the "muted audio" output at pin 8. This allows us to take advantage of the SA605's built-in muting circuit, which works by using comparator stage U2b to compare U1's RSSI output from pin 7 (proportional to the logarithm of signal strength) with an adjustable DC control voltage from muting pot VR2. When the RSSI voltage rises above the voltage from VR2, U2b's output switches high and this fed to pin 5 of U1 via a 2.2 k ohm series resistor to un-mute the audio. D2, a 5.1V Zener diode, limits the swing on pin 5 of U1 to less than 6V.

Transistor Q2 and LED1 form a simple signal strength indicator. This also uses the RSSI output from U1. In operation, the voltage across the 390 k ohm resistor and 100 nF capacitor rises from about +0.26V under no-signal conditions to about +5V with a very strong input signal. So with Q2 connected as an emitter follower and LED1 in its emitter load, the LED current and brightness are made to vary quite usefully with signal strength.

The demodulated APT signal from pin 8 of U1 is first fed through op-amp U2a, which is configured as an active low-pass filter. This has a turnover frequency of 5 kHz and is used for final de-emphasis and noise reduction. From there, the signal is fed to audio gain control VR3 and then to audio amplifier stage U3. This is a standard LM386 audio amplifier IC, configured for a gain of about 40 times. Its output is fed to both the monitor speaker socket and to a line output socket for connection to your PC's sound card.

Most of the receiver's circuitry operates from +12V, with the exception of U1 which needs +6V. As a result,

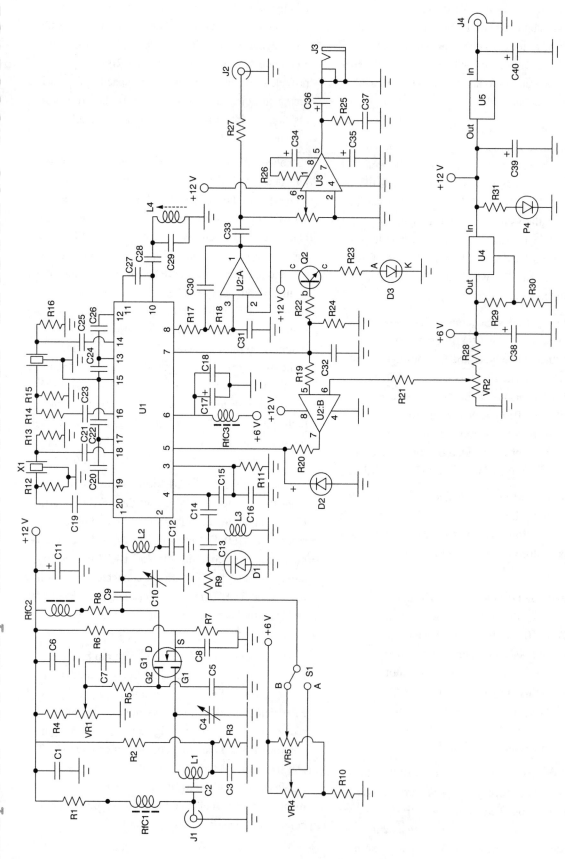

Figure 23-4 *Weather satellite receiver schematic. Courtesy of Silicon Chip Magazine*

the power supply circuitry includes U5 to provide a regulated and smoothed +12 V supply from an external supply such as a 14.5–18 V "wall-wart." This is followed by 5V regulator U4 which has its output "jacked up" using 300 ohm and 47 ohm resistors to provide close to +6 V for the SA605D (U1). Note that if you want to run the receiver from a 12 V battery, this can be quite easily done by replacing REG1 with a wire link. In addition, the 2200 μF capacitor should be replaced with a 16 V Zener diode (ZD2) for over-voltage protection.

Construction

Before beginning the project make sure that you have a clean well lit and well ventilated work space for constructing your satellite receiver. Locate all the diagrams and place them in front of you on the table or work bench. Locate a 27 to 30 watt pencil-tip soldering iron, along with some "Tip Tinner," which is used to prepare and clean your soldering iron tip. Locate a roll of 60/40 tin/lead rosin core solder as well as a few small hand tools such as a pair of small end-cutters, a needle-nose pliers, a pair of tweezers, and some small screwdrivers.

Construction of the weather satellite receiver is straightforward with virtually all of the parts mounted on a small PC board. The board is double-sided but the top copper pattern is used mainly as a ground-plane. This means that the board doesn't need to have plated-through holes but there are quite a few component leads which do have to be soldered on both sides of the board. There are also a number of "pin-through" or "hole-through" wires which have to be soldered to both sides of the PC board around U1. These connect the Earth patterns on both sides of the board and ensure that this high-gain chip operates in a stable manner.

The various input and output connectors are mounted along the rear edge of the circuit board, while the controls and indicator LEDs mount along the front edge. The only component not actually mounted on the board is S1, the channel select toggle switch. This mounts on the front panel, with its three connection lugs wired to PC board terminal pins directly underneath using very short lengths of insulated hookup wire.

Start the assembly by mounting these three terminal pins first, they are the only pins used in the receiver project. Next, mount the four connectors CON1-CON4 along the rear edge, followed by the resistors. Table 23-1 shows the resistor color codes but it's also a good idea to check each value using a digital multi-meter before

Table 23-1

Resistor color code chart

Color Band	1st Digit	2nd Digit	Multiplier	Tolerance
Black	0	0	1	
Brown	1	1	10	1%
Red	2	2	100	2%
Orange	3	3	1,000 (K)	3%
Yellow	4	4	10,000	4%
Green	5	5	100,000	
Blue	6	6	1,000,000 (M)	
Violet	7	7	10,000,000	
Gray	8	8	100,000,000	
White	9	9	1,000,000,000	
Gold			0.1	5%
Silver			0.01	10%
No color				20%

soldering it in position. All resistors are mounted to the top of the PC board but note that some of them have one lead soldered to the top copper as well as the bottom copper. Once the resistors are in place, you can use excess resistor leads to allow connection from one side of the board to the other side, since this project uses a two-sided circuit board. Each "pin-through or through-hole" is fitted by simply passing a wire through the hole in the board, then soldering it on both sides and trimming off the excess lead lengths.

The small ceramic capacitors can now all be installed on the left-hand side of the board. Note that some of these also have their "cold" leads soldered on both sides of the board, as indicated by the red dots. Once they're in, install the MKT capacitors and the electrolytic capacitors, making sure that the latter are all correctly orientated. Use the chart shown in Table 23-2 to help locate the small capacitors from the parts pile. Often small capacitors do not have their actual value printed on them but instead use a three-digit code to represent the actual value.

Finally go ahead and mount the trimmer capacitors VC1 and VC2. These should be mounted so that their adjustment rotors are connected to earth (this makes it much easier to align the receiver later). It's simply a matter of orientating them on the board.

Next, we will move on to installing the RF chokes. RFC2 and RFC3 are both pre-wound (10 μH and 68 μH respectively) but RFC1 needs to be wound on an F29 ferrite bead. It's very easy to wind though, because it requires only three turns of 0.25 mm enameled copper wire.

Winding coils

At this point in time, it's a good idea to wind and mount the remaining coils. Table 23-3 gives the winding details of the coils. As shown, L1–L3 are air-cored types, each consisting of five turns of 0.8 mm diameter wire wound on a 5 mm mandrel. Note, however, that L1 is wound using tinned copper wire, while L2 and L3 are both wound using enameled copper wire. Don't forget to scrape off the enamel at each end, so they can be soldered to the board pads.

Coils L1–L3 should all be mounted so that their turns are about 2 mm above the board. After you've fitted L1,

don't forget to fit its "tap" connection lead as well. This can be made from a resistor lead off-cut, since it's very short. It connects to a point ⅓ of a turn up from the "cold" (earth) end of the coil, i.e. just above half-way up the side of the first turn.

The final coil to wind is quad detector coil L4. Unlike the others, this is wound on a 4.83 mm OD former with a base and a copper shield can. It's wound from 20 turns of 0.25 mm enameled copper wire and tuned with an F16 ferrite slug. Coil formers with ferrite slugs are difficult to locate, check the Appendix. You could also experiment with a small "junk" IF transformer can. You could try and unwind the existing coil and re-winding your own coil over the plastic former. You must make sure that there is a ferrite slug in the center of the coil former. Once L4 is wound, and installed onto the board, cover it with its shield can. You can next mount the two ceramic filters X1 and X2. These devices can be fitted either way around but make sure that their pins are pushed through the board holes as far as they'll go before you solder them underneath.

The next step is to construct and install the small metal shield around coil L2, C10 and most of the components in the drain circuit of Q1. This shield is U-shaped and measures 20 mm high, with the front and back "arms" 36 mm long and the side section 20 mm long. The bottom edges of all three sides are soldered to the board's ground-plane in a number of places, to hold it firmly in position and to ensure it stays at earth potential.

The two 10-turn trimpots (VR4 and VR5) can now be soldered in position at the front-center of the board. They can then be followed by the three main control pots, which are all 16 mm diameter types. Trim each pot's spindle length to about 9 mm before fitting it and make sure you fit each one in its correct position as they are all different. In particular, note that VR1 and VR3 both have a value of 50 k ohm but VR1 is a linear pot while VR3 is a log type.

After installing the potentiometers, it's a good idea to connect their metal shield cans together and then run a lead to the board's top copper to earth them. This is done using a length of tinned copper wire, with a short length of insulated hookup wire then connecting them to the board copper at front right.

Table 23-2

Capacitance code information

This table provides the value of alphanumeric coded ceramic, mylar and mica capacitors in general. They come in many sizes, shapes, values and ratings; many different manufacturers worldwide produce them and not all play by the same rules. Some capacitors actually have the numeric values stamped on them; however, many are color coded and some have alphanumeric codes. The capacitor's first and second significant number IDs are the first and second values, followed by the multiplier number code, followed by the percentage tolerance letter code. Usually the first two digits of the code represent the significant part of the value, while the third digit, called the multiplier, corresponds to the number of zeros to be added to the first two digits.

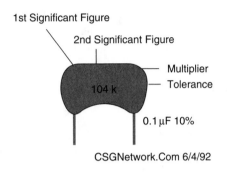

Value	Type	Code	Value	Type	Code
1.5 pF	Ceramic		1,000 pF /.001 µF	Ceramic / Mylar	102
3.3 pF	Ceramic		1,500 pF /.0015 µF	Ceramic / Mylar	152
10 pF	Ceramic		2,000 pF /.002 µF	Ceramic / Mylar	202
15 pF	Ceramic		2,200 pF /.0022 µF	Ceramic / Mylar	222
20 pF	Ceramic		4,700 pF /.0047 µF	Ceramic / Mylar	472
30 pF	Ceramic		5,000 pF /.005 µF	Ceramic / Mylar	502
33 pF	Ceramic		5,600 pF /.0056 µF	Ceramic / Mylar	562
47 pF	Ceramic		6,800 pF /.0068 µF	Ceramic / Mylar	682
56 pF	Ceramic		.01	Ceramic / Mylar	103
68 pF	Ceramic		.015	Mylar	
75 pF	Ceramic		.02	Mylar	203
82 pF	Ceramic		.022	Mylar	223
91 pF	Ceramic		.033	Mylar	333
100 pF	Ceramic	101	.047	Mylar	473
120 pF	Ceramic	121	.05	Mylar	503
130 pF	Ceramic	131	.056	Mylar	563
150 pF	Ceramic	151	.068	Mylar	683
180 pF	Ceramic	181	.1	Mylar	104
220 pF	Ceramic	221	.2	Mylar	204
330 pF	Ceramic	331	.22	Mylar	224
470 pF	Ceramic	471	.33	Mylar	334
560 pF	Ceramic	561	.47	Mylar	474
680 pF	Ceramic	681	.56	Mylar	564
750 pF	Ceramic	751	1	Mylar	105
820 pF	Ceramic	821	2	Mylar	205

Table 23-3
Coil winding details

L1   5 turns of 0.8 mm (20 ga.) tinned copper wire, 7 mm long (wound on 5 mm diameter mandrel). Tap at 0.3 turns. Air core, mounted 2 mm above board.

L2   5 turns of 0.8 mm (20 ga.) enameled copper wire, 9 mm long (wound on 5 mm diameter mandrel). Air core, mounted 2 mm above board.

L3   5 turns of 0.8 mm (20 ga.) enameled copper wire, 8 mm long (wound on 5 mm diameter mandrel). Air core, mounted 2 mm above board.

L4   20 turns of 0.25 mm (30 ga.) enameled copper wire, tight wound on 4.83 mm OD former with F16 tuning slug and shield can.

RFC1   3 turns of 0.25 mm (30 ga.) enameled copper wire wound on an F29 ferrite bead.

Mounting the semiconductors

Before we go ahead and install the semiconductors, let's take a brief look at the semiconductor pin-out diagram illustrated in Figure 23-5, which will assist you in orienting the integrated circuits. Now you can go ahead and mount the semiconductors. Begin by installing the Varicap diode at D1, and the 5.1V Zener diode at D2 and transistor Q2. Next, install regulator U5 and its associated heatsink. These parts are secured to the board using a 6 mm-long M3 screw, nut and lock-washer.

Note that U5's center pin should be soldered on both sides of the board but take care not to touch either of the two large adjacent electrolytic capacitors with the barrel of your soldering iron.

Go ahead and mount regulator U4, followed by U22 (TL072) and U3 (LM386). Note that pin 4 on both these devices should be soldered to the copper on the top of the board as well as the bottom. The two D3 and D4 are both mounted horizontally, so that they later protrude through matching 3 mm holes in the front panel. Note that they are both fitted with their cathode leads towards the left. Bend their leads down through 90° about 5 mm from the LED bodies, then solder them in position so that the axis of each LED is 5 mm above the board.

The final components to be mounted on the PC board include the Dual-Gate MOSFET at Q1 and the SA605D IC at (U1), U2 and transistor Q2. These components should be handled very carefully, using an anti-static wrist-band to avoid static charges building up on the components. The two integrated circuits should be mounted in IC sockets. IC sockets are an insurance policy against circuit failure at a later date. It is much easier to simply unplug a faulty IC, rather than trying to un-solder 14 or 16 pins from the circuit. Most people cannot un-solder 16 pins without damaging a circuit board. When mounting the IC sockets, observe a small cut-out at one edge of the socket. The cut-out will identify the top of the socket. Usually pin of the IC will be located to the left of the cut-out. Solder the sockets in their respective locations. Finally insert the ICs into their respective sockets, noting pin 1 is to the left of the small indented circle or cut-out on the top of the IC package. When handling the MOSFET be extremely

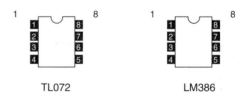

Figure 23-5 *Semiconductor pin-outs*

careful to avoid damage from static electricity. Do not move around when installing these static sensitive components and be sure to use an anti-static wrist band to avoid damage.

The completed PC board is housed in a suitable low-profile plastic instrument case.

First, you can mount switch S1 on the front panel and connect three 30 mm lengths of insulated hookup wire to the three pins on the PC board. That done, the front panel can be mated with the PC board by positioning it on the three pot ferrules and doing up the nuts. The three leads from the PC board pins can then be soldered to the switch lugs.

The rear panel is not attached to the board assembly. Instead, it simply slips over connectors CON1 and CON2 and is then slid into the rear slot when the assembly is fitted into the bottom half of the case. Now go ahead and mount the antenna connector onto the rear panel. Finally, the completed assembly is fastened in place using eight 6 mm-long 4-gauge self-tapping screws.

After completing the circuit board assembly, you should take a short rest and when we return we will inspect the circuit board for any possible "short" circuits or "cold" solder joints. Pick up the circuit board with the foil side of the circuit facing upwards toward you. First inspect the circuit board for possible "cold" solder joints. All of the solder joints should look clean, bright and shiny. If any of the solder joints look dark, dull or "blobby," then you should un-solder that particular joint, remove the solder then re-apply some solder and re-solder the joint all over again. After inspecting all of the solder joints, we can move on to looking for "short" circuits. Often sticky rosin core solder will leave a residue on a circuit board. When excess component leads are cut from the board during assembly, many times the "cut" leads will stick to the circuit board and they can "bridge" across the circuit pads and cause a "short" circuit which could damage the circuit when first powered-up. Another potential cause for a "short" circuit is a small solder ball which may have stuck to the circuit board, this too can cause a "short" circuit.

Once you have inspected the circuit board to your satisfaction, you can connect up the power supply leads to the power input jack and you will be ready to test the weather satellite receiver.

Circuit test

First, turn all three front-panel pots fully anticlockwise, connect up your antenna to the antenna input connector, and then apply power from a suitable 14–18V DC "wall-wart" power supply or a 12V battery. Check that the green power LED immediately begins glowing.

If it does, check the voltage on REG1's output lead (i.e. the right-hand lead) with your multi-meter, it should read close to +12V with respect to ground. Similarly, the voltage at REG2's (right-hand) output pin should measure very close to +6V.

If you now plug an 8 ohm speaker into CON3 terminals, and then turn up audio gain control VR3, you should hear a small amount of hiss and noise. When you turn up the RF gain control VR1 as well, this noise should increase a little further but LED1 shouldn't begin glowing except only very faintly when VR1 is turned fully clockwise.

Now turn VR1 fully anticlockwise again and use your multi-meter to measure the DC voltage at the top of the 390k resistor located just behind transistor Q2 (i.e. to the right of 5.1V Zener diode). The voltage across this resistor should be less than 0.30V and preferably about 0.26V. If it's any higher than 0.30V, the IF amplifier in IC1 may be unstable. Assuming that your receiver has passed all these tests, it should be working correctly and is now ready for alignment.

Receiver alignment

The final alignment of the weather satellite receiver is best done with a satellite signal. However, you need to give it a basic alignment first so that you can at least find the signal from a satellite when it's within range. For the basic alignment, you'll need access to a frequency counter capable of measuring up to 150 MHz and an RF signal generator which can be set to give an output at 137.50 MHz and at 137.62 MHz. It should be able to provide either unmodulated (CW) output or frequency modulation, with a modulating frequency of 2.4 kHz and a deviation of ±25 kHz or thereabouts. If the generator can't be accurately set to the above frequencies, you'll need to use the frequency counter to help set its frequency. You'll also need your multi-meter during the "tuning-up" process, to monitor received signal level.

The first step is to set the local oscillator frequencies for the two reception channels. This is done by adjusting trimpots VR4 and VR5 respectively, while measuring the oscillator's frequency with the frequency counter. The oscillator signal is coupled to the counter via a "sniffer" coil which is connected to the end of a coaxial cable. The other end of this cable is then connected to the counter's input. Note that there is no direct physical connection between the oscillator coil and the counter's sniffer coil. Instead, the sniffer coil is placed about 9 mm in front of oscillator coil (L3) and roughly on-axis (i.e. just in front of the 10 pF capacitor).

The sniffer coil can be made by winding four turns of 0.8 mm enameled copper wire on a 5 mm drill shank. Its ends can then be soldered to a BNC socket which is then connected to the end of the counter input cable. This arrangement picks up enough oscillator energy to give reliable counter readings, without needing to be any closer to L3 (to avoid "pulling" the frequency).

Assuming you want to receive the NOAA satellites, set the oscillator frequency for channel A to 132.0 MHz (using VR4), and the frequency for channel B to 132.12 MHz (using VR5). These correspond to reception frequencies of 137.5 MHz for NOAAs 12 and 15 and 137.62 MHz for NOAA 17. If you want to try for other satellites, you'll need to find out their APT frequency and set the oscillator frequency to 5.5 MHz below that figure instead, and adjust the tuning controls.

Once the oscillator frequencies have been set, the next step is to peak up the RF stage tuned circuits. This is done by setting your RF signal generator to produce an unmodulated (CW) signal at 137.5 MHz, initially with a level of about 30 mV. That done, connect the generator's output to the antenna input of the receiver, using a series DC blocking capacitor if the generator doesn't have one (so that the generator doesn't short out the +12V phantom power for the masthead amplifier).

Next, connect your multi-meter (set to the 5V DC range) across the 390k ohm resistor just behind Q2 and make sure switch S1 is set to the channel A position.

Now turn up RF gain control VR1 to about midway and use an alignment tool or a very small screwdriver to adjust trimmer cap C10 until you find a peak in the voltage reading on a multi-meter. If you can't find a peak, you may have to pull the turns on coil L2 slightly further apart to reduce its inductance. Once the peak is found, adjust VC2 carefully to maximize the meter reading (the multi-meter is reading the RSSI voltage from IC1, so it's essentially showing the received signal strength).

When you're certain that the L2/C10 circuit is tuned to 137.5 MHz, check the actual voltage reading of the meter. If it's more than 2.5V, reduce the output level from the RF generator until the multi-meter reading drops to about 2.0V. Now you're ready to peak the receiver's input tuned circuit—i.e. L1 and VC1. This is done in exactly the same way as for L2 and C10. Just adjust C4 slowly until the multi-meter indicates a peak and then carefully set C4 for the maximum peak reading. If you can't find another peak, you may need to pull the turns of L1 slightly further apart as before.

The final alignment step is to set the slug in quadrature detector coil L4 to the correct position for optimum FM demodulation of the 5.5 MHz IF signals. This is done by first switching the signal generator so that it's still producing a 137.5 MHz signal but this should now be frequency modulated—preferably with a 2.4 kHz tone and a deviation of about ±25 kHz.

Now, connect an 8 ohm speaker to the receiver's speaker socket (CON3) and turn up the audio gain control (VR3) to about the 10 o'clock position. You may not be able to hear the 2.4 kHz modulating signal at this stage but in any case, slowly and carefully adjust the slug in L4 using a non-magnetic alignment tool. Sooner or later you'll start to hear the 2.4 kHz tone and you should also be able to tune the coil for maximum audio level and minimum distortion and noise.

Once this has been done, the basic alignment of your weather satellite receiver is finished and it's ready for final alignment using the signals from a weather satellite. But before you'll be able to do this, you'll need to build or buy a suitable antenna and perhaps a small masthead pre-amplifier to boost the signal if necessary. The best antenna for APT weather satellite reception is either a turnstile or quadrifillar antenna.

Figure 23-6 *VHF turnstile antenna*

Figure 23-7 *Quadrifillar antenna*

APT antenna considerations

The crossed dipole or turnstile antenna is shown in Figure 23-6; it is recommended for nearly all APT weather satellite applications. It is made from aluminum and is strong and durable. It can be fixed onto a vertical mast and should be positioned as high as possible. The satellite signal will not go through trees or buildings and to get the best possible continuous signal it needs to be mounted onto a mast as high as possible. On the end or top of a building is preferred although, for test purposes, it can simply be stuck into the grass at ground level.

Another good choice for APT satellite reception is the quadrifilar helix antenna, see Figure 23-7. It offers slightly better performance than the crossed dipole but it is considerably more costly.

The success of weather satellite reception is very much dependent on having a good antenna. The ARRL *Weather Satellite Handbook*, is a good source of information on building antennas and low-cost antenna pre-amplifiers. You can either build a good antenna or readily purchase one from one of the vendors listed in the Appendix. Good quality, low-loss 50. coax should be used between the antenna and the satellite receiver, since satellite signals are generally weak. The coax should be plugged into the BNC jack on the rear of

the receiver. If you are using large diameter coax, it may be necessary to use a UHF-BNC adapter. Another alternative is to use a short length of smaller diameter, more flexible, coax between the main cable and the receiver. That cable can have a BNC plug on one end and a UHF plug and double female adapter on the other. If you use a 137 MHz pre-amplifier, you can overcome some cable losses by installing the pre-amp right up at the antenna. This establishes a low noise figure before going through the loss of the cable. If you find you do need a 137 MHz mast-head pre-amplifier, you can build one or buy one from Hamtronics or Ramsey Electronics. We do not recommend trying to feed B+ to power the pre-amp through the coax as some enthusiasts do. It doesn't pay to try to save the small cost of lightweight hook-up wire and de-grade the performance of the pre-amp in the process. Run a separate wire pair to power the pre-amplifier if needed.

Receiver operation

Operation of the receiver is fairly obvious, but we will comment on several features of interest. Normally, when used to drive a computer's sound card, you will want to listen on a speaker to hear what the signal sounds like. The VOLUME control can be adjusted to whatever level is comfortable for listening and proper for the computer's sound card. This is especially true during the initial acquisition period. However, most times you probably want the ability to mute the speaker; so installing a switch in line with the receiver output to the speaker is handy.

To adjust the SQUELCH control, turn it clockwise just a little beyond the point where the squelch closes with no signal (just noise). The squelch circuit has adjustable sensitivity—the further clockwise you set the control, the stronger the signal must be to open the squelch. This is handy to set how strong you want the satellite signal to be before the tape recorder turns on. Be careful not to set it too far clockwise, though, because the squelch might not open even with a strong signal. The weather satellite receiver will allow you to select two different channels. Channel selection is done with switch S1. Use the selector switch to choose channel A or B, then use potentiometer VR4 or VR5 to "tune-in" the desired satellite frequency. The satellite receiver's audio is fed to the speaker output at J3 and also to a low-level output at J2 which can be fed to either a tape recorder or decoder card in a personal computer.

Setting up you APT receive station

In order to set up your own weather satellite receiving station you will have to acquire a few more items, as seen in Figure 23-8. In addition to the satellite receiver and antenna, you will need a personal computer, some sort of interface and a software package to display the satellite picture and a satellite tracking program if desired. A tracking program is not required, but it is fun to be able to know in advance when the satellite is overhead.

How to see satellite weather pictures on your personal computer, you must follow the steps below:

1. Track the desired satellite to find when it will appear and how long it will be in view (i.e. how high in elevation it will rise). (See NASA's online J-Track program for an example of NOAA satellite tracking.)

2. Tune the receiver to the right frequency, e.g. 137.620 MHz for NOAA-14 and 137.500 for

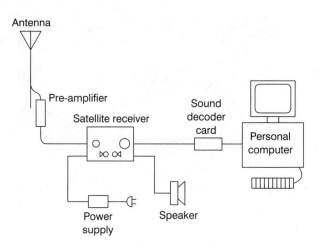

Figure 23-8 *Weather satellite receiving system diagram*

NOAA-15 and pick off the AM signal that carries the image data. The 2400 Hz AM sub-carrier signal will produce a distinctive "tick-tock" sound from the speakers.

3. Receive the signal for as much of the satellite pass as possible.

4. Decode the signal into the image. (Depending on the system, this is done in real-time or the signal is saved to file that is decoded later.) Most new computers (i.e. running at 800 MHz to 3 GHz with WIN2000 or XP software) with a good sound card can be used to decode the software directly from the satellite receiver audio.

With your 137-138 MHz turnstile or quadifillar satellite antenna connected to your weather satellite receiver, and your WX receiver connected to your sound card in your PC, via a shielded audio cable, you will be able to get stunning color images directly from weather satellites.

In order to decode an APT weather satellite signal from your WX satellite receiver, you will need to load some WX decoding software into your personal computer.

There are number of choices for APT satellite decoding software, some of which are free for the download, such as WXtoImg which is a fully automated APT and WEFAX weather satellite (wxsat) decoder. The software supports recording, decoding, editing, and viewing on all versions of Windows, Linux, and

Table 23-4

137 MHz Weather satellite receiver frequencies

Country	Satellite	Format	Frequency
USA	NOAA - 12	APT	137.500 MHz
USA	NOAA - 14	APT	137.620 MHz
USA	NOAA - 15	APT	137.500 MHz
USA	NOAA - 17	APT	137.620 MHz
USA - 2008 launch	NOAA - 18	APT	137.9125 MHz
Russia	METEOR	APT	137.300 MHz
Russia	METEOR	APT	137.400 MHz
Russia	METEOR	APT	137.850 MHz
Russia	OKEAN	APT	137.400 MHz

MacOS X. The WXtoImg software supports real-time decoding, map overlays, advanced color enhancements, 3-D images, animations, multi-pass images, projection transformation (e.g. Mercator), text overlays, automated web page creation, temperature display, GPS interfacing, and control for many weather satellite receivers, communications receivers, and scanners. Check the download site at: http://www.wxtoimg.com/downloads/.

Another good source for the WXSAT satellite decoding software is the HF-FAX web-site located at http://www.hffax.de/html/hauptteil_wxsat.htm. Once your software is loaded and your system check completed, you can refer to the chart in Table 23-4, for the most popular satellite frequencies.

Additional APT weather satellite source information

137 MHz quadrifillar antenna construction info:
http://www.askrlc.co.uk/

137 MHz turnstile antenna construction info:
http://www.qslnet.de/member/pa1pj/turnstile.pdf

TimeStep – http://www.time-step.com

Hamtronics – http://www.hamtronics.com/

Ramsey Electronics –
http://www.ramseyelectronics.com/

Dartcom – http://www.dartcom.co.uk

Swagur Enterprises, USA www.swagur.com

Vanguard Electronic Labs, USA www.cyberpad.com

Chapter 24

Analog To Digital Converters (ADCs)

Parts list

ADC - 1

C1 .1 µF, 35 volt capacitor

C2 10 µF, 35 volt electrolytic capacitor

C3 5 µF, 35 volt electrolytic capacitor

U1 LM78L05 5 volt regulator IC

U2 MAX187 A/D converter IC

ADC - 2

R1-R8 100 k ohm, $\frac{1}{4}$ watt, 5% resistor

C1 1 µF, 35 volt electrolytic capacitor

C2,C5 .1 µF, 35 volt disk capacitor

C3 47 µF, 35 volt electrolytic capacitor

C4 4.7 µF, 35 volt electrolytic capacitor

D1 1N4001 silicon diode

U1 MAX186 A/D converter IC (Maxim Semiconductor)

U2 LM7805 - 5 volt regulator IC

P1 DB-25 - 25-pin parallel port connector

Misc PC board, wire, enclosure, terminal strip, etc.

A number of radio receiver projects in this book are based on collecting data from receivers rather than listening to sound coming from a radio. In this chapter we will discuss four different approaches to buying and/or building analog to digital converters which you can use to collect or save data for analyzing at a later time. By using an analog to digital converter, you can collect, save and later playback and analyze signals that were recorded over days or over weeks and then, using special software, you can analyze the signals to determine trends of specific signals. We will present two different A/D or analog to digital converter projects which you can build with inexpensive integrated circuits and connect to your personal computer's parallel port. Next, we will show you how you can purchase a low cost commercial analog to digital converter with software and finally we will describe a high quality A/D system which is used by members of the Public Seismic Network to record earthquake signals around the country. The PSN A/D converter is a true 16-bit multi-channel assembled A/D system which connects to your personal computer's serial port. These four A/D systems are great choices for collecting radio signals over an extended period of time. Examine each of these A/D solutions and see which one is best for your particular applications. Most of the A/D collections systems described below come with free software for recording data.

ADC – 1: single channel analog to digital converter

This single channel 12-bit analog to digital converter, shown in Figure 24-1, can be built with the minimum of

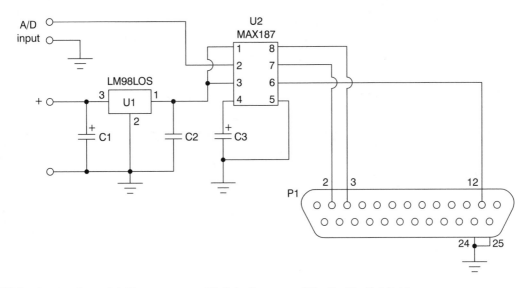

Figure 24-1 *Single channel A/D converter – ADC 1. Courtesy of Radio-Sky Publishing*

components and the software for it is free. The single channel A/D converter is built around the $14 low cost MAX187 A/D converter from MAXIM Semiconductors. The circuit consists of the A/D converter chip, a regulator chip, and a few capacitors; it can't get more simple than this! The single channel A/D converter readily connects to your computer's parallel port through a BD-25 connector. The analog input voltage can be from zero (0) to 4.095 volts as a maximum signal level. A 9 volt battery or small 9 to 12 volt DC "wall-wart" power supply can be used to power the A/D circuit, which can be housed in a small plastic case.

The A/D input from your signal source is fed to pin 2 of the MAX187 chip. The 5 volt regulator's output powers the MAX187 chip via pin 1. The MAX817 outputs to your personal computer are on pins 6, 7 and 8. Only three capacitors are needed for this A/D, two of which are used in the regulated power supply section of the circuit.

You will need a small 30 watt soldering iron, some 60/40 rosin core solder for building this project. A small jar of "Tip Tinner" from your local Radio Shack is used to clean and dress the soldering tip. Locate a few small hand tools, including a pair of small end-cutters and a needle-nose pliers and a flat-blade and Phillips screwdriver and we will begin building the A/D converter.

Heat up your soldering iron and let's get building. The ADC-1 circuit can be built on a small perf-board or direct prototyping board available at Radio Shack or

you could make your own circuit board for the project. The circuit only utilizes capacitors, both polar and non-polar types. Non-polarized capacitors are usually smaller capacitor and sometimes they do not have their actual value printed on them. Often a three-digit code is used to represent their value. For example, a . 001 µF capacitor will have (102), while a .01 µF capacitor will have a (103) marked on it. This project also uses large polarized or electrolytic capacitors, these capacitors have polarity markings on them which must be observed if you want the circuit to work properly. Use the chart in Table 24-2 to help identify the project capacitors. Identify all of the capacitors before installing them on the circuit board. Once you have identified all of the capacitors, you can go ahead and install them on the circuit board. Trim the excess component leads flush to the edge of the circuit board with your end-cutters.

Take a quick look at the semiconductor pin-out diagram shown in Figure 24-2. This diagram will help you install the integrated circuits in the A/D projects. An integrated circuit socket for this project is highly recommended, as an insurance from a possible circuit failure if it ever happens. It is much easier to simply replace the IC rather than trying to un-solder the IC from the circuit board. The IC socket will have a cut-out or notch at one end of the package. Solder the IC socket onto the PC board. Pin one (1) will be assigned to the pin just to the left of the notch or cut-out. Pin one (1) of the IC socket is connected to the

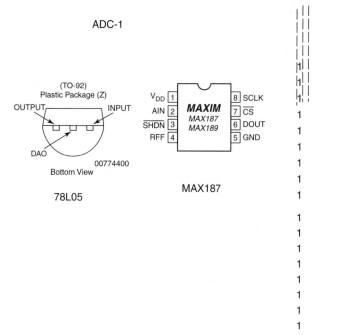

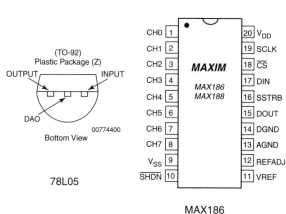

Figure 24-2 *Semiconductor pin-outs*

output of the regulator U1. When inserting the IC into its socket pay particular attention to the orientation. Pin one (1) of the IC must be inserted into pin one (1) of the IC socket. Note that there will be a notch or cut-out or small indented circle on the top of the IC package. Pin one (1) of the IC will be just to the left of the cut-out or notch.

One neat way to package the units is inside a two way parallel port manual switch box. These can be bought for about $10 at computer shows. One of the DB25 connectors is removed from the back of the switch and replaced with a power and a analog input jack. Leads 2, 3, 12, 24 and 25 from the removed DB25 are soldered directly to the appropriate places on the A to D board. The switch can then be used to pass through to the printer when the A to D is not in use. You could also elect to mount the A/D converter in a small plastic box if desired with a 25 pin connector on the rear of the enclosure.

Before applying power to the A/D circuit, take a moment to inspect the foil side of the PC board. Pick up the PC board with the foil side of the board facing upwards towards you. We are going to examine the PC board for possible "cold" solder joints and "short" circuits before applying power for the first time. Look carefully at the solder joints, they should all look clean,

shiny and bright. If any of the solder joints look dark, dull or "blobby" then remove the solder from the joint and then re-solder the joint all over again. Next, we will examine the PC board for possible "short" circuits which are often caused by the sticky rosin residue left on the circuit board from the solder. Often "cut" components or small solder balls can adhere to the foil side of the board and "short" between the circuit traces. Once you have examined the PC board carefully, you can go ahead and power-up the circuit for the first time. Install the free A/D collection software, either Data Collect Lite or Radio-Sky Pipe from the Radio-Sky web-site, or see Appendix, and start collecting data from one of your new receivers. In the event the circuit does not work, check over the installation of the electrolytic capacitors and the polarity of the power supply connections. The MAX87 chip should be available from Digi-Key Corporation, see Appendix.

ADC – 2: 8 channel analog to digital converter

An 8 channel 12-bit analog to digital converter (ADC) project is shown in Figure 24-3, and connects to the parallel port of your personal computer. This second low cost analog to digital converter circuit is one you can readily build yourself in a few hours from scratch. This A/D converter works with free software from a company known as Radio-Sky, which specializes in products for Radio Astronomy. Their free software, which works with the low cost ADC, is called Radio-Sky Pipe. The ADC is based on a single low cost integrated circuit from MAXIM Semiconductors. Pins 1 through 8 on the MAX186 A/D converted chip correspond to Channels 1 through 8. A 100 kilo-ohm resistor is placed across each input channel pin and is connected to ground. The A/D converter circuit connects to a personal computer via the parallel port. You will need to choose the appropriate parallel port address which is selected in the free Radio-Sky Pipe software. Note that all of the .1 µF (microfarad) capacitors are non-polarized types, typically disk ceramic capacitors. They should be rated for at least 15 volts. The A/D circuit may be powered by a battery or any well filtered external DC power supply. (You may

order the MAX186 IC from Radio-Sky Publishing, or from Digi-Key or Mouser Electronics, see Appendix.)

The A/D circuit can be built on a small breadboard or on a printed circuit board. A circuit board and partial kit of parts (no case, connectors, or power supply) is available from Radio-Sky. Before we begin building the single channel A/D converter you will need to secure a clean well lit work area. You will need to locate a small 27 to 33 watt pencil-tipped soldering iron, a roll of 60/40 rosin core solder and small jar of "Tip Tinner," a soldering iron tip cleaner/dresser. Collect together a few small hand tools such as a pair of end-cutters, a pair of needle-nose pliers, a flat-blade and a Phillips screw-driver and a magnifying glass. Grab the schematic, parts layout diagram and the resistor and capacitor identifications charts below, and we will begin.

This project employs eight 100,000 ohm resistors at the input to each of the A/D channels. Take a look at the resistor identification chart in Table 24-1. Each resistor will have three or four color bands on the body of the resistor, which start at one end of the resistor body.

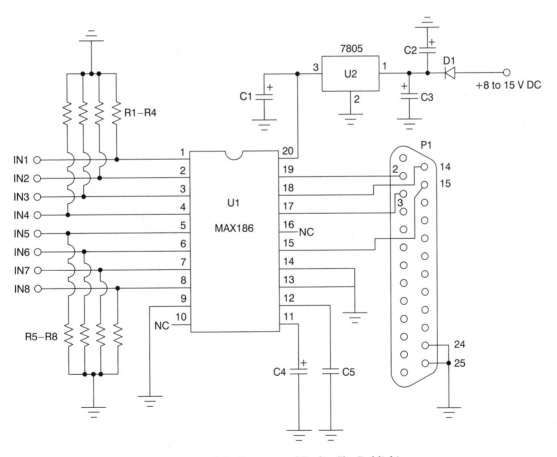

Figure 24-3 *8 channel A/D converter – ADC 2. Courtesy of Radio-Sky Publishing*

Table 24-1

Resistor color code chart

Color Band	1st Digit	2nd Digit	Multiplier	Tolerance
Black	0	0	1	
Brown	1	1	10	1%
Red	2	2	100	2%
Orange	3	3	1,000 (K)	3%
Yellow	4	4	10,000	4%
Green	5	5	100,000	
Blue	6	6	1,000,000 (M)	
Violet	7	7	10,000,000	
Gray	8	8	100,000,000	
White	9	9	1,000,000,000	
Gold			0.1	5%
Silver			0.01	10%
No color				20%

The first color band will represent the first digit of the resistor code, the second color band denotes the second digit of the code, and finally the third color code represents the resistor's multiplier value. Resistors will generally have a tolerance code which is the fourth color band. The absence of a fourth color band indicates a 20% tolerance value, while a silver color band denotes a 10% tolerance and a gold band represents a 5% tolerance resistor. A 100k ohm resistor's first band should be brown (1), its second color band is black (0) and its third band will be yellow (0000), so (1)(0) times (0000) = 100,000 ohms or 100k ohms. Place the eight resistors on the circuit board and solder them in place. Next, trim the excess component leads flush to the edge of the PC board with your end-cutters.

Capacitors are generally identified as polar and non-polar types. The polarized capacitors are also called electrolytic capacitors and they have polarity, as each of the two leads are labeled as plus (+) or minus (−) and this polarity must be observed when installing these capacitors in order for the circuit to work correctly. Non-polarized capacitors are usually small in physical size and in their actual value. Take a look at the capacitor identifier chart in Table 24-2, it illustrates the capacitor codes used to identify smaller sized capacitors. Many times small capacitors do not have room to print their entire details on the body of the capacitor, so a three-digit code is utilized to label them.

For example, a .001 µF is labeled (102), while a .01 µF capacitor is labeled as (103). Identify all of the capacitors in the ADC-2 project and separate them into piles polarized vs. non-polarized and place them in front of you. Once identified you can go ahead and install the capacitors on the PC board in their respective locations. Pay particular attention to installing the electrolytic capacitors correctly. Once you have placed the capacitors on the board, you can solder them in place, then cut the extra component leads flush to the edge of the board.

The analog to digital converter contains a single silicon diode at D1. Remember that diodes have polarity and they must be mounted correctly for the circuit to work. Diodes will generally have an anode and cathode lead, they will have a white or black band closer to one end of the diode. The end closest to the colored band is the cathode lead, look at the schematic or part layout diagram and then install the diode.

This circuit employs two integrated circuits, the main A/D converter chip and a regulator chip. An integrated circuit socket for U1 is highly recommended, so in the event of a possible IC failure, you can simply plug-in a new chip without having to un-solder 20 pins or more from the foil side of the PC board. Integrated circuit sockets will usually have a notch or cut-out at one end of the package. The pin just to the left of the notch or

Table 24-2

Capacitance code information

This table provides the value of alphanumeric coded ceramic, mylar and mica capacitors in general. They come in many sizes, shapes, values and ratings; many different manufacturers worldwide produce them and not all play by the same rules. Some capacitors actually have the numeric values stamped on them; however, many are color coded and some have alphanumeric codes. The capacitor's first and second significant number IDs are the first and second values, followed by the multiplier number code, followed by the percentage tolerance letter code. Usually the first two digits of the code represent the significant part of the value, while the third digit, called the multiplier, corresponds to the number of zeros to be added to the first two digits.

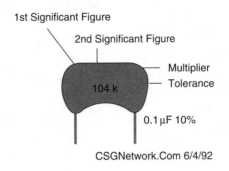

CSGNetwork.Com 6/4/92

Value	Type	Code	Value	Type	Code
1.5 pF	Ceramic		1,000 pF /.001 µF	Ceramic / Mylar	102
3.3 pF	Ceramic		1,500 pF /.0015 µF	Ceramic / Mylar	152
10 pF	Ceramic		2,000 pF /.002 µF	Ceramic / Mylar	202
15 pF	Ceramic		2,200 pF /.0022 µF	Ceramic / Mylar	222
20 pF	Ceramic		4,700 pF /.0047 µF	Ceramic / Mylar	472
30 pF	Ceramic		5,000 pF /.005 µF	Ceramic / Mylar	502
33 pF	Ceramic		5,600 pF /.0056 µF	Ceramic / Mylar	562
47 pF	Ceramic		6,800 pF /.0068 µF	Ceramic / Mylar	682
56 pF	Ceramic		.01	Ceramic / Mylar	103
68 pF	Ceramic		.015	Mylar	
75 pF	Ceramic		.02	Mylar	203
82 pF	Ceramic		.022	Mylar	223
91 pF	Ceramic		.033	Mylar	333
100 pF	Ceramic	101	.047	Mylar	473
120 pF	Ceramic	121	.05	Mylar	503
130 pF	Ceramic	131	.056	Mylar	563
150 pF	Ceramic	151	.068	Mylar	683
180 pF	Ceramic	181	.1	Mylar	104
220 pF	Ceramic	221	.2	Mylar	204
330 pF	Ceramic	331	.22	Mylar	224
470 pF	Ceramic	471	.33	Mylar	334
560 pF	Ceramic	561	.47	Mylar	474
680 pF	Ceramic	681	.56	Mylar	564
750 pF	Ceramic	751	1	Mylar	105
820 pF	Ceramic	821	2	Mylar	205

cut-out will be designated as pin 1. Be sure to align pin one (1) of the IC socket with IN1 and pin twenty (20) with the output of the regulator at U2. The actual IC itself will also have a notch or cut-out and pin one (1) will also be just to the left of the notch. Make sure that when you place the IC into its socket that you align pin one (1) of the IC with pin one (1) of the socket. Failure to observe the correct orientation of the IC will result in damage to the IC and perhaps damage to the circuit when power is first applied.

Once the circuit board is built, you can wire a DB-25 connector to the A/D chip and ground and your analog to digital converter will be just about finished. Find a suitable small plastic or metal enclosure for the circuit board. The 10 inputs can be wired to a 10-position screw terminal strip or you could use the second 25 pin connector as the input connector to the circuit. Use whatever is convenient.

Before applying power to the circuit for the first time, you will need to inspect the circuit board for possible "cold" solder joints and "short" circuits. Pick up the circuit board, with the foil side facing upwards towards you. Examine the solder joints carefully. Each of the solder joints should look clean, shiny and bright. If any of the solder joints look dark, dull or "blobby", then remove the solder from the joint and then re-solder the joint over again so that it looks good. Next, we will inspect the board for possible "short" circuits. Many times sticky residue from the rosin core solder will cause "cut" component leads and small solder balls to adhere to the foil side of the board, thus "shorting-out" circuit traces. Remove anything that looks foreign.

Now you can connect the A/D circuit to the parallel port of your computer and apply power to the circuit. You can power the A/D converter from a small DC "wall-wart" power supply. Install the free Radio-Sky software and you are ready to begin collecting data. The A/D circuit will work in the Standard version of Radio-Sky Pipe, but only Channel 1 will be available. The Pro-Upgrade allows you to use any number of channels up to 8. The channels used must be sequential. Analog inputs should be restricted to the 0 to 4.095V range.

If the circuit doesn't work on power-up, disconnect power and re-examine the circuit board to make sure that you didn't make an error when constructing

the circuit. Most common errors are installing the capacitor and diodes incorrectly. Also take note of the IC installation if the circuit does not initially work.

ADC – 3: Dataq DI-194RS/ DI-154RS starter kits

The Dataq DI-194RS is a low cost ready-made 10-bit ADC starter kit available for the low price of $24.00, while the **DI154RS** is a 12-bit ADC system priced at $39.00, see Figure 24-4. This A/D system has many features and includes a copy of WinDaq/Lite chart recorder software. This is a great budget ADC system for readers who want to begin collecting data immediately. These ADCs would go well with any of the receiver projects in this book.

4-Channel, 10-Bit, data acquisition starter kit - DI194RS

4-Channel, 12-Bit, data acquisition starter kit - DI154RS

- Compact Data Acquisition Starter Kit
- 4-Channel, 10-Bit, ±10V ADC
- Provided with a Serial Port Interface Cable

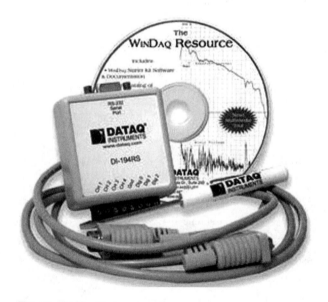

Figure 24-4 *Dataq data-logger*

- Two Digital Inputs for Remote Start/Stop and Remote Event Marker Control

- Includes WinDaq/Lite Chart Recorder Software, WinDaq Waveform Browser Playback and Analysis Software, and Documentation

- Provided Active X Controls allow you to program the DI-194 from any Windows programming environment

- Kids - Get a free Data Acquisition Starter Kit for your Science Fair Project.

The DI-194RS and DI-154RS, DATAQ's four-channel data acquisition starter kits, are a low-cost way to experience WINDAQ software, now with more features and capabilities than ever before. With these starter kits, you can digitize virtually any transducer's analog output signal and record it to your PC's hard disk. At the same time, view the transducer's output on your PCs monitor in a triggered sweep (oscilloscope-like) or scrolling (chart recorder-like) display format. The starter kits provide a taste of the exceptional power and speed possible with WINDAQ software. They provide 12-bit (DI-154RS) and 10-bit (DI-194RS) measurement accuracy, a ±10V analog measurement range, up to 240 samples/second throughput, and four analog input channels. A CD demonstrates WINDAQ Waveform Browser (WWB), our playback and analysis software, but to get a hands-on illustration of the data recording and display capabilities of WINDAQ/Lite, you need a DI-154RS or DI-194RS starter kit. When connected to your PC's serial port, these starter kits allow you to record, display, and analyze data using your own signals. The kits ship with WINDAQ/Lite (Recording Software and Playback and Analysis software). Data acquisition rates up to 240 samples per second are supported for Windows 2000 and XP.

The DI-154RS and DI-194RS derive their power directly from the RS-232 serial port to which it is connected. So there are no batteries to replace or external power supplies to connect. The DI-154RS and DI-194RS starter kits are equipped with two digital inputs for remote start/stop and remote event marker control. Both starter kits are provided with an ActiveX Control Library that allows you to program the starter kits from any Windows programming environment.

The included WINDAQ/Lite data acquisition software offers real time display and disk streaming for the Windows environment. Their real time display can operate in a smooth scroll or triggered sweep mode of operation, and can be scaled into any unit of measure. Event markers with comments allow you to annotate your data acquisition session with descriptive information as you're recording to disk. Raise your productivity to new heights with WINDAQ's unique multi-tasking feature. Record waveform data to disk in the background while running any combination of programs in the foreground—even WINDAQ Waveform Browser playback software to review and analyze the waveform data as it's being stored! The DI-194RS Module with the WINDAQ Resource CD, screwdriver, and cable—all included with each starter kit. DI-154RS module case is identical.

DI-194RS Starter Kit:

http://www.dataq.com/products/startkit/di194rs.htm

DI-154RS Starter Kit:

http://www.dataq.com/products/startkit/di154rs.htm

ADC – 4: PSN true 16-bit serial output analog to digital converter

The Public Seismic Network analog to digital board is a true 16-bit analog to digital converter for serious researchers who want high speed and high accuracy and excellent resolution. This ADC is used by the Public Seismic Network, a group of dedicated researchers based in California who collect earthquake events. Many researchers world-wide use this equipment to collect data. PSN ADC's are used to collect seismic, magnetic, radio and weather information. A free data collection software package is also available. The PSN board costs about $180, see Figure 24-5.

True 16-bit PSN A/D board

16-Bit ADC board features:

- Low Cost Analog to Digital Card for Seismic and Other Low Frequency Data Recording

- True 16-Bit Resolution with Oversampling (more information)

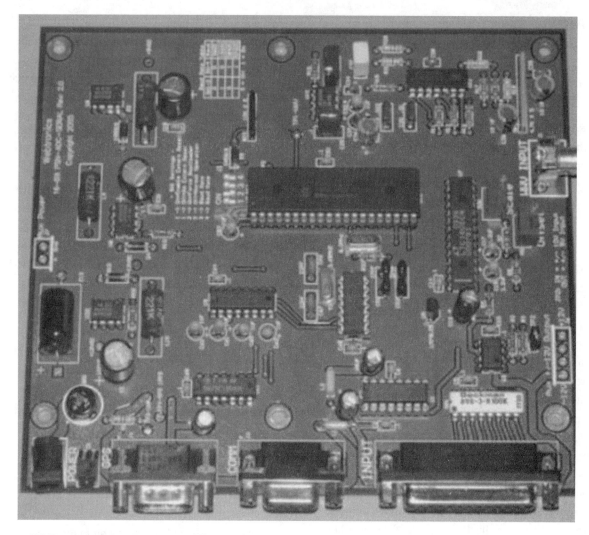

Figure 24-5 *16-bit analog to digital converter*

- Standalone ADC Board with RS-232 Serial Output

- 8 Single-Ended Analog Input Channels

- Compatible with WinSDR Datalogging Software

- Time Accuracy < 3 Millisecond with GPS Timing System

- Uses GPS, WWV, WWVB and Other Time References for Accurate Time Keeping

- Low power Requirement: < 1 Watt without Timing Reference; < 1.5 Watts with GPS Timing

- Uses Low Power High Speed PIC18F452 Microprocessor

Specifications

- Analog Inputs:

  - Channels: 8 Single-Ended Inputs

- Resolution: 16-Bit Successive Approximation

- Accuracy: 16 Bits

- Dynamic Range: 96 dB

- Input Range: ± 5 or ±10 Volts, Jumper Selectable

- A/D Converter Chip: Linear Technology LTC1605CN

- Input Impedance: 4.7k or 100k (Note 1)

- Sample Rates:

  - 200, 100, 50, 20 and 10 Samples Per Second

- WWV Time Correction Option:

  - Tone Detector and Bandpass Filter to Detect 800 ms Top of the Minute Tone Mark

  - ±20 Millisecond Time Accuracy when Locked to WWV Time Signal

- GPS Support:
  - Supports these GPS Receivers: Garmin GPS 16/18 Sensor or the Motorola ONCORE series of GPS receivers
  - ±3 Millisecond Time Accuracy when Receiver is Locked to GPS Time
- Board Size:
  - 6.500 × 5.750 inches or 16.51 × 14.605 centimeters
- Power Supply Requirements:
  - 16 to 28 VDC or 11 to 18 VAC at 100 MA or more
- PSN-ADC-SERIAL Version II Board without WWV Timing Circuitry and Power Supply - Cost: $185.00
- WWV Timing Option - Cost: $25.00
- Garmin GPS 18 Timing System - Cost: $98.00
- Power Supply Options:
  - 110 VAC 60 Hz Wall Mounted Power Supply - Cost: $10.00
  - Universal 110 VAC to 220 VAC 50 / 60 Hz Power Supply with one Adapter Plug - Cost: $25.00
- Six foot 9-pin Male / Female RS-232 Cable - Cost: $5.00
- Enclosures for the PSN-ADC-SERIAL Board.
- Plastic Box with Metal Front/Rear Panels - Cost: $15.00
  Box Size: L: 8.0 W: 7.8 H: 2.5 inches. This box can be used to hold the A/D and Amp/Filter boards. Note: This is just the box, you will need to drill any front/rear panel and mounting holes.

True 16-bit performance

The Analog to Digital converter chip on the PSN-ADC-SERIAL board has an internal noise level of ± 3 counts. With the use of over-sampling, the internal noise level of the chip can be averaged out to produce a true 16-bit

(96 dB) dynamic range. The amount of oversampling depends on the sample rate. The number of samples that are averaged together at 200 samples per second is 16 and at sample rate below 200 sps the number of samples averaged together is 32.

Board differences

There are two versions of this ADC board. The boards are identical except for the Analog Input connector and location on the board. The board with the DB25 right angle connector is for customers who only need the ADC board. This version allows access to all of the connectors at the edge of the board. The board with the 26 pin ribbon connector header can be used with the PSN-ADC-EQAMP Amplifier/Filter board with mating 26 pin header connector. This arrangement produces a module that can be mounted in an off-the-self box.

Software

WinSDR (Download) can be used to record data using the PSN-ADC-SERIAL Version I or II ADC boards. This program is included with the purchase of the board. This Windows DLL Driver (pdf version) can be used by programmers who would like to write their own data-logger program or add the A/D board to an existing program. The DLL Driver zip file includes an example, written in VC++ 6.0, showing how to use the PSN A/D Board DLL file.

Earthworm, a powerful seismic software package that can run under various operating systems, can now receive data using the PSN-ADC-SERIAL board or WinSDR. See this link for more information. A device driver for the Windows version of Seislog is now available. Contact me for more information.

To order a board contact Larry at: 1.650.365.7162

16-Bit ADC system board: http://psn.quake.net/serialatod.html#16-Bit%20Performance

WinSDR software: http://psn.quake.net/winsdr/

Appendix

Electronic Parts Suppliers

We have listed some electronic parts suppliers. We cannot guarantee any particular level of service from these companies and supply this list only as a convenience to our readers.

Allegro Electronic Systems - Semiconductor manufacturer
3 Mine Mountain Rd. Cornwall
Bridge, CT. 06754
Phone: 1-203-672-0123
http://www.allegromicro.com/

AmaSeis Seismic Software - Free seismic software available from: Alan Jones
www.binghamton.edu/faculty/jones/AmaSeis.html

Amidon Associates Inc. - Toroidal cores
240 Briggs Ave
Costa Mesa, CA 92626
Phone: 714-850-4660 1-800-898-1883
http://www.amidoncorp.com/
Email: sales@amidoncorp.com

Analog Devices Inc. - Now selling their parts to individual semiconductors on the Internet
3 Technology Way
Norwood, MA - 02062
http://www.analog.com/

Centerpointe Electronics, Inc. - I-F can type inductors/transformers
5241 Lincoln Ave. Unit #A-7
Cypress, CA 90630
Email: cpcares@sbcglobal.net
http://www.cpcares.com/

Circuit Specialist Inc. - Good parts selection, test equipment, kits
220 South Country Club Drive #2
Mesa, AZ 85210
Phone: (800) 528-1417 +1-480-464-2485
http://www.cir.com/

C & S Sales - Some kits and test equipment
Wheeling, IL 60090
Phone: 1-800-292-7711
http://www.cs-sales.com/
sales@cs-sales.com

Dan's Small Parts and Kits - Low tech site, but nice parts selection
PO Box 3634
Missoula, Montana. 59806-3634
Phone: 406 258 2782
http://www.danssmallpartsandkits.net/

DATAQ Instruments, Inc. - A/D data-loggers
241 Springside Drive
Akron, OH 44333
Phone: 330-668-1444
http://www.dataq.com/
E-mail: info@dataq.com

Debco Electronics Inc. - Sometimes has some good bargains
4025 Edwards Road
Cincinnati, Ohio, USA, 45209
Phone: (513) 531-4499
http://www.debco.com/

Digi-Key Corporation - Extensive selection of electronic parts on-line
PO Box 677
Thief River Falls, Minnesota 56701
http://www.digikey.com/

Downeast Microwave Inc - Microwave components and kits for amateurs
954 Rt. 519 Frenchtown
NJ 08825 USA
Phone: 908-996-3584 (Voice)
http://www.downeastmicrowave.com/

Edmund Scientific Company - Science items, parts and kits
101 E. Gloucester Pike
Barrington, New Jersey 08007
Phone: 800-728-6999
http://scientificsonline.com/

Electronic Goldmine - Many unique surplus parts
9322 N. 94th Way
Suite 104
Scottsdale, AZ 85258
http://www.goldmine-elec.com/

Electronix Express - Parts and kits
365 Blair Road
Avenel, New Jersey 07001
Phone: 1-800-972-2225
http://www.elexp.com/

Electus Distribution - Ferrite coil formers
100 Silverwater Road, Silverwater
Sydney, Australia
http://www.electusdistribution.com.au/productResults.asp?form=CAT&CATID=34&SUBCATID=83

Elenco Electronics Inc. - Radio and electronic kits
150 Carpenter Avenue
Wheeling, IL 60090 USA
http://www.elenco.com
email: sales@elenco.com
(847) 541-3800
(800) 533-2441

Fat Quarters Software - FM-1, FM-3 magnetometer sensors
24774 Shoshonee Dr.
Murieta, CA 92562
Phone: 909-698-7950
http://www.fatquarterssoftware.com

FAR Circuits - Circuit boards for many projects.
18N640 Field Court
Dundee, Illinois 60118
Phone: (847) 836-9148 Voice/Fax
http://www.farcircuits.net/

Freescale Semiconductor - Semiconductor manufacturer
7700 West Parmer Lane
Mailstop PL-02
Austin, TX 78729
http://www.freescale.com/

Hamtronics - Radio kits and parts
65 Moul Road
Hilton, NY 14468-9535
http://www.hamtronics.com/

Kits and Parts - Semiconductors and Toroids
http://partsandkits.com/

Linear Technology Corporate Headquarters - Manufacturer of high quality op-amps
1630 McCarthy Blvd.
Milpitas, CA 95035-7417
Phone: 408-432-1900
http://www.linear.com/index.jsp

Loadstone Pacific - Ferrite coil formers, inductors
4769 East Wesley Drive
Anaheim, CA 92807
Phone: 800-694-8089 714-970-0900
http://www.lodestonepacific.com/coilforms

Maplin Electronics - AM Radio chip ZN414/MK484
Maplin Electronics Ltd
National Distribution Center
Valley Road
Wombwell
Barnsley
South Yorkshire
S73 0BS
www.maplin.co.uk

Maxim Integrated Circuits - IC manufacturer - great
documentation samples
120 San Gabriel Drive
Sunnyvale, CA 94086
Phone: 408-737-7600
http://www.maxim-ic.com/

Mincircuits - RF parts, MMICs, mixers, etc.
PO Box 350166,
Brooklyn, NY 11235 U.S.A.
Phone: (718) 934-4500
http://www.mini-circuits.com/

Mouser Electronics - Extensive electronics
components, no minimum order
1000 North Main Street
Mansfield, TX 76063-1514
http://www.mouser.com/

National Semiconductor - Semiconductor manufacturer
2900 Semiconductor Dr.
PO Box 58090
Santa Clara, California
USA 95052-8090
Phone: (408) 721-5000
http://www.national.com/

Newark Electronics - Extensive inventory of
electronic parts
4801 North Ravenswood Avenue
Chicago, Illinois 60640-4496
Phone: 1-800-342-1445
http://www.newark.com/

Ocean State Electronics - AM Radio chip
ZN414/MK484
PO Box 1458
6 Industrial Drive
Westerly, RI 02891
Phone: 1-401-596-3080
http://www.oselectronics.com

Onset Computers - HOBO Data Loggers
PO Box 3450
Pocasset, MA 02559
Phone: 800-564-4377
www.onsetcomp.com

Public Seismic Network - 16-bit Analog to Digital
converter
Larry Cochrane
http://psn.quake.net/serialatod.html
email: lcochrane@webtronicsc.oom

Radio Jove Project - Radio Jupiter project
c/o Dr. James Thieman
Code 633
NASA's GSFC
Greenbelt, MD 20771
Phone: 1-301-286-9790

Radio Shack - The Internet version of the omnipresent
store
Riverfront Campus World Headquarters
300 Radio Shack Circle
Fort Worth, TX 76102-1964
Phone: (817) 415-3700
http://www.radioshack.com/home/index.jsp

Radio-Sky Publishing - Amateur radio astronomy,
parts and publications
PMB 242
PO Box 7063
Ocean View, Hawaii 96737
http://www.radiosky.com/

Ramsey Electronics - Electronic kit supplier
590 Fishers Station Dr.
Victor, NY 14564 USA
Phone: 800-446-2295
http://www.ramseyelectronics.com/

RF Parts Company - Small RF parts
435 South Pacific Street
San Marcos, California 92078
http://www.rfparts.com/

RP Electronics
2060 Rosser Ave.
Burnaby B.C.
Canada V5C 5Y
Phone: (604) 738-6722
http://www.rpelectronics.com

Speake & Co Limited - Manufacturer of FM1, FM-3
    magnetometer Elvicta Estate
Crickhowell
Powys NP8 1DF
United Kingdom

Texas Instruments Inc. - Semiconductor manufacturer
12500 TI Boulevard
Dallas, TX 75243-4136
Phone: 800-336-5236
http://www.ti.com/

Toko America, Inc.
1250 Feehanville Drive
Mt. Prospect, IL 60056
Phone: (847) 297-0070
1-800-PIK-TOKO
http://www.tokoam.com/

Vectronics - Radio kits
300 Industrial Park Road
Starkville, MS 39759
http://www.vectronics.com/

Vishay Intertechnology, Inc. - Dual matched
    N-Channel FETs
63 Lincoln Highway
Malvern, PA 19355-2143
http://www.vishay.com/

Interesting Web-sites

Amateur Radio Relay League - Ham radio organization
225 Main Street
Newington, CT, 06111-1494 USA
Phone: 860-594-0200
email: http://www.arrl.org/

ELFRAD Group - Low frequency research group
114 Artist Ln
Statesville, NC 28677
http://www.elfrad.com/
email: cplyler@elfrad.com

Radio Waves Below 22 kHz -Interesting low frequency
    web-site
http://www.vlf.it/

Appendix: Electronic Parts Suppliers

Index

Headings in *italics* indicate projects. Page numbers in *italics* indicate illustrations.